POLYMER PROCESS ENGINEERING

Richard G. Griskey Ph.D., P.E.

Institute Professor (Emer.)
Chemistry and Chemical Engineering
Stevens Institute of Technology

Chapman & Hall

New York • Albany • Bonn • Boston • Cincinnati • Detroit • London • Madrid • Melbourne
Mexico City • Pacific Grove • Paris • San Francisco • Singapore • Tokyo • Toronto • Washington

Cover design: Edgar Blakeney

Copyright © 1995
By Chapman & Hall

Printed in the United States of America

For more information, contact:

Chapman & Hall
115 Fifth Avenue
New York, NY 10003

Chapman & Hall
2-6 Boundary Row
London SE1 8HN
England

Thomas Nelson Australia
102 Dodds Street
South Melbourne, 3205
Victoria, Australia

Chapman & Hall GmbH
Postfach 100 263
D-69442 Weinheim
Germany

Nelson Canada
1120 Birchmount Road
Scarborough, Ontario
Canada M1K 5G4

International Thomson Publishing Asia
221 Henderson Road #05-10
Henderson Building
Singapore 0315

International Thomson Editores
Campos Eliseos 385, Piso 7
Col. Polanco
11560 Mexico D.F. Mexico

International Thomson Publishing - Japan
Hirakawacho-cho Kyowa Building, 3F
1-2-1 Hirakawacho-cho
Chiyoda-ku, 102 Tokyo
Japan

1 2 3 4 5 6 7 8 9 10 XXX 01 00 99 98 97 96 95

Library of Congress Cataloging-in-Publication Data

Griskey, Richard G.
  Polymer process engineering / Richard G. Griskey.
    p.   cm.
  Includes bibliographical references and index.
  ISBN 0-412-98541-1
  1. Polymers.   I. Title.
TP1087.G75   1995
668.9—dc20   94-31630
                                       CIP

British Library Cataloguing in Publication Data available

To Dr. Pauline B. Griskey
my inspiration, my best friend,
and my dear wife

CONTENTS

PREFACE

Polymers are ubiquitous and pervasive in industry, science, and technology. These giant molecules have great significance not only in terms of products such as plastics, films, elastomers, fibers, adhesives, and coatings but also less obviously though none the less importantly in many leading industries (aerospace, electronics, automotive, biomedical, etc.). Well over half the chemists and chemical engineers who graduate in the United States will at some time work in the polymer industries. If the professionals working with polymers in the other industries are taken into account, the overall number swells to a much greater total.

It is obvious that knowledge and understanding of polymers is essential for any engineer or scientist whose professional activities involve them with these macromolecules. Not too long ago, formal education relating to polymers was very limited, indeed, almost nonexistent. Speaking from a personal viewpoint, I can recall my first job after completing my Ph.D. The job with E.I. Du Pont de Nemours dealt with polymers, an area in which I had no university training. There were no courses in polymers offered at my alma mater. My experience, incidentally, was the rule and not the exception.

Since that time, formal education in polymers has grown to the extent that most universities now offer relevant course work. One important aspect of polymers has lagged behind, that of polymer processing. There are probably a number of reasons why this has occurred. One of many is the lack of a suitable textbook.

The present text attempts to deal with this problem by providing a book that starts with first principles and then moves through the semiempirical and empirical approaches needed in polymer processing. While rigorous, the text is not excessively mathematical or theoretical but rather a blend of quantitative and, when needed, semiquantitative approaches. The readership for the text is seniors and first-year graduate students in engineering and science as well as industrial practitioners.

The material in the text is an outgrowth of personal experience in industry, both full-time and as a consultant, and in academe, teaching and doing research. Approaches and treatments used in the text were developed in courses taught at the University of Cincinnati (required courses in interdisciplinary graduate program in materials science); at the Virginia Polytechnic Institute (required courses in the graduate materials engineering science program; and senior chemical engineering courses); at New Jersey Institute of Technology (required graduate courses in polymer engineering and science); and senior and graduate courses at the University of Denver, at the University of Wisconsin–Milwaukee, and at universities in Poland, Brazil, Australia, and Algeria.

Further, the relevance of the text to professional practitioners is clearly demonstrated in the use of the material in successful short courses taught for the American Institute of Chemical Engineers and the American Society for Mechanical Engineers.

A suggested syllabus for a university course would be to start with the first chapter, which gives a concise but thorough grounding in polymer classifications, structural characteristics, and methods of characterization. Next, Chapters 2 through 6 should be covered as an interrelated unit. This set of chapters gives the student required background in the relation of thermodynamics, rheology, heat transfer, mass transfer, and chemical kinetics to polymers. Chapters 7 through 11 treat the polymer-processing operations, such as extrusion and injection molding. Finally, Chapter 12, which treats the important linkage between processing, polymer structural characteristics, and polymer properties, is an important capstone to the text.

Over the years, it has been apparent that students profit immensely from worked examples. For this reason, a large number of such relevant solved problems have been included. Likewise, illustrations have been used to aid the learning process as much as possible.

Also included in the text are a large number of problems at the end of Chapters 2 through 11. These problems are designed to emphasize the practice of polymer processing. Indeed, many of them were taken from industrial situations. Additionally, sets of Mini Projects appear at the end of most of the chapters. These open-ended Mini Projects require more time and effort on the student's part than the problems do. As such, they can be assigned as team or group efforts. Further, one or two of these could be a term-long project for individual students. In any case, the combination of the problems and projects should further enhance the text's ability to optimize a sound approach to polymer processing.

All books are our companions, but textbooks fill a special role in that regard. They not only guide us into new areas of knowledge and understanding but can also motivate us to go beyond what is known to what has to be known. It is my strong wish that the present text will do just that.

CHAPTER 1

POLYMER BASICS, STRUCTURAL CHARACTERISTICS, CHARACTERIZATION

1.1 INTRODUCTION

Man has marked the epochs of his history by the materials that he has used to progress—the Stone Age, the Bronze Age, the Iron Age. If we were to name our own time, it could truly be called the Polymer Age. Each passing year sees unbelievable surges in the production and use of plastics, fibers, films, adhesives, elastomers, and other polymer products. The plastic industry itself continues to grow at a rapid pace not only in the United States but also worldwide (see Fig. 1-1). What has brought this about? What is the basis for the importance of polymers in our lives?

Some Definitions

In order to develop a better understanding of the impact of polymers, let us first define a *polymer*. The word itself comes from two Greek words: *poly-* meaning "many" and *-mer* from *meros*, meaning "part" or "member." Hence, a *polymer* is a large molecule built by the repetition of many small repeating units. A polymer can also be characterized as a giant molecule, or *macromolecule*.

As an example of a polymer, take cellulose. A chemical analysis of the substance shows that cellulose has an empirical chemical formula of $C_6H_{16}O_5$ (molecular weight of 162). Yet, a typical cellulose sample might have a molecular weight of 486,000. This indicates that such a cellulose molecule would be made up of 486,000/162, or 3000, repeating units. This number of repeating units is called the *degree of polymerization*.

1

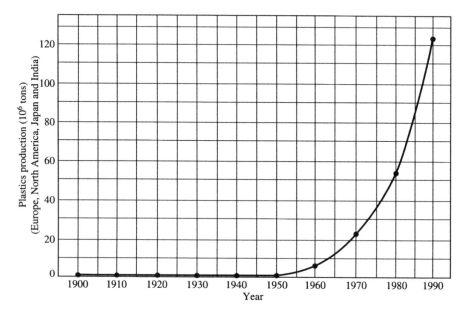

Fig. 1-1   Plastics production [data from ref. 1].

Polymers can be categorized by a number of methods. One technique is to separate them broadly into classifications of natural, or biological, polymers and nonbiological, or synthetic, materials. Whereas synthetic polymers are of fairly recent vintage, natural polymers go back into prehistory. Cellulose, the chief component of plant cell walls, and protein, an essential component of all living cells, are both polymers. Hence, plants and animals are in large measure polymeric in nature.

History of Polymers

From the earliest recorded times, man has utilized natural polymers. Such materials as asphalt, amber, and gum mastic are referred to in the written records of very early civilizations and, later, of the ancient Greeks and Romans. The pattern of use followed the classic human approach, namely, putting something to work without understanding its fundamental nature. Unfortunately, this failure to study delayed the effective utilization of synthetic polymers until the twentieth century.

Consider the climate of scientific opinion in the 1800s. Although many scientists recognized the unusual properties of natural polymers, the question that preoccupied them was: What are these materials? Some said that they were giant molecules, while other subscribed to the notion that they were colloids. The scientists of that day tended to accept the latter concept.

This unexplored acceptance of the colloidal concept derived from the notion that the indicated high molecular weights found for polymers must be in error-since other colloids (i.e., nonpolymers) had reasonable molecular weights. In addition, molecular associations and complexes that were supposed to be physical aggregates of small molecules were used to explain the ''erroneous'' values of molecular weight. Another weakness in human nature may account for this conclusion: it is sometimes more appealing to attribute something to mysterious forces—in this case, colloidal forces—than to consider it in a serious analytical light.

The Rise of Polymer Technology. Although acceptance of colloidal theory unquestionably delayed the development of synthetic polymers, progressive and increased use of modified natural materials continued. The early part of the nineteenth century, for example, saw the development of vulcanization of natural rubber of Goodyear in the United States and McIntosh and Hancock in Great Britain. In 1868, there occurred what many consider to be the start of the plastics industry. At that time, billiards and pool were increasingly popular sports. However, a shortage of elephant tusks, the prime source of the ivory used to make billiard balls, sparked a search for a substitute. This search led John Wesley Hyatt, a Newark, New Jersey, entrepreneur, to produce celluloid, a cellulose nitrate plasticized with camphor. Ultimately, the material was used not only for billiard balls but also for explosives, photographic and motion picture film, and the once popular celluloid collar.

Other materials were gradually introduced. The first of the totally synthetic plastics was the family of phenol-formaldehyde resins discovered initially in 1907 by Baekeland (i.e., Bakelite). Other modified natural polymers (rayon, cellulose acetate) were introduced by the 1920s.

Polymer Science—The Gateway to the Polymer Age. In 1920, an event occurred that set the stage for the rapid development of polymeric materials. In that year, a German chemist, Hermann Staudinger, proposed that polymers were truly macromolecules. His successful efforts to prove this hypothesis were the catalyst for a burst of scientific endeavor that ultimately brought us into the Polymer Age. It is somewhat ironic that Staudinger finally received a Nobel Prize in chemistry in 1953, many years after his outstanding efforts.

The efforts of Staudinger led others into the study of polymers. If there were such a thing as a Polymer Hall of Fame, it would record the efforts of Mark, Flory, Meyer, and others, who labored in the 1920s and 1930s to open up the body of knowledge.

One of the most interesting of these polymer pioneers was Wallace Carothers. Originally a professor at Harvard, he joined E. I. du Pont de Nemours & Company as a research chemist. His work on polymerization resulted in the discovery not only of nylon but also of many other commercially important polymers.

World War II—Into the Polymer Age.   The onset of the World War II brought the impetus needed for significant breakthroughs in both polymer science and engineering. Nations faced with the loss of essential materials had to search for acceptable substitutes. Germany, for example, cut off early from supplies of natural rubber, initiated large-scale programs for the development of a synthetic rubber. For the United States, the entry of the Japanese into the war soon eliminated sources of natural rubber, silk, and many metals. The response was a massive scientific and engineering program that led to large-scale production of nylon (as a substitute for silk), synthetic rubber, silicones, latex-based paints, and many plastics. Progress was not limited to technological applications but included major scientific gains such as Debye's work on light scattering of polymer solutions, Flory's research on viscosity of polymer solutions, Harkins' development of a theory of emulsion polymerization, and Weissenberg's research into normal stresses in polymer flow. The combination of the pioneering efforts of Staudinger, Mark, Flory, Meyer, Carothers, and others, together with the massive technological breakthroughs resulting from World War II, placed us firmly in the Polymer Age.

The Polymer Age—Present and Future.   In the years following World War II, important polymer engineering and scientific discoveries continued. New polymeric materials such as epoxies, polyesters, polypropylene, polycarbonate, polyimides, and many others have found their way into the marketplace. Coincidentally, new fundamental breakthroughs have been made not only in polymer science but also in the engineering aspects of polymers. The result is an area of scientific engineering and technological endeavor that continues to develop without signs of abatement.

The future is not always easy to discern. One thing, however, appears certain: the Polymer Age will continue for as long as our imaginations can take us.

1.2   CLASSIFICATION OF POLYMERS

Classification by Structure and Processibility

We have seen how polymers can be broadly classified into categories of natural or biological polymers and synthetic or nonbiological polymers. It is also possible to use other classifications such as *polymer structure*, or *processibility*.

Polymers can be subdivided into three and possibly four structural groups. These structures are *linear*, *branched*, and *cross-linked* (possibly subdivided into two- or three-dimensional cross-linked types). Schematic representatives are given in Fig. 1-2, where the letters represent mers.

As seen in Fig. 1-2, the *linear* polymer is essentially a long chain consisting of many mers joined together. The concept of a linear or chain polymer is somewhat oversimplified. While it is true that the structure has length without appreciable thickness, it is also true that the bonds joining the mers are not like chain linkages since they are not freely rotating. Perhaps a better way of looking at the linear polymer is to consider it as a number of units joined by the yokes and crosses of universal joints. As these yokes become larger and clumsier, the angle of swing between the mers decreases until a semirigid structure is formed. Typical examples of linear polymers are polyethylene, polyvinylalcohol, or polyvinylchloride.

Branched polymers are those that have side chains growing out of the chain molecule itself. These branches can vary in number and length. Branching can be caused by impurities or the presence of monomers (the small molecules used for polymerization reactions), which have several reactive groups. It is important to distinguish between such polymers as polypropylene or polystyrene and those with branched structures. In the former cases, the side groups (methyl or benzyl)

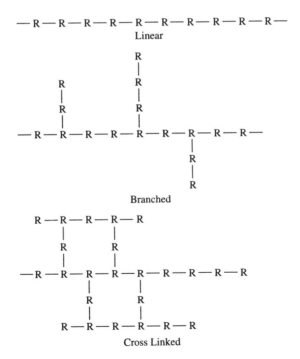

Fig. 1-2 Polymer structures.

are part of the reacting monomer. The latter instances represent actual cases in which a side chain forms by reaction.

Cross-linked structures are those in which two or more chains are joined together by side chains. When cross-linking is not of too great a degree, we obtain a kind of loose network that is essentially two-dimensional. High degrees of cross-linking give a tight structure that is three-dimensional. This cross-linked polymer is also characterized as a space polymer.

Cross-linking is usually brought about by chemical reaction. A typical process is the vulcanization of rubber by which cross-links are formed to join together the polymer molecules in a two-dimensional structure. Thermosetting plastics, on the other hand, represent the three-dimensional case.

It should be understood that there is necessarily no hard and fast classification of polymers by structure. The spectrum that is formed is a gradual one with some overlaps. Nevertheless, the concept of such structures is a useful one.

If we return to our structural classifications and isolate the two ends of the spectrum—linear and highly cross-linked polymers—we find that we can establish still another method of classification based on *processibility*. Linear polymers can be melted or thermally softened and as such can be termed *thermoplastic*. On the other hand, the highly cross-linked material cannot be melted or appreciably softened. In fact, such a polymer will thermally decompose or burn before it melts. The highly cross-linked polymers are classified as *thermosetting polymers*, or *thermosets*. One further distinction between *thermoplastics* and *thermosets* is that the former are generally somewhat soluble whereas the latter are insoluble. Table 1-1 summarizes the two classes.

Methods of Synthesis as a Classification

Still another technique for polymer classification is the method of synthesis. Generally speaking, there are two broad methods of synthesis: addition and condensation polymerization. *Addition polymerization* involves the formation of a polymer from a monomer (a small molecule which is converted into a mer) without the loss of an atom. *Condensation polymerization*, on the other hand, takes place between two monomers and yields a by-product small molecule.

Addition polymerization takes place by a multistep reaction, that is, a chain

TABLE 1-1   Polymer Classification by Processibility

	Thermoplastic	*Thermoset*
Structure	Linear, branched	Highly cross-linked
Solubility	Soluble or partly soluble	Insoluble
Fusibility	Can be melted or softened	Infusible

reaction. This involves initiation, propagation, and termination or chain transfer reactions. A more detailed consideration will be delayed for the time being. If a simplistic look is taken at *addition polymerization,* we find that we can form such polymers by combining monomers with no loss of atoms, involving only simple bonding as with polyethylene:

$$X\ CH_2{-}CH_2 \rightarrow H{-}[CH_2{-}CH_2]_x{-}H \qquad (1\text{-}1)$$

Another way to form addition polymers is by rearrangement, again without loss of an atom, as with a polyurethane:

$$X\ O{-}C{-}N{-}R{-}N{-}C{-}O\ +$$

$$\qquad\qquad\qquad\qquad\overset{\text{H}\ \ \text{O}}{\underset{|\ \ \ \|}{X\ H{-}O{-}R{-}O{-}H \rightarrow O{-}C{-}N{-}[R{-}N{-}C{-}O{-}R]_x{-}OH}}\qquad (1\text{-}2)$$

Typical addition polymers are polyethylene, polypropylene, polystyrene, polyvinylacetate, and polytetrafluoroethylene (Teflon).

A typical condensation polymerization is that between a diamine and a dibasic acid to form a polyamide (i.e., a nylon):

$$X\ H_2N(CH_2)_6NH_2 + X\ HO{-}\overset{O}{\overset{\|}{C}}(CH_2)_4{-}\overset{O}{\overset{\|}{C}}{-}OH \rightarrow$$

$$H[{-}\overset{H}{\overset{|}{N}}{-}(CH_2)_6{-}\overset{H}{\overset{|}{N}}{-}\overset{O}{\overset{\|}{C}}{-}(CH_2)_4{-}\overset{O}{\overset{\|}{C}}]{-}OH + X\ H_2O \qquad (1\text{-}3)$$

Typical condensation polymers are polyamides (the nylons), polyesters, and certain polyurethanes.

Table 1-2 summarizes the differences between condensation and addition polymers.

Copolymerization

In many instances, polymers are synthesized with only one monomer or mer repeating unit. These are called *homopolymers.* On the other hand, if more than one monomer is used, the result is a *copolymer.* Systems in which two monomers combine to form a repeating structure such as a nylon [see Eq. (1-3)] are considered homopolymers.

The purpose of copolymerization is to synthesize a material with special

TABLE 1-2   Comparison Addition and Condensation Polymerization

Addition	Condensation
1. Chain reaction	Stepwise reaction
2. Only growth reaction adding one unit at a time	Any two molecular species can react
3. Monomer decreases steadily	Monomer disappears early
4. High polymer at once	Polymer DP rises steadily
5. Long reaction times give high yields	Long reactions times give high DP
6. Only polymer and monomer present	All molecular species (dimers, etc.) present
7. No by-product	By-product

properties. For example, by copolymerizing acrylonitrile, butadiene, and styrene, we obtain a new and more versatile material, ABS. Notice that the term *copolymerization* covers all combinations of two or more monomers.

Copolymers are categorized by the placement of the mers, as Figure 1-3 shows. For example, in a random copolymer, mers can appear anywhere in the chain whereas, in the block copolymer, regular runs of the mers appear. The graft copolymer attaches the other mers as branches.

Polymeric materials offer the practitioner a wide range of potential products that can be utilized effectively to meet the pressing needs of modern technology. The possibilities of physical and chemical modifications, as well as of copolymerization, mean that a polymer can literally be tailor-made for given end uses.

1.3   STRUCTURAL CHARACTERISTICS OF POLYMERS

Polymers as giant molecules have a number of important structural characteristics that determine their end-use mechanical, electrical, thermal, chemical, and optical properties. These include:

1. Molecular weight
2. Molecular weight distribution
3. Degree of branching
4. Degree of cross-linking
5. Polarity of polymer chains
6. Flexibility of polymer chains
7. Macrocrystalline structure
8. Fine crystalline structure
9. Orientation of polymer chains

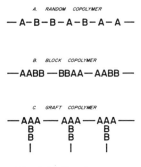

Fig. 1-3   Copolymer types.

The principal influence of these characteristics is on the intermolecular forces in a given polymer. Since such intermolecular forces are additive, they give polymers their peculiar end-use properties. Also, since the processing of the polymer can and does change the structural characteristics, they constitute a means of tracking and or controlling property development. The importance of polymer structural characteristics cannot be understated. For this reason, a discussion of each of the characteristics and the methods used to measure them (polymer characterization techniques) will be dealt with in succeeding sections.

1.4   POLYMER STRUCTURE AND BEHAVIOR—EFFECT OF TEMPERATURE

Changes in temperature can and do influence the behavior of polymers. Such changes can, in some cases, bring about a change in phase (e.g., from a solid to a molten form). Such a change would involve a melting point, or melting temperature. Transitions that result in a change of state or phase are classified as first-order transitions.

Another form of transition can take place in polymers. This type of transition (second-order) results in the type of behavior shown in Fig. (1-4), which plots the logarithm of a mechanical modulus for a polymer (stress divided by strain) as a function of temperature.

As can be seen, the polymer goes from a ''glass'' to a ''leather'' to a ''rubber'' and ultimately to ''liquid flow.'' At the same time, mechanical moduli of the polymer also change from a very high value to a plateau value and ultimately to a very low value.

The temperature at which the polymer goes from a glass to a leather is termed the *glass transition temperature,* or *Tg.* Glass temperature is a second-order

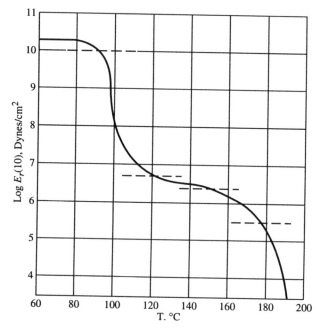

Fig. 1-4 Five regions of viscoelastic behavior [67]

transition as contrasted to a first-order transition as, for example, the melting point of a semicrystalline polymer.

The molecular bases for each of the regions shown in Figure 1-4 can be explained in a simplified manner.

Glassy region: The segments of the polymer chains are literally frozen in fixed positions, which form a disordered structure resembling a lattice. These segments vibrate around a fixed position as do molecules of a molecular lattice.

Leathery or transition region: Here the polymer chain segments undergo short-range diffusional motion. A typical diffusion time from one site to another is on the order of 10 s.

Rubbery region: The short-range diffusional motions of the segments become very rapid. Now many chain segments become involved, which increases retardation by chain entanglement. In fact, these entanglements resemble temporary cross-links.

Rubbery flow: The motion of polymer molecules as a whole becomes important. Now major configurational changes take place on the order of 10 s.

Liquid flow: Long-range configurational changes occur very rapidly (<10 s).

All polymers have a glass transition temperature. These can be determined experimentally from dilatometric data (see Fig. 1-5). Here the inflection point

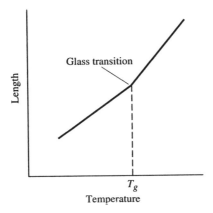

Fig. 1-5   Glass temperature inflection point [68]

shows the glass temperature. Other techniques such as sonic velocity can also be used.

It is also important to see that the glass transition and melting temperatures are related through Beaman's rule (8).

$$0.56 < \frac{T_g}{T_{melt}} < 0.76 \tag{1-4}$$

In the case of semicrystalline polymers, glass temperature is not easily determined with the usual measurements. For such polymers, an indirect technique can be used. Basically, this involves taking two semicrystalline homopolymers and preparing a group of amorphous copolymers that have varying mass fractions of each of the homopolymer materials. Then, according to the equation

$$\frac{1}{T_g} = \frac{m_1}{T_{g_1}} + \frac{m_2}{T_{g_2}} \tag{1-5}$$

a plot of T_g for the copolymers vs the mass fraction will result. Extrapolation of the plot to $m_1 = 0$ and $m_2 = 1.0$ gives the values for the two components. This technique can, of course, also be used if one of the homopolymers has a known T_g.

Values of T_g, together with additional discussion, will be deferred until Chapter 2.

1.5 MOLECULAR WEIGHT

One of the most important of a polymer's structural characteristics is its molecular weight. As described earlier, polymers are giant molecules made up of repeating units (mers). The result is that we can describe this situation by quoting a degree of polymerization or an appropriate molecular weight.

The former is the number of repeating units (mers) in the polymer chain. Molecular weights are average values defined in various ways. As, for example,

$$\overline{M}n = \frac{\text{Number average}}{\text{molecular weight}} = \frac{\Sigma n_i \, M_i}{\Sigma n_i} = \Sigma N_i \, M_i \qquad (1\text{-}6)$$

where n_i is the number of molecules of molecular weight M_i, and N_i is the number of moles of each species.

Likewise the weight average molecular weight $\overline{M}w$ is given as

$$\overline{M}w = \frac{\Sigma \, N_i \, M_i^2}{\Sigma \, N_i \, M_i} = \Sigma W_i \, M_i \qquad (1\text{-}7)$$

where W_i = weight fraction of species i.

Still another molecular weight is the Z average $\overline{M}_Z$:

$$\overline{M}_Z = \frac{\Sigma \, N_i \, M_i^3}{\Sigma \, N_i \, M_i^2} = \frac{\Sigma W_i \, M_i^2}{\Sigma W_i \, M_i} \qquad (1\text{-}8)$$

One additional molecular weight is the viscosity average value M_v:

$$\overline{M}_v = (\Sigma \, W_i \, M_i^a)^{1+a} \qquad (1\text{-}9)$$

where a is a constant related to the solvent used to dissolve the polymer.

Various techniques can be used to determine average molecular weight. These can be subdivided into absolute and indirect methods. In turn, the absolute method breaks down into categories of physical and chemical methods. The methods are listed in Table 1-3.

One of the most widely used and popular of these techniques is the measurement of viscosity. This is done by measuring the viscosity behavior of a polymer in a dilute solution. These measurements are usually made (at constant temperature) in capillary viscometers of the kind shown in Fig. 1-6.

Basically, the viscosity of the pure solvent and of solutions at various concentrations is measured. The terms used in the computation are shown in Table 1-4.

In Table 1-4, t and t_0 are the viscometer efflux times for solutions and solvent and C is the appropriate concentration. Figure 1-7 graphs the determination of

·TABLE 1-3 **Methods of Determining Molecular Weights**

Absolute Methods	*Indirect Methods*
Physical:	
Vapor pressure changes	Viscosity
Cryoscopic measurements	Solubility
Boiling measurements	Electron microscope
Osmotic tests	Sound velocity
Sedimentation tests	Chromatographic adsorption
Light scattering tests	
Chemical:	
End-group analyses	
Tracer techniques	

$[\eta]$. The $[\eta]$ is in turn related to the viscosity average molecular weight by the relation

$$[\eta] = K \, [\overline{M}_v]^a \qquad (1\text{-}10)$$

Table 1-5 gives some constants for various polymer-solvent systems.

Other widely used techniques are osmotic and light-scattering methods. The former are carried out in units of the type shown in Fig. 1-9, whereas the latter

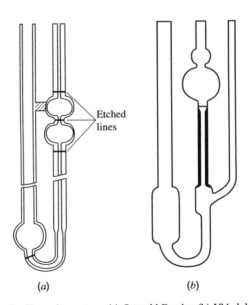

Etched
lines

(a)   (b)

Fig. 1-6   Capillary viscometers: (a) Ostwald-Fenske, (b) Ubbelohde [10].

TABLE 1-4   Viscosity Nomenclatue

Solvent viscosity	η_0
Solution viscosity	η
Relative viscosity	$\eta_r = \dfrac{\eta}{\eta_0} = \dfrac{t}{t_0}$
Specific viscosity	$\eta_{sp} = \eta_r - 1$
Reduced viscosity	$\eta_{red} = \dfrac{\eta_{sp}}{c}$
Inherent viscosity	$\eta_{inh} = \dfrac{(\ln \eta r)}{c}$
Intrinsic viscosity	$[\eta] = \left(\dfrac{\eta sp}{c}\right)_{c=0}$
	$= \left[\dfrac{\ln \eta r}{c}\right]_{c=0}$

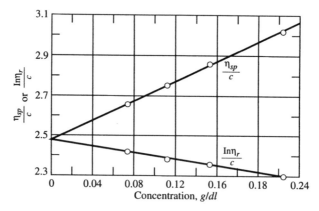

Fig. 1-7   Determination of intrinsic viscosity for polystyrene in benzene [11,12].

TABLE 1-5   Constants for viscosity average molecular weight

Polymer	Solvent	$T, °C$	$K \times 10^4$	a
Polycarbonate	Dioxane	25	4.8	0.67
Polyvinyl alcohol	Water	50	5.9	0.67
Polyvinyl acetate	Acetone	50	2.8	
Polyisobutylene	Cyclohexane	30	2.6	0.70
Polystyrene	Methyl ethyl ketone	25	3.9	0.58
Polymethyl methacrylate	Benzene	20	0.84	0.73
Cellulose acetate	Acetone	25	1.49	0.82

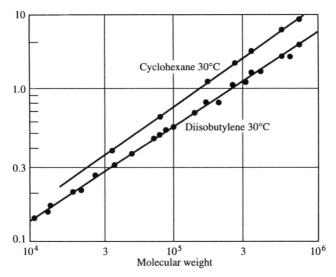

Fig. 1-8   Intrinsic viscosity–molecular weight relations for polyisobutylene [13–15].

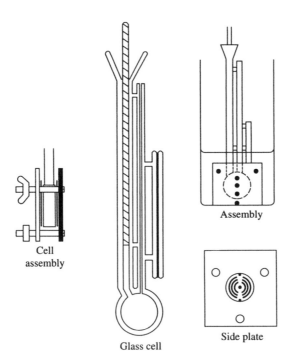

Fig. 1-9   Zimm–Myerson Osmometer [16].

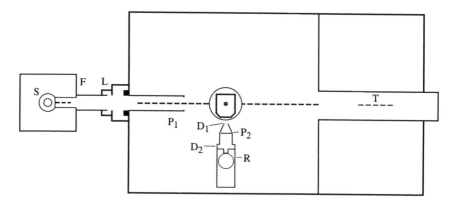

Fig. 1-10   Light-scattering photometer [17].

are measured in a light-scattering photometer (Fig. 1-10). The procedure used to determine molecular weight is to carry out tests at various concentration levels and then to extrapolate the data to zero concentration (see Figs. 1-11 and 1-12). The intercept for the osmotic pressure data (osmotic pressure π divided by concentration c) gives $RT/\overline{M}n$, where $\overline{M}n$ is the number average molecular

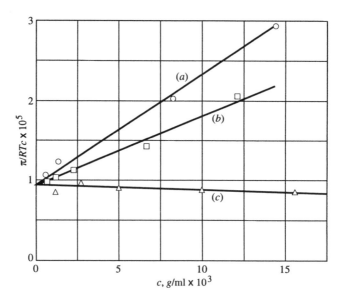

Fig. 1-11   Plot for molecular weight determination by osmotic pressure for nitrocellulose in (a) acetone, (b) methanol, and (c) nitrobenzene [18, 19].

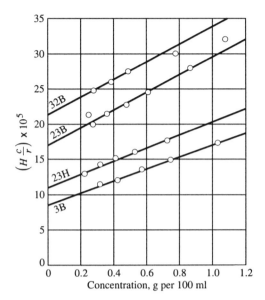

Fig. 1-12 Plot for molecular weight determination of cellulose acetate fractions by light scattering [20, 21].

weight. Likewise, the HC/τ intercept, where H is a constant, c the concentration, and τ the turbidity, gives a value of $1/\overline{M}w$, where $\overline{M}w$ is weight average molecular weight.

Sedimentation studies are carried out in an ultracentrifuge (Fig. 1-13). Basically, the molecular weight relates to the concentration (C_1, C_2) at two radii (r_1, r_2), temperature T, the polymer's partial specific volume $\bar{v}$, and the angular velocity W.

$$M = \frac{2\ RT\ \ln\ (c_2/c_1)}{(1 - \overline{vp})w^2\ (r^2_2 - r_1^2)} \tag{1-11}$$

Three types of molecular weights ($\overline{M}n$, $\overline{M}w$, and $\overline{M}z$) can be determined from sedimentation data.

End-group studies involve chemically measuring the number of end groups or polymer chains. This in turn yields the number of molecules in a given mass and hence $\overline{M}n$. The technique is used mainly for condensation polymers. Some specific cases are shown in Table 1-6.

The remaining chemical method, the tracer technique, measures a given group or groups purposely incorporated into the growing polymer chain. An example

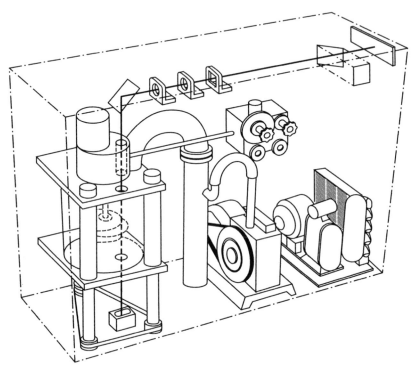

Fig. 1-13   Analytical ultracentrifuge [22].

TABLE 1-6   End-Group Methods

Polymer	End Group	Method
Hydroxy acid polyester	—COOH	Acid determination
Amino acid polyamides		Salt formation
Amino acid polyamides	—NH$_2$	Amino group determination
Glycol polyethers	—OH	Acid take-up during acetylation
Ethylene	—CH$_3$	Infrared spectra
Polymers termination with acetic acid	$-O-\overset{\displaystyle O}{\overset{\displaystyle \|}{C}}-CH_3$	Acetate determination

is the use of *p*-bromobenzoyl peroxide as a tracer initiator for polystyrene. Bromine analysis gives a good indication of molecular weight.

Measuring vapor pressure changes, cryoscopic points, and boiling points are adaptations of the much used physical chemistry technique of determining colligative properties. Chromatographic adsorption can yield molecular weight data. It is, however, used more often to determine molecular weight distribution and, as such, will be discussed in Section 1-6. Number average molecular weights for low molecular liquid polymers can be measured by sound velocities. The electron microscope has also been used to determine the molecular weight of polymers precipitated from solution. This requires that the density of the polymer be known.

Still another technique that can be used is the rheological behavior of the polymer. Since polymers are non-Newtonian (see Fig. 1-14), they have continually varying viscosity (see Fig. 1-15). Further, if the shear stress exerted is plotted against the shear rate (or strain rate), a flow curve results on a logarithmic plot. This contrasts with a Newtonian fluid, which gives a straight line with a slope of 1, that is, a 45° line. Molecular weight will influence the flow curve behavior (see Fig. 1-16). Some typical instruments for making such measurements are shown in Figs. 1-17 and 1-18.

Table 1-7 gives a comparison of some at the principal methods used to determine polymer molecular weights.

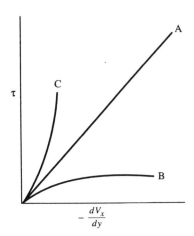

Fig. 1-14   Comparison of shear stress vs shear rate for (A) Newtonian, (B) shear thinning, and (C) shear thickening fluids.

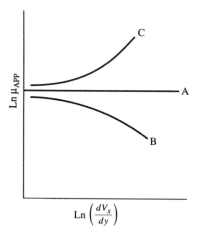

Fig. 1-15   Logarithm of apparent viscosity vs logarithm of shear rate for (A) Newtonian, (B) shear thinning, and (C) shear thickening fluids.

1.6   MOLECULAR WEIGHT DISTRIBUTION

Polymers are materials that have a variety of molecular weights present in any given mass. The average molecular weights, previously discussed, characterize the *average* degree of polymerization for that mass. It is also important in many instances to know the distribution of the moleculear weights in the sample.

Such distributions give the frequency of the occurrence of a certain degree of polymerization. Typical distribution curves for two polymers are shown in Fig. 1-19. The curve for cellulose acetate shows a narrow distribution, whereas

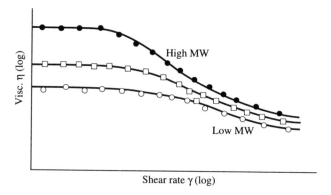

Fig. 1-16   Effect of polymer molecular weight.

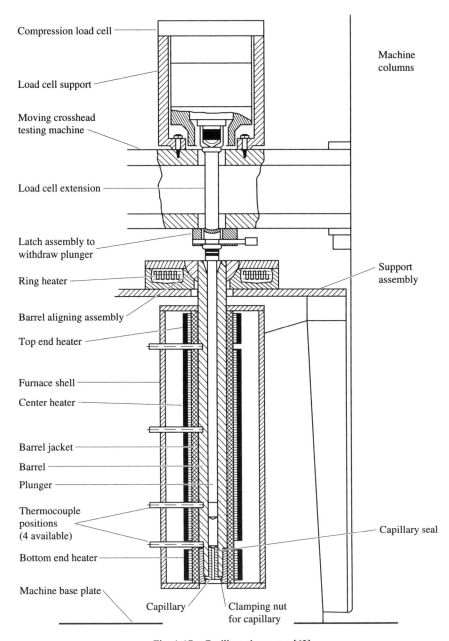

Compression load cell

Load cell support

Moving crosshead
testing machine

Load cell extension

Latch assembly to
withdraw plunger

Ring heater

Barrel aligning assembly

Top end heater

Furnace shell

Center heater

Barrel jacket

Barrel

Plunger

Thermocouple
positions
(4 available)

Bottom end heater

Machine base plate

Capillary

Clamping nut
for capillary

Machine
columns

Support
assembly

Capillary seal

Fig. 1-17   Capillary rheometer [63].

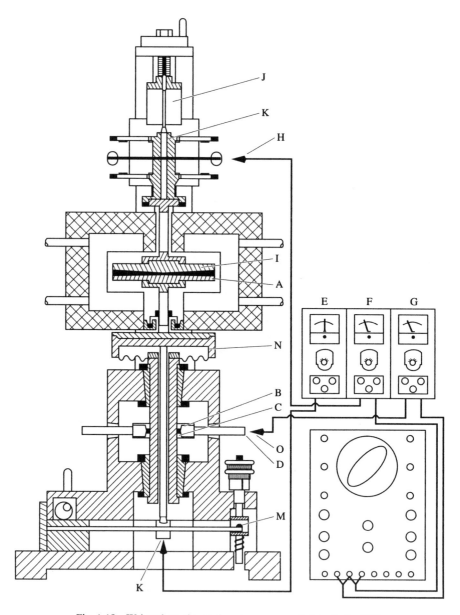

Fig. 1-18   Weissenberg rheogoniometer a cone and plate device [63].

TABLE 1-7   Comparison Polymer Molecular Weight Methods

Method	Molecular Weight Measured	When Insensitive	High-Concentration Effect
Osmotic pressure	$\overline{Mn}$	High molecular weight	Gives low molecular weight reading
Light scattering	$\overline{Mw}$	When solute refractive index approaches solvent refractive index	Gives low molecular weight reading
Sedimentation	$\overline{Mn}, \overline{Mw}, \overline{Mz}$	—	—
End-group	$\overline{Mn}$	High molecular weight	—
Viscosity	$\overline{Mw}$	Low molecular weight	Molecular weight high

Source: Ref. 24.

the plot for polystyrene indicates a broad distribution. Another method of presenting molecular weight distributions is by a cumulative curve (see Fig. 1-20).

There are a number of methods for determining the distribution of molecular weights. The principal methods are:

1. Fractionation
2. Sedimentation
3. Gel permeation chromatography
4. Rapid estimates

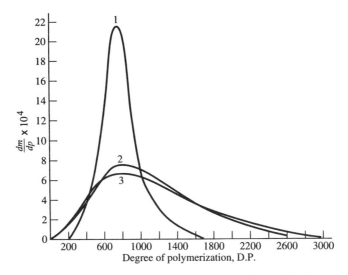

Fig. 1-19   Molecular weight distribution curves for (1) cellulose nitrate, (2) polystyrene, and (3) theoretical case [25].

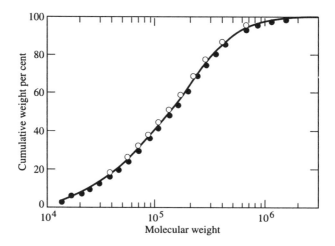

Fig. 1-20   Cumulative molecular weight distribution [26,27].

The first technique, fractionation, consists of separating a polymer solution into separate groups by fractional precipitation and then measuring the mass and average molecular weight of the fractions. The method is time-consuming and none too precise.

The sedimentation studies mentioned earlier use the ultracentrifuge to determine a number of different molecular weights ($\overline{M}n$, $\overline{M}w$, $\overline{M}z$, etc.). It is then possible to construct the distribution curve from these data.

A method that has won general favor is gel permeation chromatography. Basically, polymer samples are separated by size of molecule. The separation is accomplished by using beads of a rigid, porous gel whose pores are about the same size as polymer molecule dimensions (see Fig. 1-21).

A typical chromatogram is given in Fig. 1-22. These data are in turn related to $[\eta]M$ as shown in Fig. 1-23. The result is the establishment of a distribution curve (Fig. 1-24). This curve compares gel permeation to fractionation data.

Methods are also available for rapidly estimating polymer molecular weight distribution. These methods include swelling, in which the amount of polymer precipitated vs nonsolvent added gives a cumulative distribution curve (i.e., higher molecular weights first precipitate out with nonsolvent addition) and turbidimetric titration, which uses turbidity induced by nonsolvent addition as a measure of molecular weight distribution. These methods are described in detail in Ref. 28.

The ratio of the weight average to number average molecular weight also gives an index of distribution. The $\overline{M}w$ is much more sensitive to high molecular weight species and, thus, is always greater than $\overline{M}n$. If the polymer is reasonably

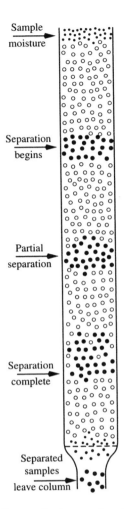

Fig. 1-21   Schematic diagram of gel permeation chromatographic unit [29].

homogeneous with a narrow distribution, the $\overline{M}w/\overline{M}n$ ratio is about 1.5 to 2.0. With a broad distribution, however, the ratios could be as high as 20 to 50. Hence, the relation of $\overline{M}w$ to $\overline{M}n$ provides a measure of polydispersity in a polymer sample.

Finally, melt rheology can be used to determine if molecular weight distributions differ. In Fig. 1-25, two melt flow curves, one for a narrow and one for a broad distribution, are shown. As can be seen, there is a considerable difference in their flow behavior.

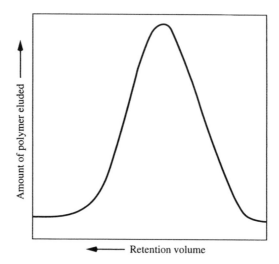

Fig. 1-22   Chromatogram [30].

1.7   BRANCHING

Branching in polymers cannot be as easily determined as either molecular weights or molecular weight distributions. Basically, in many instances, only the relative effects of branching or certain measurements can be indicated. For example, branching will generally depress the glass temperature of a polymer. Hence, if a polymer has not been affected by other possible T_g depressants, such as a copolymer, additives, or a change in morphology, then branching could have occurred.

Another situation involving a degree of branching is the effect on intrinsic viscosity. The greater the branching, the more compact the packing of the polymer in solution, with the result that, at equal molecular weights, a branched polymer will have a lower intrinsic viscosity. End-group measurements used in conjunction with another molecular measurement can also give some insight into branching. A comparison of actual end groups to those predicted from molecular weight will reveal that branching occurs.

There are other indicators as well. For example, since branched molecules tend to become still more highly branched, they concentrate at the high molecular weight end of the distribution. The resulting presence of a long, high molecular weight tail shows that the possibility of branching can increase the solubility of a polymer [35].

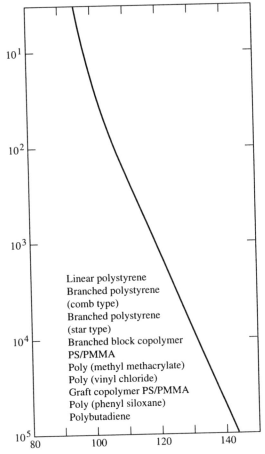

Fig. 1-23   Chromatographic data [31,32].

1.8   CROSS-LINKING

Cross-linking is not an easily determined characteristic. Because cross-linking moves a polymer toward insolubility and infusibility, one possible indication of cross-linking is a change in the polymer's fusibility or solubility. Also, a change in a polymer property itself can indicate cross-linking. Another method for determining cross-linking is to measure a property before and after the addition of varying amounts of a chemical that will destroy cross-linkages. The relative charge of the property then indicates the relative amount of cross-linking. Spectraphotometric techniques can also be used to determine cross-linking. The pres-

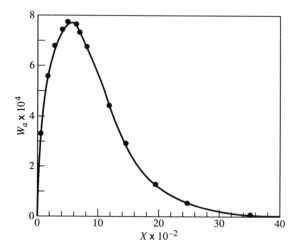

Fig. 1-24   Molecular weight distribution curve (circles); determined by GPC compared to theoretical (solid curve) [33,34].

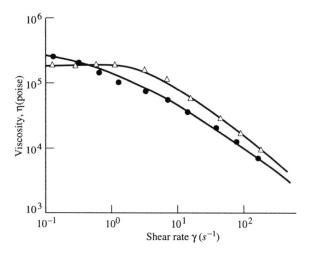

Fig. 1-25   Effect of molecular weight distribution on flow curves for polystyrene; narrow distribution has open symbols; broad distribution has solid symbols. [ ].

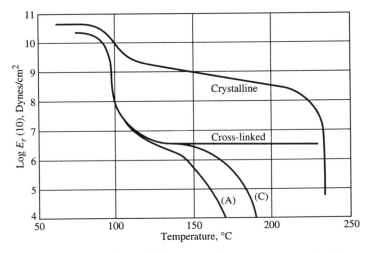

Fig. 1-26   Effect of cross-linking on modulus-temperature behavior [36].

ence of cross-linking affects the modulus-temperature behavior as shown in Fig. 1-26.

1.9   POLARITY

Molecules containing strongly polar groups give rise to large secondary forces. This is shown in Table 1-8, which lists relative molar cohesions for various polymers.

TABLE 1-8   Relative Polymer Molar Cohesions

Polymer	Mer Group	Relative Molar Cohesion per 5 A˙ Length
Polyethylene	$-CH_2-CH_2-$	1.0
Polyisobutylene	$-CH_2-\underset{\underset{CH_3}{\mid}}{\overset{\overset{CH_3}{\mid}}{C}}-$	1.2
Polyvinyl chloride	$-CH_2-CHCl-$	2.6
Polystyrene	$-\underset{\underset{C_6H_5}{\mid}}{CH}-CH_2-$	4.0
Polyamides	$-\overset{\overset{O}{\parallel}}{C}-(CH_2)_x-\overset{\overset{O}{\parallel}}{C}-NH(CH_2)_y-NH-$	5.8

Source: Refs. 37, 38.

As can be seen, polyethylene, which is made up of essentially nonpolar units ($-CH_2-$), has a much lower relative cohesion than polyamides, which have strongly polar mers

$$-\overset{\overset{\displaystyle O}{\|}}{C}-(CH_2)_x \qquad -\overset{\overset{\displaystyle O}{\|}}{C}-NH-(CH_2)_y-NH-.$$

The larger secondary forces so developed are then reflected in the properties.

Polarity is generally measured by using some thermodynamic property as an index. For example, the polymer's melting point can be used since this is a measure of molar cohesion. Another useful index is the polymer's cohesive energy. This function is defined in terms of cohesive energy density (C.E.D.):

$$C.E.D. = \frac{\Delta H_v - RT_v}{V_v} \tag{1-12}$$

where H_v is the heat of vaporization, T_v the vaporization temperature, and V_v the molar volume at vaporization.

Values of C.E.D. cannot be directly determined for polymers. It is, however, possible to get effective measurements by swelling or solubility studies. These involve testing a given polymer in a series of solvents whose solubility parameter δ is known:

$$\delta = (C.E.D.)^{1/2} \tag{1-13}$$

The peak value for swelling or solubility is taken as the δ value for the polymer. In this regard, see Fig. 1-27, which shows such a determination for natural

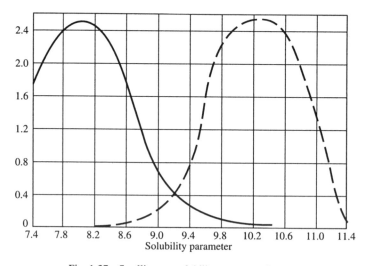

Fig. 1-27   Swelling vs. solubility parameter [42,43].

rubber and a synthetic rubber (Monsanto's N-5400 Vyram rubber). The δ values are 8.0 for natural rubber and ~10.3 for the synthetic.

Table 1-9 gives solubility parameter values for a group of polymers.

1.10   FLEXIBILITY

The most basic approach to measuring polymer chain flexibility is to use the concept of a completely flexible chain molecule and then introduce the ways in which structures can restrict rotation and hence alter flexibility. This, however, presents some problems. The first is that poor results are obtained unless all the ways that structures can reduce flexibility are considered.

For example, polyacrylonitrile, which is theoretically 100% flexible, is actually stiff because of the steric hindrance of its large nitrile groups. Likewise, polyamides, also theoretically 100% flexible, are quite stiff because of hydrogen bonding between the chains.

The stiffness of the chain can be obtained by measuring a value for $\bar{r}^2$ (which is a mean square ends to distance for a chain) by angular light scattering of a dilute polymer solution or simultaneous stress and birefringence measurements on a solid polymer in its rubbery state. The stiffness z is then

$$z = \left(\frac{\bar{r}^2}{N l_0^2}\right)^{1/2} \tag{1-14}$$

where N is the number of links in the polymer chain and l_0 is the length of each link.

Another useful method for measuring flexibility is to use the material's sonic modulus. The sample used should be unoriented and measured below the glass temperature. If this is done, the chain stiffness will correlate with sonic velocity; that is, the greater the sonic velocity, the greater the chain stiffness.

TABLE 1-9   Polymer Solubility Parameters

Polymer	Solubility Parameter δ
Polyethylene	7.88
Polyisobutylene	8.06
Polystyrene	8.56
Polyvinyl acetate	9.38
66 Nylon	13.6

Source: Refs. 39–42.

1.11 CRYSTALLINITY

The question of crystallinity in polymers can be a confusing one. This is best understood by comparing polymers to such crystalline materials as metals or even smaller organic molecules. These latter materials can be obtained in forms that are essentially almost 100% crystalline. Polymers, on the other hand, can range from purely amorphous, that is, with no order, to materials with varying degrees of semicrystallinity (up to a maximum of about 90%). In addition, crystals of organics such as benzene are very large, macroscopic, with respect to molecular size. The polymer, however, has regions of crystallinity that are much smaller. Although these may be large in terms of molecular size, they are nonetheless small compared to ordinary organic crystals.

Crystallinity in polymers, involving as it does significant amorphous regions, causes difficulties in establishing an appropriate description. The earliest mechanism proposed was the fringed micelle theory (see Fig. 1-28), which postulated that a polymer chain could pass through both amorphous and crystalline regions, that is, ordered and disordered areas. Later studies indicated that it was possible to form a polymer microcrystal by folding a single polymer chain in such a manner that a folded lamella type of structure was formed (see Fig. 1-29). This would seem to be a more sophisticated approach. However, it is difficult to relate this completely to the semicrystalline nature of polymers. Efforts to do this are covered in other sources [48,49]. At present, it can be said that no completely accepted definitive model for polymer ordered–disordered (crystalline–amorphous) regions exists.

Although insight into polymer crystallinity is not completely satisfactory, there are certain factors affecting it that are generally known and recognized. For example, the effect of such parameters as branching, stereoisomerism (tac-

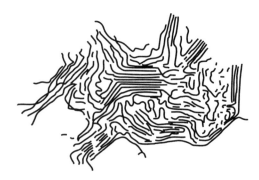

Fig. 1-28   Schematic diagram of the fringed micelle model of polymer amorphous-crystalline structure [44,45].

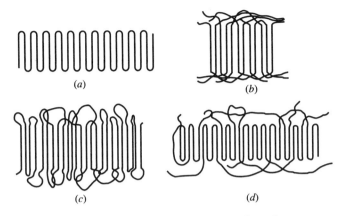

Fig. 1-29   Various folded chain models [46,47].

ticity), cross-linking, and copolymerization can be described at least qualitatively.

Branching can, in certain instances, disrupt crystallinity. This is especially true for short, irregular branches. However, if the side chains are over 5 to 14 atoms in length, then crystallization of these side chains is favored. Very long side chains have little influence because such branches tend to resemble entire polymers. For this last case, only the branch points and ends are disruptive.

Cross-linking can also cause local imperfections in crystalline structure. Still another factor that can affect polymer crystallinity is copolymerization. For example, the homopolymers polyvinylidene fluoride,

$$- (CH_2 \; CF_2)_n^- \tag{1-15}$$

and polytrifluorochloroethylene,

$$- [CF_2 \; CF \; (Cl)]_n^- \tag{1-16}$$

are both highly crystalline. Yet, if a random copolymer is synthesized around a 1:1 ratio, we obtain an essentially completely amorphous material. A partial thermodynamic explanation [50] of the effect of branching, cross-linking, and copolymerization is furnished by the expression

$$\left(\frac{1}{T_m} - \frac{1}{T_m^\circ}\right) = \frac{R}{\Delta H_u} \ln X_A \tag{1-17}$$

where T_m° is the melting point of the pure homopolymer; T_m the melting point of the polymer with branching, cross-linking, or copolymerization; R the gas constant; ΔH_u the heat of fusion per repeating unit; and X_A the mole fraction of crystallizable units. Actually, a number of refinements have to be made to use the equation in certain cases. There are described in Refs. 51 and 52.

The other factor that can seriously affect crystallinity is stereoisomerism, which occurs in polymers that contain asymmetric carbon atoms, that is, carbon chain atoms having both a hydrogen atom and another group R:

$$
\begin{array}{c}
\text{H} \\
| \\
\text{—C—} \\
| \\
\text{R}
\end{array}
\qquad (1\text{-}18)
$$

In such cases, the positioning of the hydrogen atom or R group affects molecular packing and, consequently, crystallinity.

Figure 1-30 illustrates the three cases that can occur: 1) isotactic, where all the R groups line up on one side of the chain; (2) syndiotactic, where R's and H's alternate in some regular fashion; and (3) atactic, where R and H positioning is completely random. In the foregoing cases, isotactic is the most crystalline, atactic is essentially amorphous, and syndiotactic somewhere between the other two.

We can illustrate the effect of tacticity by considering density. The more crystalline the polymer, the higher its density. As shown in Table 1-10, distinct

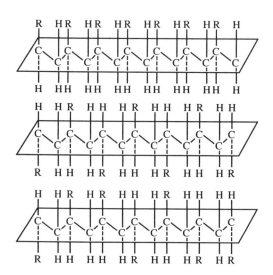

Fig. 1-30   Tacticity in polymers: (a) isotactic, (b) syndiotactic, and (c) atactic [53,54].

TABLE 1-10   Densities of Crystalline and Amorphous Material

Polymer	Crystalline, g/cm^3	Amorphous, g/cm^3
Polyethylene	1.000	0.852
Nylon 66	1.24	1.09
Polypropylene	0.936 (isotactic)	0.85

values of density occur for both purely crystalline and amorphous material. A particularly good example for tacticity is polymethyl methacrylate, which has densities of 1.22 for isotactic, 1.19 for syndiotactic, and 1.188 for atactic polymer.

The crystallines in polymers can form aggregates known as *spherulites*. The spherulites represent an ordered structure and can range in size from submicron to a few hundredths of a millimeter. These aggregates can be easily seen with cross polarizers under a light microscope. The spherulites are spherical in shape and have the aspect of Maltese crosses (see Fig. 1-31). In many instances, the size of the spherulites influences polymer properties.

A number of techniques exist for measuring crystallinity in polymers. These include X-ray diagrams, infrared techniques, density, calorimetric and differential thermal analysis, and nuclear magnetic resonance. The X-ray diffraction pattern is a widely used technique for determining polymer crystallinity. There are significant differences between amorphous and semicrystalline polymers (Figs. 1-32 and 1-33). In essence, the semicrystalline polymer has sharp diffraction rings as a result of the crystalline material. Although X-ray diffraction patterns can give excellent results, they require, a careful analysis by a highly skilled scientist. Additionally, of course, an excellent X-ray machine must be available, and such devices are rather expensive.

The infrared technique is also popular. It involves measuring a sample and then comparing it on a linear scale to values for amorphous and crystalline polymers. Sample preparation is difficult and time-consuming. Also, a highly skilled analyst is required.

In contrast to the X-ray and infrared techniques, the measurement of density is an economical, rapid, and rather precise method of measuring the degree of crystallinity of a polymer. Such densities can be readily measured in a density gradient tube. This device can be set up in a graduated cylinder by mixing together two fluids so that the resultant liquid density varies from top to bottom. Calibrated beads of known density will then yield a calibration curve for the tube (density vs scale reading). Polymer samples (pellets, powders, fibers, film bits, etc.) can then be dropped into the tube and their densities read. Degree or

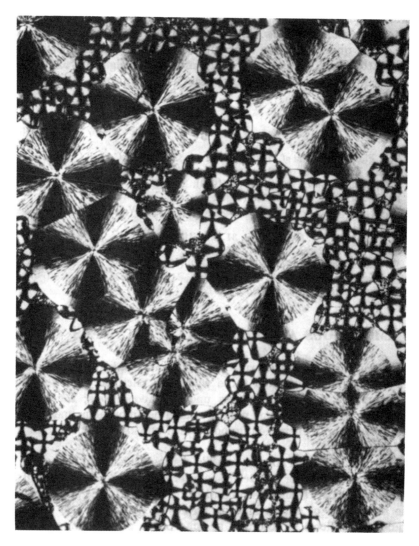

Fig. 1-31   Optical photograph with crossed polarizers of spherulites [55].

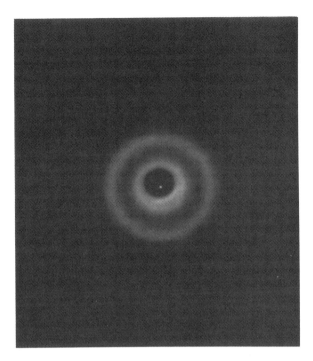

Fig. 1-32   X-ray diffraction pattern for an amorphous polymer [56].

percent crystallinity is then

$$\% \text{ crystallinity} = \left(\frac{\rho_{sample} - \rho_{amorphous}}{\rho_{crystalline} - \rho_{amorphous}} \right) \times 100 \qquad (1\text{-}19)$$

The calorimetric or differential thermal analysis methods indicate the amount of crystallinity by the size of the area associated with the peak that occurs in the scans (see Fig. 1-34 and 1-35). These areas can be compared to those for a polymer with known crystallinity. The technique is rapid and quite precise. It does, however, require a first-class analytical device, which is not inexpensive.

Nuclear magnetic resonance spectroscopy is a highly sophisticated analytical technique that can also be used to measure polymer crystallinity. Figure 1-36 shows proton resonance line width, and the second moment shows the crystalline behavior as an area.

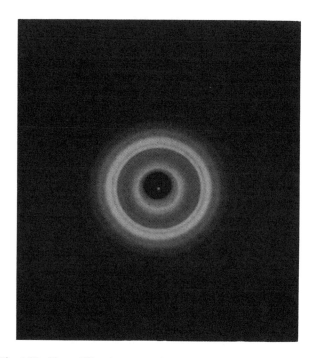

Fig. 1-33   X-ray diffraction pattern for a semicrystalline polymer [56].

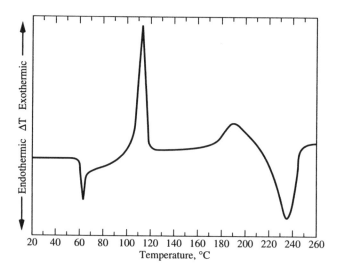

Fig. 1-34   Differential thermal analysis scan [68,59].

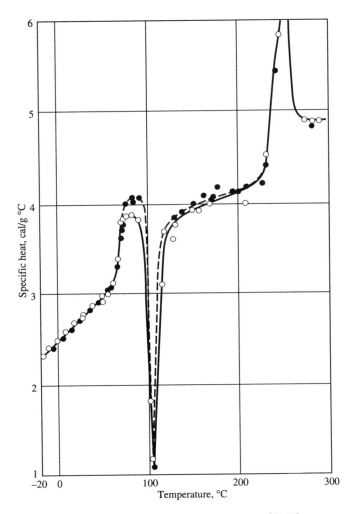

Fig. 1-35   Polymer specific heat vs temperature [57,58].

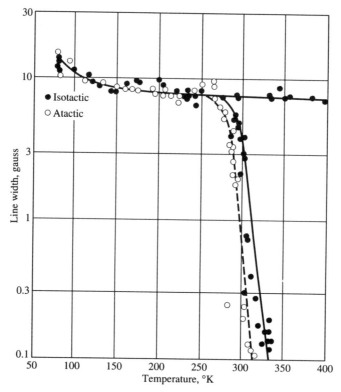

Fig. 1-36   Nuclear magnetic resonance determination of crystallinity [60,61].

Tacticity is determined principally by either nuclear magnetic resonance or infrared absorption. Figure 1-37 compares proton magnetic resonance spectra for three different tacticities of polymethyl methacrylate.

In Fig. 1-38, infrared absorption spectra are shown for two different tacticities.

1.12   ORIENTATION

Orientation represents the alignment of the polymer molecules. Although usually associated with crystalline structures, orientation can and does occur in amorphous materials. If mechanical stress is applied, either the amorphous polymer molecule or the crystallite will be oriented. It is possible for orientation to bring

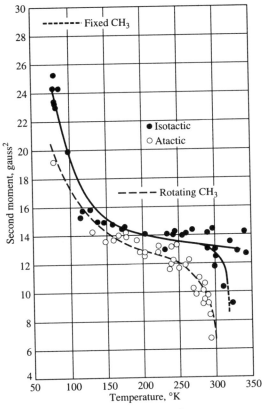

Fig. 1-36   (*Continued*)

about crystallization. This occurs when crystalline polymers have been melted into an amorphous state and then are oriented.

There are three principal methods for measuring orientation in polymers. One technique is to use X-ray diffraction patterns for crystalline polymers. As can be seen in Fig. 1-39, the oriented crystallites give distinctive characteristics. The technique suffers from the same problems as does X-ray crystallinity measurement: cost, time, and the requirement of a highly skilled analyst.

Orientation can also be determined by birefringence. This is done by measuring with an optical polarized microscope the refractive indices of the polymer in the directions parallel and perpendicular to the orientation. Then,

$$\text{Birefringence} = \Delta n = r_x - r_y \tag{1-20}$$

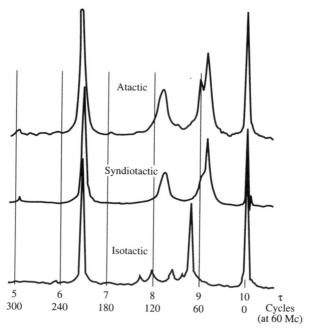

Fig. 1-37   Proton magnetic resonance for polymethyl methacrylate stereoisomers in CDCl [ ].

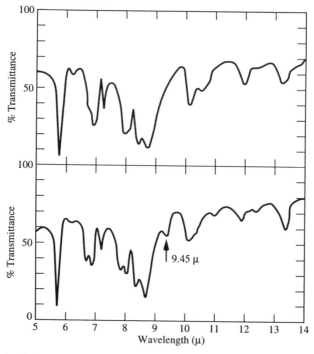

Fig. 1-38   Infrared absorption spectra (polymethyl methacrylate): top, isotactic; bottom, atactic [9].

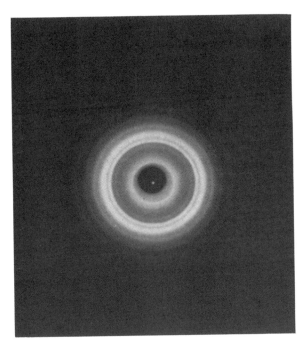

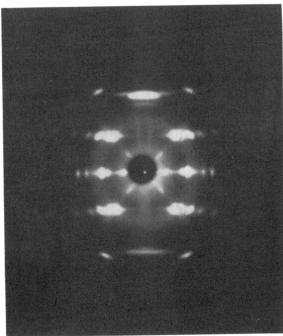

Fig. 1-39   Comparison of (a) unoriented and (b) oriented Polymers [62].

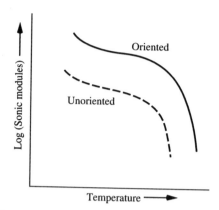

Fig. 1-40   Relation of sonic modulus to orientation.

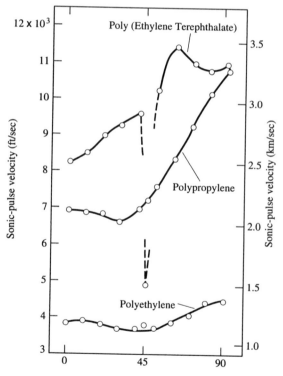

Fig. 1-41   Sonic velocity for three polymer films plotted against orientation. The poly(ethylene terephthalate) film had a double ($\theta = 0$, $90°$) orientation, the polypropylene film had only transverse ($\theta = 90°$ orientation), and the polyethylene film had a light transverse orientation [64].

where r_x = the refractive index in the direction of orientation and r_y the refractive index perpendicular to orientation. Unfortunately, the procedure is extremely time-consuming and is also limited to small samples.

Sonic modulus measurements can also give orientation data (see Figs. 1-40, 1-41).

The third method for determining orientation is to measure the sonic modulus. This is done by transmitting a sound wave through the polymer sample. Elastic modulus E is related to the density ρ, Poisson ratio W, and the sound wave's velocity v by

$$E = \pi (1 - W^2)v \qquad (1\text{-}21)$$

REFERENCES

1. SCHMIDT, A.K., and MARLIES, C.A., *Principles of High Polymer Theory and Practice*, McGraw-Hill, New york (1948), pp. 27, 33, 34.
2. GILMAN, H., *Organic Chemistry*, Wiley, New York (1943).
3. PAULING, L, *Nature of the Chemical Bond*, Cornell Univ. Press, Ithaca, NY, (1939).
4. Margenau, H., *Rev. Mod. Phys.* **11**, 1 (1939).
5. HUGGINS, M.L., *J. Organic Chem.* **1**, 407 (1936).
6. WINDING, C.C., and HIATT, G.D., *Polymeric Materials*, McGraw-Hill, New York (1961), p. 66.
7. DEANIN, R.D., *Polymer Structure Properties and Application*, Cahners Books, Boston, (1972), p. 89.
8. NIELSEN, L.E., *Mechanical Properties of Polymers*, Reinhold, New York (1962), pp. 171–172.
9. MEARES, P., *Polymers Structure and Bulk Properties*, Van Nostrand, New York (1965), pp. 261–263.
10. BILLMEYER, F.W. JR., *Textbook of Polymer Science*, 2nd Ed. Wiley-Interscience, New York (1971), p. 85.
11. Ibid., p. 86.
12. EWART, R.H., "Significance of Viscosity Measurements on Dilute Solution of Polymers," *Advances in Colloid Science, Vol. II*, H. MARK and G.S. WHITBY, eds., Interscience, New York (1946), pp. 197–251
13. BILLMEYER, *Textbook of Polymer Science*, p. 87.
14. FLORY, P.J., *J. Am. Chem. Soc.* **65** 372 (1943).
15. KRIGBAUM, W.R., and FLORY, P.J., *J. Am. Chem. Soc.* **75**, 1775, (1953).
16. BILLMEYER, *Textbook of Polymer Science* p. 71.
17. Ibid., p. 82.
18. Ibid., p. 74.

19. GEE, G., *Trans. Faraday Soc.* **40**, 261 (1944).

19A. Dobry, *J. Chem. Phys.* **32**, 46 (1935).

20. SCHMIDT AND MARLIES, *Principles*, p. 254.

21. STEIN, R.S., and DOTY, P., *J. Am. Chem. Soc.* **68**, 159 (1946).

22. BILLMEYER, *Textbook of Polymer Science*, p. 91.

23. SCHMIDT AND MARLIES, *Principles*, p. 236.

24. GOLDING, B., *Polymers and Resins*, Van Nostrand, New York (1959), p. 86.

25. SCHMIDT AND MARLIES, *Principles*, p. 7.

26. BILLMEYER, *Textbook of Polymer Science*, p. 52.

27. CROUZET, P., FINE, F., and MAGNIN, P., *J. Appl. Polymer Sci.* **13**, 205 (1969).

28. GRISKEY, R.G., and FOK, S.Y., *Tech. Doc. Report* No. ASD-TDR-62-879 (1962).

29. BILLMEYER, F.W. JR., *J. Paint Technology* **41**, 3 (1969).

30. BILLMEYER, *Textbook of Polymer Science*, p. 54.

31. Ibid., p. 55.

32. GRUBISIC, Z., REMPP P., and BENOIT, H., *J. Polymer Sci.* **B5**, 753 (1967).

33. BILLMEYER, *Textbook of Polymer Science*, p. 56.

34. MAY, J.A., and SMITH, W.B., *J. Phys. Chem* **72**, 216 (1968).

35. SCHMIDT AND MARLIES, *Principles*, p. 77.

36. TOBOLSKY, A.V., *Properties and Structure of Polymers*, Wiley, New York (1960), p. 75.

37. SCHMIDT AND MARLIES, *Principles*, p. 46.

38. MARK, H., *Ind. Eng. Chem.* **34**, 1343 (1942).

39. TOBOLSKY, *Properties and Structures*, p. 66.

40. BURRELL, H., *Interchem. Rev.* **14**, 3 (1955).

41. BRISTOW, G.M., and WATSON, W.F., *Trans. Faraday Soc.* **54**, (1951).

42. WILT, M.H., ''Use of Cohesive Energy Density,'' Monsanto Chemical Co., Tech. Bull, St. Louis, MO (July 15, 1955).

43. TOBOLSKY, *Properties and Structures*, p. 65.

44. BILLMEYER, *Textbook of Polymer Science*, p. 160.

45. BRYANT, W.M.D., *J. Polymer Sci.* **2**, 547, (1947).

46. BILLMEYER, *Textbook of Polymer Science*, p. 159.

47. INGRAM, P., and PETERLIN, A., *Encyclopedia of Polymer Science and Technology*, H. MARK, N.G. GAYLORD, and N. BIKALIS, ed., Vol. 9, Wiley, New York (1968), pp. 204–274.

48. GEIL, P.H., *Polymer Single Crystals*, Wiley-Interscience, New York (1963).

49. PETERLIN, A., *Man Made Fibers, Science and Technology*, H. MARK and E. CERNIA ed., Wiley, New York (1967), pp. 283–340.

50. FLORY, P.J., *Principles of Polymer Chemistry*, Cornell Univ. Press, Ithaca, NY (1953), pp. 565–576.

51. WUNDERLICH, B., and Bodily, *J. Polymer Sci.* **A2**, **4**, 25 (1966).

52. GRISKEY, R.G., and FOSTER, G.N., *J. Polymer Sci.* **A1**, **8**, 1623.

53. BILLMEYER, *Textbook of Polymer Science*, p. 142.

54. NATTA, G., and CORRADINI, P., *J. Polymer Sci.* **39**, 29 (1959).

55. DEANIN, *Polymer Structure*, p. 353.

56. TOBOLSKY, *Properties and Structures*, pp. 45, 46.

57. SMITH, C.W., and DALE, M., *J. Polymer Sci.* **20**, 37 (1965).
58. BILLMEYER, *Textbook of Polymer Science*, pp. 122, 123.
59. KE, B., *Organic Analysis, Vol. IV*, J. MITCHELL, JR., I.M. KOLTHOFF, E.S. PROS-KAUER, and A. WEISSBERGER, eds. Interscience, New York (1960).
60. BILLMEYER, F.W., JR. *Textbook of Polymer Science*, 1st Ed., Interscience, New York (1962), pp. 101, 102.
61. MANDELL, E.R., and SCHLICHTER, W.P., *J. Appl. Phys.*, 1438 (1958).
62. BILLMEYER, *Textbook of Polymer Science*, 2nd Ed., p. 179.
63. MCCALL, D.W., and SCHLICHTER, W.P., in *Newer Methods of Characterization*, B. Ke, ed., Wiley, New York (1964).
64. MEARES, *Polymers Structure.*
65. PRICE, H.L., *SPE Journal* **24**, No. 2, 54 (1968).
66. SAMUELS, R.J., *J. Polymer Sci* **20C**, 253 (1967).
67. TOBOLSKY, *Properties and Structures*, p. 73.
68. ROSATO, D.V. and ROSATO, D.V., *Injection Molding Handbook*, p. 727. Van Nostrand Reinhold, New York (1986).

CHAPTER 2

THERMODYNAMICS OF POLYMER SYSTEMS SOLID, MOLTEN, AND THERMALLY SOFTENED

2.1 INTRODUCTION

Polymers are generally processed with large changes in temperature and pressure. Because of this, it becomes important to determine as accurately as possible the energy demands for a given process. Such determinations are best made by applying the science of thermodynamics. This subject treats the behavior of systems (in equilibrium states) with their surroundings.

The thermodynamic properties needed for the application of the science emanate from the first two laws of thermodynamics. In word form, the first law is: "Energy can change form but be neither created nor destroyed". For a nonflow static system that is not elevated, the first law is

$$dU = \bar{d}Q - \bar{d}W \tag{2-1}$$

where dU is the internal energy change of the system, dQ the heat transferred to the system, and dW the work performed by the system. The crosses on the work and heat terms indicate that they are inexact differentials. If the system is not at rest or if it is elevated, then the first law is

$$dE = \bar{d}Q - \bar{d}W \tag{2-2}$$

Here dE includes kinetic and potential energy as well as internal energy.

For a system involving expansion work (i.e., *P-V* work), we can define a system property enthalpy (*H*) as

$$H = U + PV \tag{2-3}$$

where *V* is the specific volume.
Furthermore,

$$C_p = \left(\frac{\partial H}{\partial T}\right)_p \tag{2-4}$$

where C_p is the specific heat at constant pressure.
Hence, for a constant-pressure process,

$$dH = C_p \, dT \tag{2-5}$$

The specific heat associated with constant volume is defined as

$$C_v = \left(\frac{\partial U}{\partial T}\right)_v \tag{2-6}$$

Also,

$$C_p - C_v = \frac{T\epsilon^2}{\rho\beta} \tag{2-7}$$

where ϵ is the coefficient of thermal expansion, defined as

$$\epsilon = -\frac{1}{\rho}\left(\frac{\partial \rho}{\partial T}\right)_p = \frac{1}{V}\left(\frac{\partial V}{\partial T}\right)_p \tag{2-8}$$

The β is compressibility, defined as

$$\beta = \frac{1}{\rho}\left(\frac{\partial \rho}{\partial P}\right)_T = \frac{1}{V}\left[\left(\frac{\partial V}{\partial P}\right)_T\right] \tag{2-9}$$

so that the change of enthalpy at constant temperature with pressure is

$$dH = \left[V - T\left(\frac{\partial V}{\partial T}\right)_p\right] dP \tag{2-10}$$

For a steady-state flow system

$$\Delta H + \Delta \text{ P.E. } + \Delta \text{ K.E. } = Q - W_s \tag{2-11}$$

where Δ P.E. and Δ K.E. are the changes in potential and kinetic energy. W_s is shaft work, that is, work realized in actual mechanical performance.

The second law of thermodynamics can be stated as: It is impossible to convert a given quantity of heat completely into work. The property associated with the second law is entropy. For a nonflow system

$$dS \geqq \frac{dQ}{T} \tag{2-12}$$

where T is the absolute temperature. Likewise, for a flow system,

$$dS \geqq \Sigma \frac{dQ}{T} + (S_{\text{in}} - S_{\text{out}}) \tag{2-13}$$

Other properties used in thermodynamics include the Gibbs and Helmholtz free energies G, A. These are defined as

$$G = H - TS \tag{2-14}$$

and

$$A = U - TS \tag{2-15}$$

The entropy change at constant pressure is given by

$$\left(\frac{\partial S}{\partial T}\right)_p = \frac{C_p}{T} \tag{2-16}$$

and at constant pressure

$$dS = \frac{C_p}{T} \, dT \tag{2-17}$$

Likewise, the entropy change at constant temperature with changing pressure is

$$\left(\frac{\partial S}{\partial P}\right)_T = - \left(\frac{\partial V}{\partial T}\right)_p \tag{2-18}$$

and

$$dS = - \left(\frac{\partial V}{\partial T}\right)_p dp \qquad (2\text{-}19)$$

From the preceding discussion, it is apparent that any calculation of thermodynamic properties such as enthalpy or entropy requires two basic pieces of information, namely, pressure-volume-temperature (*P-V-T*) data and calorimetric (C_p) data. It is, of course, also possible to use appropriate equations of state and calorimetric equations relating C_p to temperature for these purposes.

Next, sources of these needed data, as well as the appropriate relations, will be discussed in detail.

2.2 POLYMER PRESSURE-VOLUME-TEMPERATURE DATA

The volumetric behavior of polymers with temperature and pressure is an important aspect of many polymer processing operations. This means that a knowledge of such behavior is important to the polymer engineer and scientist.

A search of the literature shows a number of available publications of such data. However, as with any area, these can vary widely in quality, and caution must be exercised in their use and application. A problem of considerable magnitude is that many of the published measurements were made almost dynamically. This meant that equilibrium was not even approached. The result is that the data cannot really be used since they do not represent correct *P-V-T* information. It therefore becomes important for the user to verify completely that any published data represent equilibrium values.

Published pressure-volume-temperature data are summarized in Table 2-1, together with ranges of temperature and pressure.

Again, the potential user is cautioned to check original references carefully as to methods of measurement. This will enable the determination of whether the data truly represent equilibrium thermodynamics data.

In addition to the foregoing, mention should also be made of the pioneering work of Bridgman [19], who made rough compression measurements of a number of commercial plastics, and of Kovacs [20], who presented coefficients of thermal expansion for a number of polymers.

Some plots of acutal pressure-volume-temperature data for various polymers are given in Figs. 2-1–2-7. Before commenting on the behavior shown, we must give some consideration to the nature of the polymers and the transition points encountered.

First of all, solid polymers can be broadly classified into two general areas by morphology: amorphous and semicrystalline. The latter types exhibit a melt-

TABLE 2-1   Polymer Pressure-Volume-Temperature Data

Polymer	Temperature Range, °C	Pressure Range, $N/(m)^2 \times 10^{-5}$	Reference
Polyethylene (ρ_0 = 0.917 g/cm³)	0–185	1–2,000	1–3
Polyethylene (ρ_0 = 0.920 g/cm³)	25–178	1–1,200	4
Polyethylene (ρ_0 = 0.920 g/cm³)	20–170.4	1–2,000	5
Polyethylene (ρ_0 = 0.920 g/cm³)	20–80	1–10,000	6
Polyethylene (ρ_0 = 0.930 g/cm³)	0–180	1–2,000	7
Polyethylene (ρ_0 = 0.958 g/cm³)	23–250	1–618	8
Polyethylene (ρ_0 = 0.973 g/cm³)	19–203	1–2,000	5
Polyethylene (ρ_0 = 0.976 g/cm³)	23–165.5	1–2,000	1–3
Ethylene-propylene Copolymer (ρ_0 = 0.933 g/cm³)	25–250	1–618	9
Polypropylene	23–250	1–618	10
Polyisobutylene	26–200	1–618	11
Polymethyl methacrylate	−60–150	1–1,000	12
Polymethyl methacrylate	20–139.3	1–2,000	5
Polymethyl methacrylate	25–175	1–1,160	4
Polyvinyl chloride	−60–100	1–1,000	12
Polyvinyl chloride	20–97	1–2,000	5
Polyvinyl chloride	−60–100	1–1,000	12
Polyvinyl chloride (plasticized with 20% diocytyl phthalate)	−50–65	1–1,000	12
Polyvinyl chloride (plasticized with 30% dioctyl phthalate)	−60–30	1–1,000	12
Polytetrafluoroethylene	10–80	1–10,000	6
Ethylene–tetrafluoroethylene copolymer	20–70	1–10,000	6
Polymonochlorotrifluoroethylene	20–80	1–10,000	6
Polymonochlorotrifluoroethylene	26–227	1–10,000	13
Polyvinylidene fluoride	20–70	1–10,000	6
Polyvinyl fluoride	20–70	1–10,000	6
Polyvinyl alcohol	20–80	1–10,000	6
Polyhexamethylene adipamide (66 nylon)	22–202	1–9,000	14

TABLE 2-1 *(Continued)*

Polymer	Temperature Range, °C	Pressure Range, $N/(m)^2 \times 10^{-5}$	Reference
Polyhexamethylene sebacamide	25–260	1–1,855	15
Polystyrene	25–178	1–12,000	13
Polystyrene	25–232	1–1,910	16
Polystyrene	20–248.9	1–2,000	5
Ethyl cellulose	25–196	1–1,190	4
Cellulose acetate butyrate	25–181	1–1,190	4
Polyester ($\rho_0 = 1.194$ g/cm³)	20–50	1–10,000	6
Polyester ($\rho_0 = 1.177$ g/cm³)	20–80	1–1,800	17
Polyester ($\rho_0 = 1.196$ g/cm³)	20–75	1–1,800	
Polyester ($\rho_0 = 1.22$ g/cm³)	20–80	1–1,800	
Polyester ($\rho_0 = 1.25$ g/cm³)	20–62	1–200	
Polyurethane ($\rho_0 = 1.16$ g/cm³)	20–260	1–1,800	
Epoxy (59.5% dodecenyl succinic anhydride hardener)	23–182	1–12,000	
Epoxy (12.6% meta-phenylene diamine hardener)	26–205	1–12,000	
Epoxy (13% tris (dimethylaminomethyl) phenol tri-2-ethyl hexaote hardener)	24–193	1–12,000	
Polysiloxane	23–165	1–10,000	
Polyester (cross-linked)	24–160	1–10,000	

ing point and actually form melts. The former, however, do not melt but only thermally soften. Better insight can be obtained from Fig. 2-8, where specific heat behavior is shown for an amorphous and a semicrystalline polymer. As can be seen, the semicrystalline material shows a definite peak at its melting point, whereas the amorphous material gradually increases.

Similarly, the semicrystalline material will undergo a significant volume

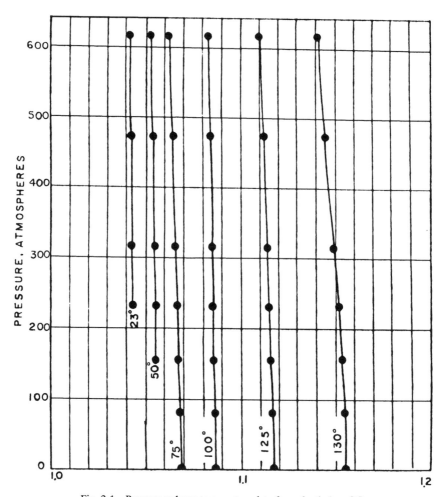

Fig. 2-1   Pressure-volume-temperature data for polyethylene [8]

change at constant temperature for a given pressure. Such an occurrence is a first-order transition point. This is illustrated in Fig. 2-5 for nylon 6-10. The amorphous polymer, on the other hand, will not have a melting point and the attendant volume change (see, for example, the polyisobutylene data of Fig. 2-7).

All polymers do, however, undergo second-order transitions, which occur when the volume-temperature curve at a given pressure undergoes a discontinuity but not a volume change at constant temperature. The most important of these transitions is the glass transition temperature T_g. Figure 2-9 shows the occurrence of such a point for atactic polypropylene. Table 2-2 [21] lists T_g

HIGH DENSITY POLYETHYLENE

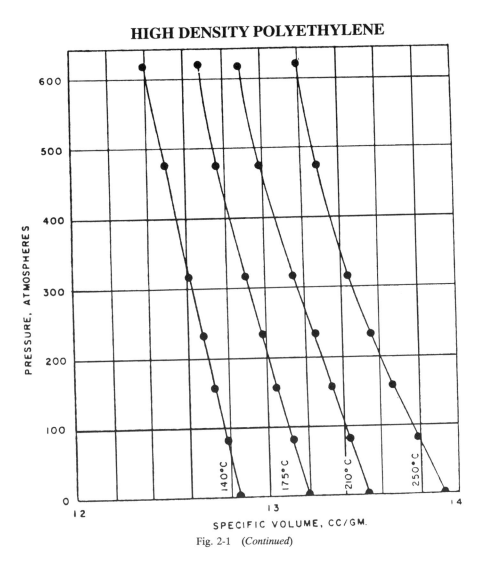

Fig. 2-1   (*Continued*)

values for various polymers at atmospheric pressure. Additionally, it has been shown that [22, 23]

$$0.56 < \frac{T_g}{T_{\text{melt}}} < 0.76 \qquad (2\text{-}20)$$

with the ratio for the majority being 0.66. Both temperatures are absolute values.

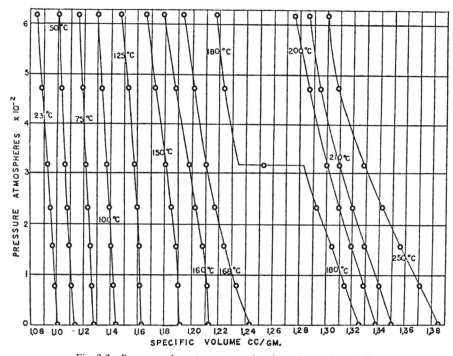

Fig. 2-2   Pressure-volume-temperature data for polypropylene [10].

The glass transition represents the point at which the polymer segments go from only vibrational motions (i.e., as "frozen" in the glassy state) to, at first, short-range and, later, long-range diffusional motion. This change is usually marked by a significant reduction in the modulus of elasticity of the polymer.

Other second-order transitions can and do occur in polymers. The data of Figs. 2-4 and 2-5 provide an example. Here the pressure-volume-temperature data show discontinuities for the solid polymer that cannot be attributed to the glass temperatures [15]. The behavior represents an unknown second-order transition. Incidentally, the shift of this discontinuity with pressure is about 0.016°C/atm compared to a general range of such shifts of 0.013–0.018°C/atm [12].

2.3 POLYMER EQUATIONS OF STATE

Actual experimental pressure-volume-temperature data represent the best means of supplying this needed information. However, such data are not always available and also cannot be easily determined. The result is that appropriate equa-

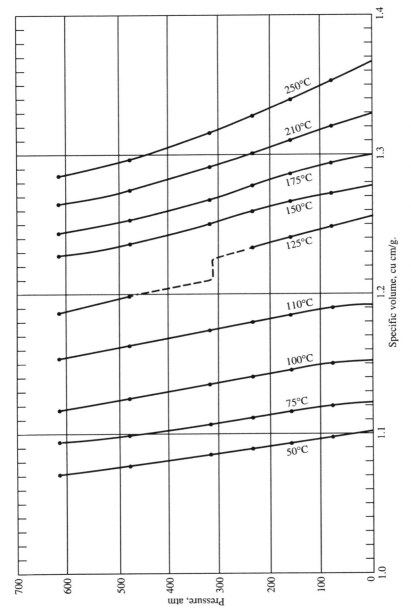

Fig. 2-3 Pressure-volume-temperature data for ethylene–propylene copolymer [9].

57

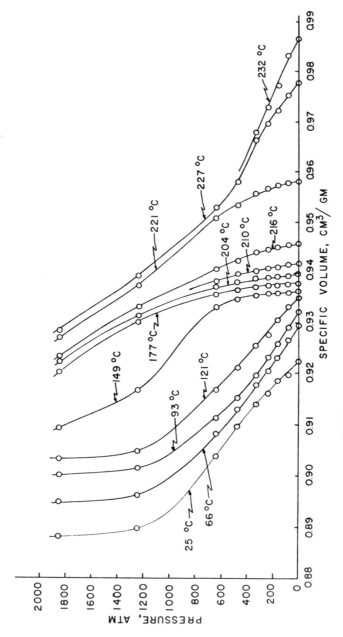

Fig. 2-4 Pressure-volume-temperature data for nylon 610 [15].

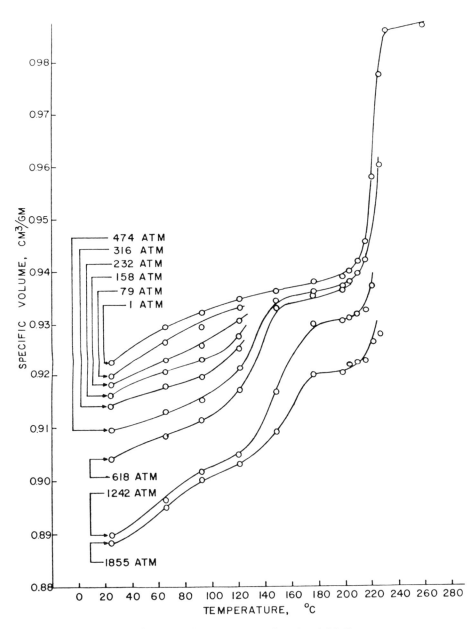

Fig. 2-5 Volume temperature for nylon 610 [15].

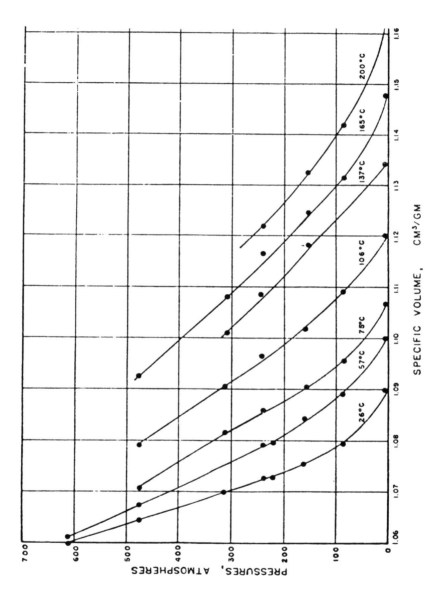

SPECIFIC VOLUME, CM³/GM

PRESSURES, ATMOSPHERES

Fig. 2-6 Pressure-volume-temperature data for polyisobutylene [11].

60

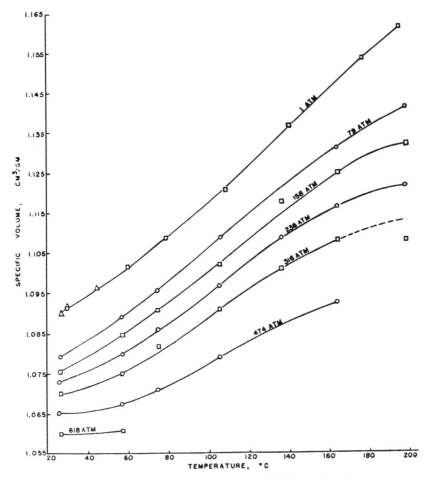

Fig. 2-7 Volume vs temperature for polyisobutylene [11].

tions of state must be developed. Incidentially, such equations also make it possible to interpolate existing data properly.

Any suitable equation of state should meet certain criteria. In the case of polymers, these are:

1. The equation must be applicable over wide ranges of temperature and pressure.
2. If possible, the equation should be generalized. In other words, the equation should apply to a large number of different polymers rather than one particular polymer such as polyethylene.

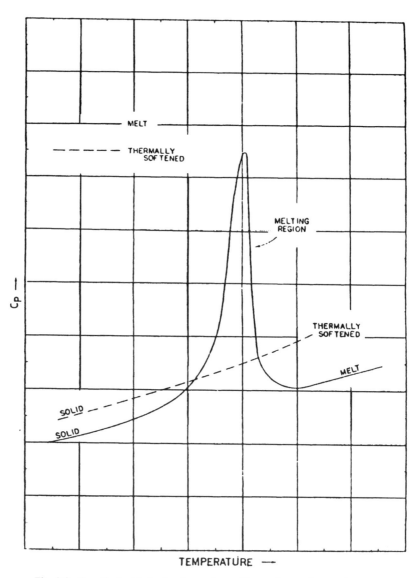

Fig. 2-8   Specific heat behavior for semicrystalline and amorphous polymers.

3. The equations should be simple and easily used. Complex forms requiring excessive mathematical manipulation should be avoided.
4. The equation should be such that it can be used to predict the behavior of polymers for which no pressure-volume-temperature data are available. Ideally, we

ASPECTS OF POLYMER PHYSICS

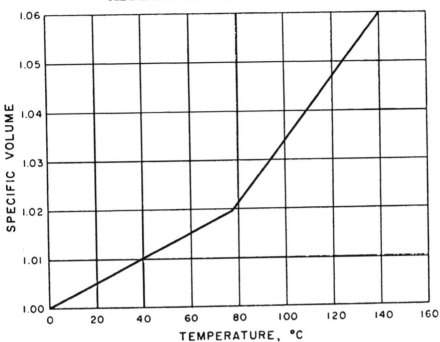

Fig. 2-9   Glass transition point [84].

should be able to calculate pressure-volume-temperature data for a polymer where only easily measured physical properties are known, such as glass temperature at atmospheric pressure, melting temperature at atmospheric pressure, or density at room temperature and atmospheric pressure.

A number of equations of state, (6), (16), and (24–37), have been developed for polymers. These are summarized in Table 2-3 with respect to the criteria listed, the basis of the given equation, and applicability.

Whereas all the equations have their merits it is recommended that the two equations based on corresponding states theory be used. The reason is that these represent generalized equations that depend only on a set of common parameters and on easily accessible data (polymer density at room temperature and atmospheric pressure, glass temperature at atmospheric pressure).

TABLE 2-2   **Experimental Values of** T_g

Polymer	$T_g,°K$ exp.
Polypropylene	253
Polystyrene	373
Polyisobutene	198
Polyvinyl fluoride	253/313
Polyvinyl chloride	356
Polyvinylidene fluoride	233/286
Polyvinylidene chloride	255/288
Poly 1,2-difluoroethylene	323/371
Polychlorotrifluoroethylene	318/325
Poly-α-methyl styrene	390/465
Polybutadiene, *cis*	171
Polybutadiene, *trans*	259
Polymethylene oxide	190/243
Polyethylene oxide	206
Polytrimethylene oxide	195
Polytetramethylene oxide	185/194
Polyparaxylylene disulphide	296
Polydecamethylene tetrasulphide	197
Polyethylene adipate	203/223
Polyethylene dodecate	202
Polydecamethylene adipate	217
Polyethylene terephthalate	343/350
Polydecamethylene terephthalate	268/298
Polydiethyleneglycol malonate	244
Polydiethyleneglycol octadecanedioate	205
Polymetaphenylene isophthalate	428
Poly 4,4'-methylene diphenylene carbonate	393
Poly 4,4'-isopropylidene diphenylene carbonate	414/423
Poly 4,4'-tetramethylene dibenzoic anhydride	319
Poly 4,4'-methylenedioxy dibenzoic anhydride	357
Polyhexamethylene adipamide	318/330
Polydecamethylene sebacamide	333
Polyheptamethylene terephthalamide	396
Polyparaphenylene diethylene sebacamide	378
Polytetramethylene hexamethylene diurethante	303
Polyphenylene dimethylene hexamethylene diurethane	329
Polyhexamethylene dodecamethylene diurea	322
Polymethylene *bi*-5 (oxydiparaphenylene) sulphone	453
Polyoxy *bis*-(oxydiparaphenylene) ketone	423
Polyisopropylidene diparaphenylene metaphenylene disulphonate	388
Poly 3,5-dimethyl paraphenylene oxide	453/515
Polydimethyl siloxane	150

TABLE 2-3 Comparison of Polymer Equations of State

Equation	Basis of Equation	General?	Simple, Easy to Use?	Reference	Applicable at High Pressure	Applicable at High Temperature	Requires Experimental pressure-volume-Temperature Data	Applies to Solid(s) or Molten (M)
Flory et al.	Quantum mechanics	No	No	24	No	?	Yes	S
DiBenedetto	Quantum mechanics	No	No	25	No	Yes	Yes	S
Spencer–Gilmore	Modified Van der Waal equation	No	Yes	16	Yes	Yes	Yes	S
Tait	Semiempirical	No	Yes	26,27	Yes	Yes	Yes	S
Murnaghan	Semiempirical	No	Yes	28	Yes	Yes	Yes	S
Birch	Semiempirical	No	Yes	29	Yes	Yes	Yes	S
Weir	Virial equation of state	No	Yes	6,30	Yes	?	Yes	S
Griskey–Whitaker	Corresponding state theory	Yes	Yes	31	Yes	Yes	No	S
Smith	Hirai–Eyring equation	No	Yes	32	Yes	Yes	Yes	S,M
McGowan	Semiempirical	No	Yes	33,34	Yes	Yes	Yes	S,M
Kamal	Virial equation of state	No	Yes	35	Yes	Yes	Yes	S,M
Griskey–Rao	Corresponding state theory	Yes	Yes	36	Yes	Yes	No	M
Olabisi–Simha	Corresponding state theory	Yes	No	37	Yes	Yes	Yes	S,M

More specifically, for amorphous and solid semicrystalline polymers, the equation of Ref. 31,

$$V = \left[\frac{(0.01205)}{(\rho_0)^{0.9421}} \right] (P)^{n-1} \left(\frac{T}{T_g} \right)^{m+1} R \tag{2-21}$$

is recommended. The volume here is in cm^3/g, pressure P in atm, the temperatures in $°K$, T_g the glass temperature at atmospheric pressure, R the gas constant in atm cm^3 $°K$ g mole, and ρ_0 the density in gm/cm^3 at room temperature and atmospheric pressure. The n and m terms are functions of pressure (see Figs. 2-10 and 2-11).

Likewise, for molten semicrystalline polymers, the relation of Ref. 36 is recommended.

$$V = K \left(\frac{T}{T_g} \right)^x P^y \tag{2-22}$$

All terms are the same as for Eq. 2-21. The x and y are functions of pressure (Figs. 2-12 and 13), while K is a function of ρ_0 (Fig. 2-14).

Both Eqs. (2-21) and (2-22) generally predict polymer pressure-volume-temperature to within 1% or less. Again, the relative ease of use of Eqs. (2-21) and (2-22) should be re-emphasized. All that is needed are T_g ρ_0 and Figs. 2-10 –2-14. Hence, it is possible rapidly to estimate pressure-volume-temperature data for a given polymer. Furthermore, the glass temperature can be estimated for any polymer from one point of pressure-volume-temperature data.

The volume change due to melting for semicrystalline polymers can be estimated [21] from the equation

$$\Delta V_m = 0.19 \frac{T_m}{298} V_w \tag{2-23}$$

where ΔV_m is the change in molar volume on melting, T_m the melting temperature in $°K$, and V_w the Van der Waals volume in $cm^3/mole$. Table 2-4 lists the mer weights (used to calculate moles) and V_w values for various polymers.

2.4 POLYMER CALORIMETRIC DATA

Polymer calorimetric data are a necessary ingredient not only for determination of thermodynamic properties but also, ultimately, for design purposes. It is also important, however, to provide appropriate correlations for estimating such data where experimental points are limited or unavailable.

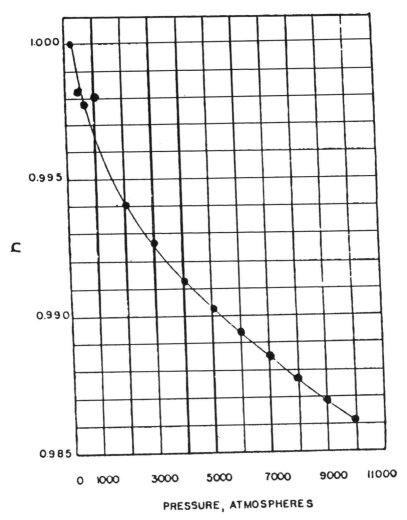

Fig. 2-10   Function n for Eq. (2-22) vs pressure [31].

The following sections will summarize existing sources of data, as well as appropriate correlations that can be used for calorimetric data.

Table 2-5 summarizes sources of data for polymers in the existing literature.

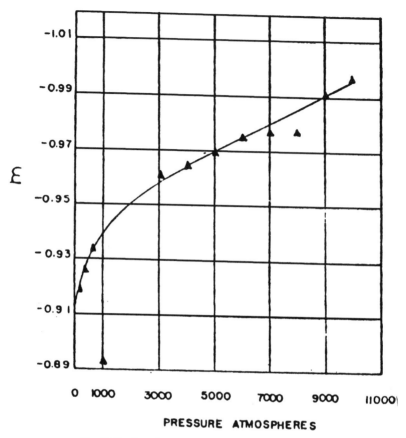

Fig. 2-11   Function *m* for Eq. (2-21) vs pressure [31].

2.5   CORRELATIONS FOR POLYMER CALORIMETRIC AND LATENT HEAT DATA

Whenever possible, actual experimental calorimetric data should be used. Frequently, however, this cannot be done. In such situations, it becomes necessary to estimate polymer calorimetric data from appropriate correlations. These will be discussed in some detail.

Van Krevelen and Hoftyzer [21] have proposed empirical relations for the specific heat behavior of both solid (crystalline) and liquid (amorphous) polymers. These relations are:

$$\frac{C_p^s\,(T)}{C_p^s\,(298\ \text{K})} = 0.106 + 0.003\ T \tag{2-24}$$

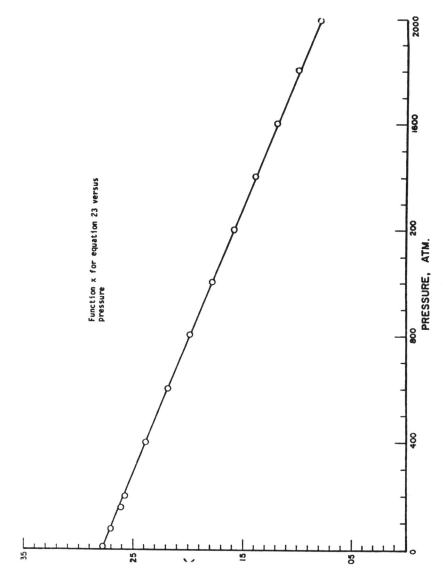

Fig. 2-12   Function x vs pressure [36].

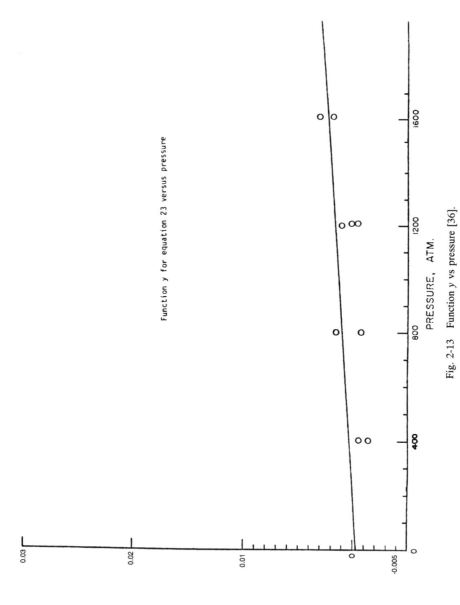

Fig. 2-13 Function y vs pressure [36].

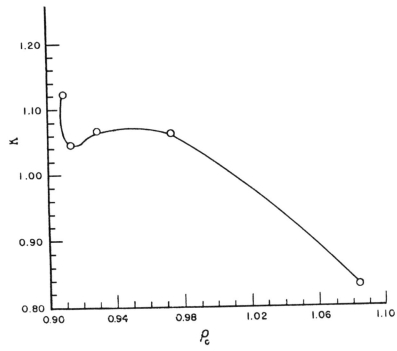

Fig. 2-14   Function K for Eq. (2-23) vs density at 25°C [36].

and

$$\frac{C_p^l\,(T)}{C_p^l\,(298\text{ K})} = 0.64 + 0.0012\,T \qquad (2\text{-}25)$$

where C_p^s (298 K) and C_p^l (298 K) are the C_p values at 298 K, calculated from the group contributions of the various organic groups to the polymer mer, and T is the temperature in °K. The C_p^s (298 K) and C_p^l (298 K) values are computed, respectively, from the studies of Shaw [84] and Satoh [85].

Specific heat values for amorphous polymers are computed directly from Eq. (25). Semicrystalline polymers are handled by a weighting procedure.

$$(C_p)_{\text{ semicrystalline}} = (C_p^s)x + (C_p^l)(1 - x) \qquad (2\text{-}26)$$

where x is degree of crystallinity.

Values of C_p^s (298 K) and C_p^l (298 K) are listed in Table 2-6 for a variety of polymers.

TABLE 2-4   Van der Waals Volumes

Polymer	M, g/mol	V_w cm³/mol
Polyethylene	28.0	20.46
Polypropylene	42.1	30.68
Poly (1-butene)	56.1	40.91
Poly (1-pentene)	70.1	51.14
Polyisobutene	56.1	40.90
Poly (4-methyl-1-pentene)	84.2	61.36
Polystyrene	104.1	62.85
Polybutadiene	54.1	37.40
Polyisoprene	68.1	47.61
Polychloroprene	88.5	45.56
Polyvinyl chloride	62.5	28.63
Polyvinylidene fluoride	64.0	25.56
Polychlorotrifluoroethylene	116.5	32.95
Polyformaldehyde	30.0	13.93
Polyethylene oxide	44.1	24.16
Polytetramethylene oxide	72.1	44.62
Polyacetaldehyde	44.1	24.15
Polypropylene oxide	58.1	34.38
Polyepichlorohydrin	92.5	42.56
Polyvinyl alcohol	44.0	25.05
Polyvinyl acetate	86.1	45.88
Polymethyl acrylate	86.1	45.88
Polyethyl	100.1	56.11
Polyisopropyl acrylate	114.2	66.33
Polybutyl acrylate	128.2	76.57
Poly-sec. butyl acrylate	128.2	76.56
Poly (2,2 dimethylpropyl acrylate)	142.2	86.78
Poly (1-ethylpropyl acrylate)	142.2	86.79
Polymethyl mechacrylate	100.1	56.10
Polyethyl methacrylate	114.1	66.33
Polypropyl methacrylate	128.2	76.56
Polyisopropyl methacrylate	128.2	76.55
Polybutyl methacrylate	142.2	86.79
Polyisobutyl methacrylate	142.2	86.78
Poly-sec. butyl methacrylate	142.2	86.78
Poly-tert. butyl methacrylate	142.2	86.77
Polyhexyl methacrylate	170.3	107.25
Poly-(2-ethylbutyl methacrylate)	170.3	107.24
Polyoctyl methacrylate	198.4	127.71
Polydodecyl methacrylate	254.4	168.63
Poly-(2-methoxyethyl methacrylate)	144.2	80.26
Poly-(2-propoxyethyl methacrylate)	172.2	100.72
Polycyclohexyl methacrylate	168.2	99.23
Polyethylene terephthalate	192.2	94.18
Polydecamethylene terephthalate	304.4	176.02
Polyethylene phthalate	192.2	94.18

TABLE 2-4 *(Continued)*

Polymer	M, g/mol	V_w cm³/mol
Polyethylene (2.6-naphthalate)	242.2	119.86
Polyethylene (2.7-naphthalate)	242.2	119.86
Polybisphenol carbonate	254.3	135.71
Nylon 6	113.2	70.71
Nylon 7	127.2	80.94
Nylon 8	141.2	91.17
Nylon 9	155.2	101.40
Nylon 10	169.3	111.63
Nylon 11	183.3	121.86
Nylon 12	197.3	132.09
Nylon 10,9	324.5	213.03
Nylon 10,10	338.5	223.26

Additional correlations have been developed from amorphous [67] and molten polymers [46]. Griskey and Hubbell [69] showed that C_p data for methacrylic polymers could be correlated using a form of the Sakiadis–Coates equation originally developed for organic liquids. Additionally, it has been shown [46] that the C_p data for molten polyolefins could be correlated as shown in Fig. 2-15.

Van Krevelen and Hoftyzer [21] give a rough approximation for latent heat of fusion:

$$\frac{\Delta H_m}{C_p^l \ (298 \ \text{K})} = 0.55 \ (T_M - T_g) \qquad (2\text{-}27)$$

where ΔH_m is the latent heat of fusion in cal/mole, C_p^l (298 K) is the quantity defined earlier, and T_M and T_g are, respectively, the melting and glass temperatures in °K.

A somewhat better value of ΔH_m can be estimated from the use of group contributions [21] Table 2-7 lists such contributions for some typical polymer mer constituents.

2.6 CALCULATIONS FOR POLYMER CALORIMETRIC DATA

Example 2-1

Estimate the latent heat of fusion for polypropylene.

TABLE 2-5   Polymer Calorimetric Data

Polymer	Temperature Range, °C	Data for Solid (S) or Molten (M)	Reference
Polyethylene (~55% crystalline)	22–167	S,M	38
Polyethylene (ρ_0 = 0.912, 0.917 g/cm³)	−23–197	S,M	39
Polyethylene (~92% and 82% crystalline)	−23–177	S,M	40
Polyethylene (~90% and 60% crystalline)	27–157	S,M	41
Polyethylene (ρ_0 of 0.917, 0.95, 0.96 g/cm³)	27–167	S,M	42
Polyethylene (ρ_0 of 0.924, 0.968 g/cm³)	−183–157	S,M	43
Polyethylene (~65% and 85% crystalline)	−143–177	S,M	44
Polyethylene (ρ_0 = 0.952 g/cm³)	25–230	S,M	45,46
Polyethylene (~99% crystalline)	−93–137	S	47
Polyethylene (ρ_0 = 0.960 g/cm³)	27–187	S,M	48
Polyethylene	127–197	S,M	49
Polypropylene (ρ_0 = 0.9103, 0.8667 g/cm³)	25–210	S,M	50
Polypropylene (~65% crystalline)	−180–177	S	51
Polypropylene (ρ_0 = 0.9050, 0.913 g/cm³)	30–180	S,M	42
Polypropylene (ρ_0 = 0.9063 g/cm³)	25–230	S,M	52
Poly(1-butene)	−53–158	S	53
Poly(4-methyl-1-pentene)	−193–267	S,M	54
Polyisobutylene	−259–107	S	55
Polystyrene (atactic)	−223–127	S	56
Polystyrene (atactic, isotactic)	−173–187	S	57
Polystyrene (atactic)	73–157	S	44
Polystyrene (atactic, isotactic)	−13–267	S	58
Polystyrene (atactic)	47–67	S	41
Polystyrene (atactic)	43–147	S	42
Polybutadine (*cis*-1,4; *trans*-1,4)	−253–67	S	59
Polyisoprene (amorphous; *cis*-1,4)	−258–147	S	60
Polyoxymethylene	27–227	S,M	48
Polyoxymethylene	27–177	S	42
Polyoxymethylene	77–192	S	61
Polyoxyethylene	−183–87	S	62
Polyoxyethylene	27–202	S,M	63
Polyoxypropylene	−193–87	S,M	62
Polytetrahyrofuran (69% crystalline)	−193–87	S,M	64
Poly (2,6 dimethyl-1,4 phemylenee ether)	−193–297	S,M	65
Poly [3,3-*bis* (chloromethyl) oxacyclobutane]	−253–33	S	66
Polymethyl methacrylate	−43–127	S	67
Polymethyl methacrylate	27–197	S	44
Polymethyl methacrylate	27 to 187	S	68
Polymethyl methacrylate	127–300	S	69
Poly (dimethylamenoethyl) methacrylate	120–300	S	69
Poly (cyclohexyl) methacrylate	120–300	S	69
Poly (allyl) methacrylate	120–300	S	69
Polyethylacrylate	120–300	S	69

TABLE 2-5 *(Continued)*

Polymer	Temperature Range, °	Data for Solid (S) or Molten (M)	Reference
Polytetrafluoroethylene	−23–127	S	70
Polytetrafluoroethylene	0–440	S,M	71
Polyvinyl chloride	−150–157	S	44
Polyvinyl chloride	−223–157	S	72
Polycaprolactam	−13–267	S	73
Polyhexamethylene adipamide	27–277	S,M	74
Polyhexamethylene sebacide	27–277	S,M	74
Polyhexamethylene sebacide	220–290	M	75
Nylon 7 [mer:$(CH_2)_6(CONH)$]	−203–67	S	76
Polymethane [(mer: $OCONH(CH_2)_4NHCOO(CH_2)_4$ and mer $OCONH(CH_2)_6NHCOO(CH_2)_2$ $0(CH_2)_2)$]	−53–107	S	77
Propene polysulfone	−253–27	S	78
1-butene polysulfone	−173–27	S	78
1-butene polysulfone	−253–27	S	78
Polycarbonate (amorphous)	−173–287	S	79
Polycarbonate (amorphous)	27–177	S	42
Polyvinyl acetate	−153–177	S	44
Polyethylene terephthalate	−13–277	S,M	80
Polyethylene sebacate	−23–137	S,M	81
Polychlorotrifluoroethylene	−3–247	S,M	82
Polyvinylidene chloride	−213–27	S	83

The polypropylene mer is

$$-CH_2-CH-$$
$$|$$
$$CH_3$$

From Table 2-7,

Group	Group Contribution (cal/mole)
$-CH_2-$	900
$-CH-$	900
$-CH_3$	600
(Total)	2,400

We arrive at an estimated value of 2400 cal/mole, compared to an actual value of 2100–2600.

TABLE 2-6 C_p^s and C_p^l **Data for Various Polymers**

Polymer	C_p^s (298), cal/mo °K	C_p^l (298), cal/mo; °K
Polyethylene	12.1	14.5
Polypropylene	17.15	21.1
Polybutene	23.2	28.3
Polyisobutene	22.3	26.6
Poly(4-methylpentene)	34.3	42.1
Polyisoprene	26.4	32.2
Polystyrene	30.2	41.7
Polymethylene oxide	10.1	15.8
Polyethylene oxide	16.1	23.0
Polytetramethylene oxide	28.2	37.5
Polypropylene oxide	21.2	29.6
Polyvinyl chloride	16.2	(21.8)
Polychlorotrifluoroethylene	(24.7)	(28.0)
Polyvinylidene chloride	20.4	(28.0)
Polytetrafluoroethylene	(23.3)	(23.5)
Polychloroprene	25.5	(32.9)
Polyvinyl alcohol	13.8	23.0
Polyvinyl acetate	(28.2)	36.6
Polymethyl methacrylate	(33.3)	42.1
Polyacrylonitrile	(15.8)	—
Nylon 6	(39.2)	57.8
Nylon 6,6	(78.5)	115.6
Nylon 6,10	(102.7)	144.6
Polyethylene terephthalate	(52.9)	72.5
Polyethylene sebacate	(82.5)	103.6
Poly(*bis*-phenol-A carbonate)	(68.9)	(94.9)
Polypropylene sulphone	(29.2)	—
Polybutylene sulphone	(35.2)	—
Polyhexene sulphone	(47.3)	—
Poly(2,6-dimethylphenylene oxide)	34.3	48.3

( ) indicates most probable value.
Source: Ref. 21.

Example 2-2

Estimate the specific heat of molten 66 nylon at 280°C.
For 66 nylon (Table 2-6), the C_p^l (298 K) is 115.6; then,

$$\frac{C_p^l \ (553 \ \text{K})}{C_p^l \ (298 \ \text{K})} = 0.64 + 0.0012 \ (553)$$

$$C_p^l \ (573 \ \text{K}) = 150.8 \ \text{cal/mole}$$

$$C_p^l \ (573 \ \text{K}) = 0.67 \ \text{cal/g}$$

The experimental value [74] is given as 0.75 cal/g.

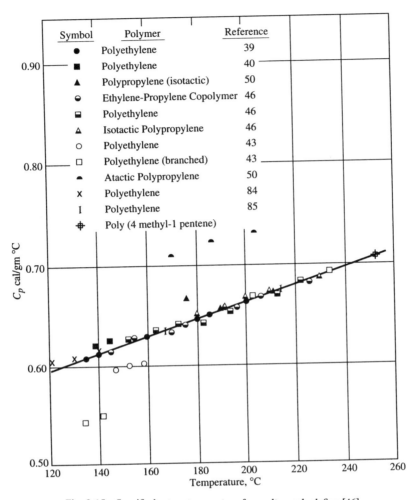

Fig. 2-15   Specific heat vs temperature for molten polyolefins [46].

Example 2-3

Estimate the specific heat of polystyrene at 200°C.
The C_p^s (298 K) for the polymer is 30.2 (Table 2-6). Then,

$$\frac{C_p^s \ (473 \ K)}{C_p^s \ (298 \ K)} = 0.106 + 0.003 \ (473 \ K)$$

$$C_p^s \ (473 \ K) = 46.9 \ \text{cal/mole}$$

TABLE 2-7   **Approximate Values for Group Contribution to the Heat of Fusion of Polymers (ΔH_m)**

Group	Group Contribution, cal/mol
—CH$_3$	600
—CH$_2$	900
—CH—	900
—CH=CH—	500
—C=CH—	500
—⟨⟩—	5300
—O—	400
—COO—	−1000
—CONH—	700

or

$$C_p^s \ (473 \text{ K}) = 0.445 \text{ cal/g}$$

Compared to an experimental value of 0.513 cal/g [58].

2.7 THERMODYNAMIC PROPERTIES FOR POLYMER SYSTEMS

The principal thermodynamic properties used in processing are enthalpy and entropy. Basically, these are functions of pressure and temperature that can be calculated from appropriate calorimetric and pressure-volume-temperature data. Another factor that must be considered is the state of the polymer (i.e., solid or molten).

For example, for a solid polymer, the enthalpy and entropy are given, respectively, by Eq. (2-28) and (2-29):

$$H_{T,P} - H_{T^*,P} = \int_{T^*}^{T} (C_p)_{\text{solid}} \, dT \tag{2-28}$$

$$S_{T,P} - S_{T^*,P} = \int_{T^*}^{T} \frac{(C_p)_{\text{solid}}}{T} \, dT \tag{2-29}$$

where H is enthalpy in Btu/lbm or cal/g, S entropy in Btu/°R lbm or cal/g °K, T temperature in absolute units (°K or °R), P pressure in psia or atm, C_p the specific heat at constant pressure (cal/g°K or Btu/lbm°R), and T^* a base temperature.

Equations (2-28) and 2-29) represent the changes in enthalpy and entropy (at constant pressure) from the base temperature T^* to any temperature T. In the calculation of these properties, the constant pressure is taken at atmospheric since most C_p data are available at that pressure. The base temperature T^* can be any number of different points (0°C, 0 K, etc.). It is set to some extent by the availability of C_p data. For example, a T^* lower than 0°C should not be chosen if specific heat data are not available below that temperature.

The end result of computing enthalpies and entropies from Eqs. (2-28) and (2-29) seem to yield data dependent on the T^* selected. However, it is important to notice that most computations involving enthalpy and entropy actually involve *changes* in these quantities. For example, suppose we want to determine the enthalpy change from a temperature T_1 to another higher temperature, T_2. Then,

$$\Delta H \text{ from } T_1 \text{ to } T_2 = (H_{T_2,P} - H_{T^*,P}) - (H_{T_1,P} - H_{T^*,P}) \tag{2-30}$$

$$\Delta H \text{ from } T_1 \text{ to } T_2 = H_{T_2,P} - H_{T_1,P} \tag{2-31}$$

Hence, the *change* in enthalpy or entropy, at constant pressure, with temperature is independent of the selected base temperature T^*.

A method frequently used to compute an enthalpy change is to solve the $C_p dT$ integral graphically. This is shown schematically in Fig. 2-16, where the area *ABCD* represents the enthalpy change. The same approach can be used for entropy by plotting C_p/T vs T and then evaluating the area under the curve.

As mentioned earlier, semicrystalline and amorphous polymers undergo somewhat different experiences as temperature is increased. The former polymers ultimately exhibit a melting point and form a true melt, whereas the latter

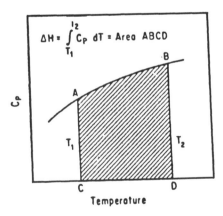

Fig. 2-16 Change of enthalpy with temperature.

soften but do not melt. The difference between the two types can be further seen in Fig. 2-8, which illustrates the C_p behavior for both semicrystalline and amorphous polymers. As can be seen, there is no discontinuity for the amorphous polymer. Hence, the solid polymer equations (2-28) and (2-29) for enthalpy and entropy are valid for high temperatures. On the other hand, the discontinuity (melting) for the semicrystalline polymer makes it necessary to change the form of the equations for these materials.

The altered form for semicrystalline polymers is shown in Eqs. (2-32) and (2-33):

$$(H_{T,P})_{\text{melt}} - H_{T^*,P} = \int_{T^*}^{T_{\text{fusion}}} (C_p)_{\text{solid}} \, dT + \Delta H_{\text{fusion}} + \int_{T_{\text{fusion}}}^{T} (C_p)_{\text{melt}} \, dT \quad (2\text{-}32)$$

$$(S_{T,P})_{\text{melt}} - S_{T^*,P} = \int_{T^*}^{T_{\text{fusion}}} \frac{(C_p)_{\text{solid}}}{T} \, dT + \frac{\Delta H_{\text{fusion}}}{T_{\text{fusion}}} + \int_{T_{\text{fusion}}}^{T} \frac{(C_p)_{\text{melt}}}{T} \, dT \quad (2\text{-}33)$$

The specific heats for the molten and solid cases are different in Eqs. (2-32) and (2-33). This can be seen by considering Fig. 2-8. Also, the heat of fusion of the material must be included. Basically, these equations involve (at constant pressure) taking material up to the melting point (T fusion), melting it (heat of fusion), and then moving from the melting point up to the final temperature.

Once again, as with the solid material, the choice of T^* becomes unimportant since the usual case involves determining changes in enthalpy or entropy. For example, a change in enthalpy of a melt from one temperature to another would be given by

$$\Delta H_{\text{melt}} \text{ from } T_1 \text{ to } T_2 = \left[\left(H_{T_2,P} \right)_{\text{melt}} - \left(H_{T^*,P} \right) \right] - \left[\left(H_{T_1,P} \right)_{\text{melt}} - \left(H_{T^*,P} \right) \right] \quad (2\text{-}34)$$

$$\Delta H_{\text{melt}} \text{ from } T_1 \text{ to } T_2 = \left(H_{T_2,P} \right)_{\text{melt}} - \left(H_{T_1,P} \right)_{\text{melt}} \quad (2\text{-}35)$$

There can also be a pressure effect on either enthalpy or entropy. Such effects are given by Eqs. (2-34) and (2-35):

$$H_{T,P} - H_{T,P^*} = \int_{P^*}^{P} \left[V - T \left\{ \frac{\partial V}{\partial T} \right\}_P \right] dP \quad (2\text{-}36)$$

$$S_{T,P} - S_{T,P^*} = - \int_{P^*}^{P} \left(\frac{\partial V}{\partial T} \right)_P dP \quad (2\text{-}37)$$

These equations are at some constant temperature T. Also the P^* represents a standard pressure (usually atmospheric). Hence, the bases usually taken for pressure and temperature are atmospheric pressure and 0 or 25°C.

The pressure correction can be applied for either a solid or molten polymer. The only stipulation is that the function $(\partial V/\partial T)_p$ or $[V - (\partial V/\partial T)_p]$ be determined in the appropriate state, that is, solid or melt.

Figures 2-17 and 2-18 show plots of both these functions vs pressure for molten high-density polyethylene. The pressure-volume-temperature data for these graphs were obtained from Refs. 5 and 8. Graphically integrating the appropriate curve gives the enthalpy or entropy correction due to pressure change. This correction applied to the value of enthalpy or entropy at atmospheric pressure will, in turn, yield the appropriate thermodynamic function at the pressure and temperature. Figures 2-19–2-29 give enthalpy and entropy data as functions of temperature and pressure for various polymers.

2.8 POLYMER THERMODYNAMIC PROPERTIES APPLIED TO POLYMER PROCESSING

Example 2-4

Molten high-density polyethylene is flowing through a tube. At one point, the average temperature of the polyethylene is 290°F, with a pressure of 1500 psia. A distance away, the pressure has dropped to atmospheric. If the tube is heavily insulated, what is the new polyethylene temperature?

For a steady-state flow system,

$$\Delta H + \Delta \text{P.E.} + \Delta \text{K.E.} = Q - W_s$$

Then, for the case given, ΔP.E., ΔK.E., W_s are all zero, and

$$\Delta H = Q$$

but, for heavy insulation, $Q \cong 0$ and

$$\Delta H = 0$$

Then, using Fig. 2-19 at the pressure of 1500 psia and 290°F, H is 275 Btu/lbm. Following a constant-enthalpy path to atmospheric pressure, we find that the temperature is 294.6°F.

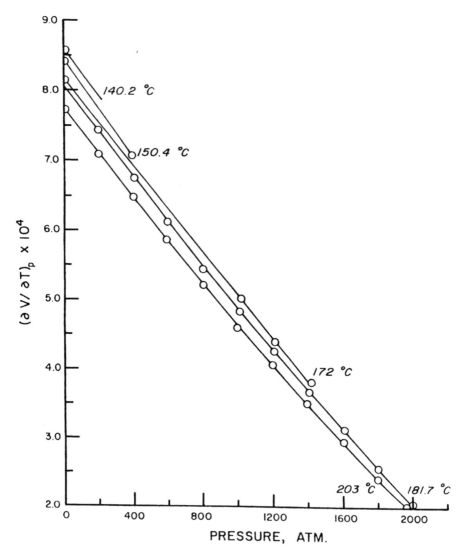

Fig. 2-17 $(\partial V/\partial T)_p$ vs pressure for polyethylene.

Example 2-5

Molten polypropylene is pumped through a screen pack with a pressure drop of 2500 psi at a flow rate of 20 lbm/min. The polymer enters the pack at 5000 psi and 400°F. What is the temperature change across the pack?

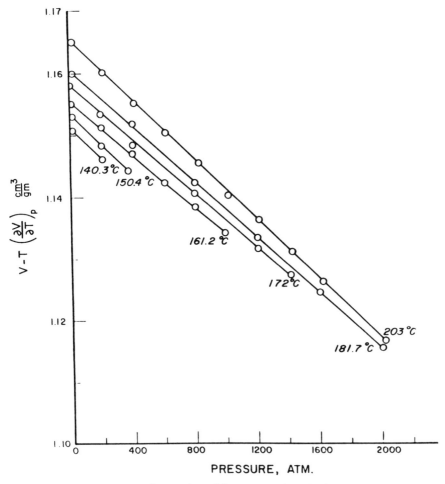

Fig. 2-18 $[V - T\,(\partial V/\partial T)_p]$ vs pressure for polyethylene.

Once again,

$$\Delta H + \Delta \text{P.E.} + \Delta \text{K.E.} = Q - W_s$$

Then, since ΔP.E. and ΔK.E. are negligible and W_s is zero,

$$\Delta H = Q$$

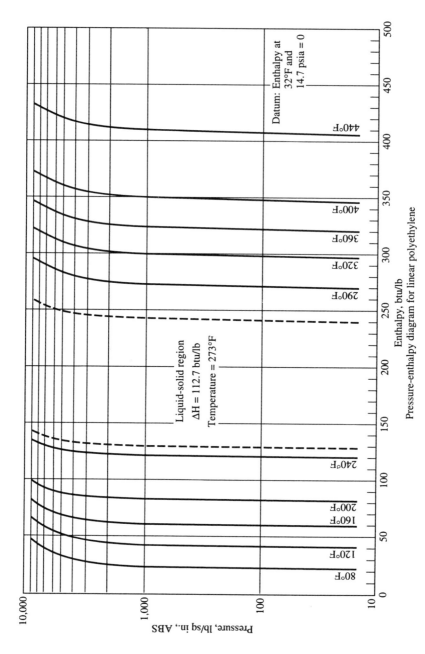

Fig. 2-19 Enthalpy and entropy data for polyethylene [45,52].

Pressure-enthalpy diagram for linear polyethylene

Enthalpy, btu/lb

Pressure, lb/sq in., ABS

Datum: Enthalpy at 32°F and 14.7 psia = 0

Liquid-solid region
ΔH = 112.7 btu/lb
Temperature = 273°F

440°F
400°F
360°F
320°F
290°F
240°F
200°F
160°F
120°F
80°F

84

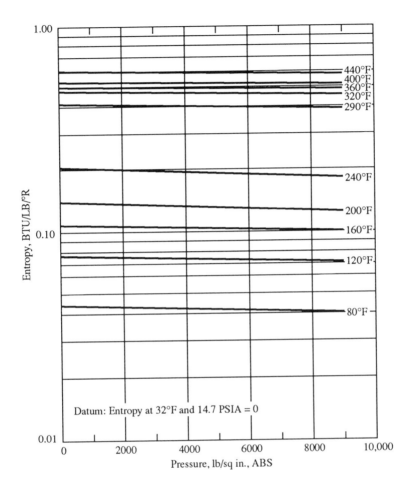

Fig. 2-20   Enthalpy and entropy data for polyethylene [45,52].

But, if the pack is heavily insulated, $(Q = 0)$ and

$$\Delta H = 0$$

From Fig. 2-21 at 5000 psi and a temperature of 400°F, the H value is 266 Btu/lbm. Following an isenthalpic path to 2500 psi, the temperature is found to be 411°F.

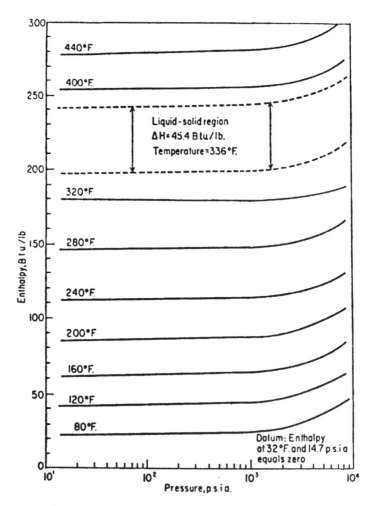

Fig. 2-21   Enthalpy and entropy data for polypropylene [85].

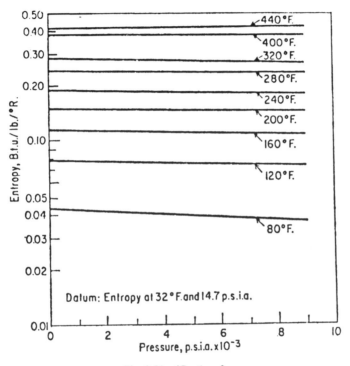

Fig. 2-21 (*Continued*)

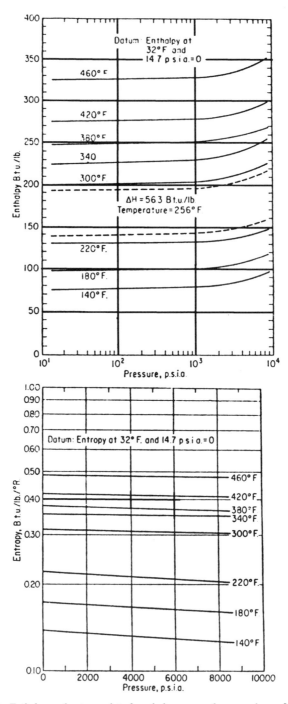

Fig. 2-22   Enthalpy and entropy data for ethylene–propylene copolymer [86].

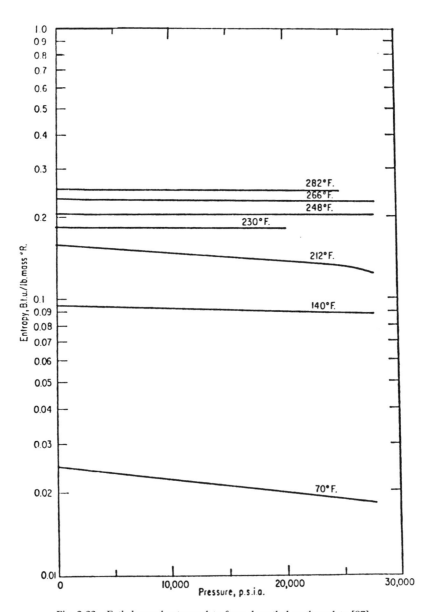

Fig. 2-23   Enthalpy and entropy data for polymethyl methacrylate [87].

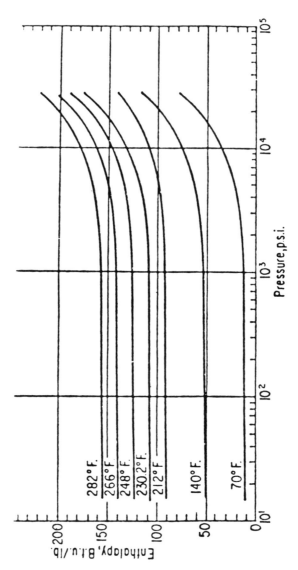

Fig. 2-23 *(Continued)*

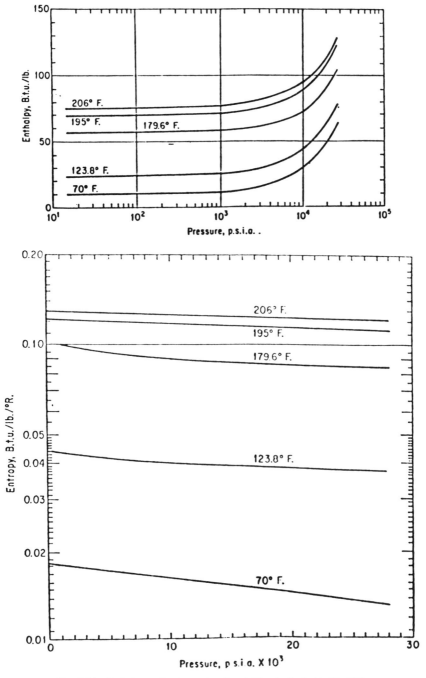

Fig. 2-24   Enthalpy and entropy data for rigid polyvinyl chloride [88].

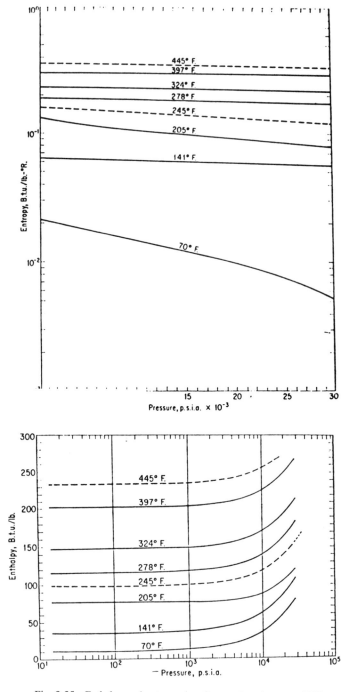

Fig. 2-25 Enthalpy and entropy data for atactic polystyrene [89].

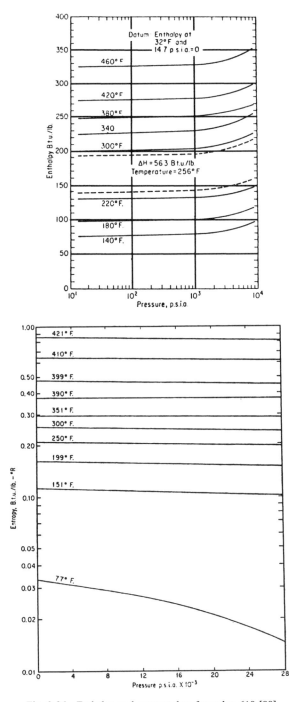

Fig. 2-26   Enthalpy and entropy data for nylon 610 [90].

93

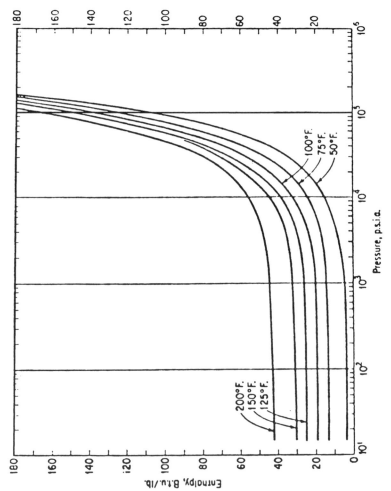

Fig. 2-27 Enthalpy and entropy data for polytetrafluoroethylene [91].

94

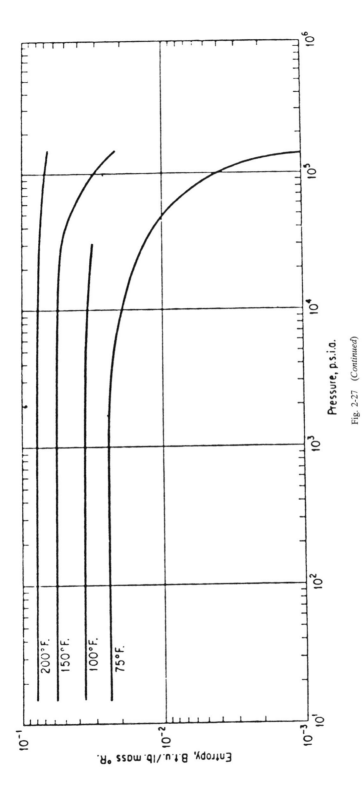

Pressure, p.s.i.a.

Fig. 2-27 *(Continued)*

95

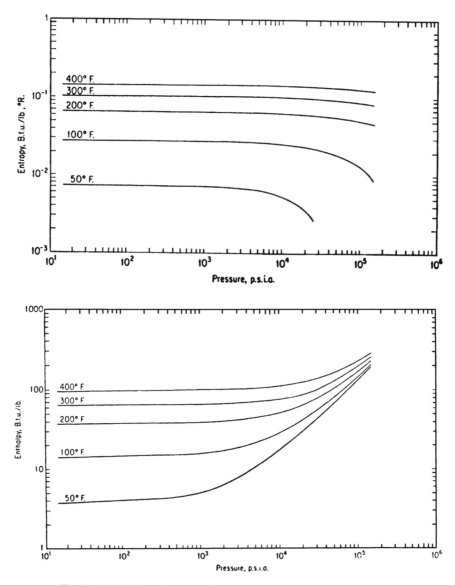

Fig. 2-28   Enthalpy and entropy data for polychlorotrifluoroethylene [92].

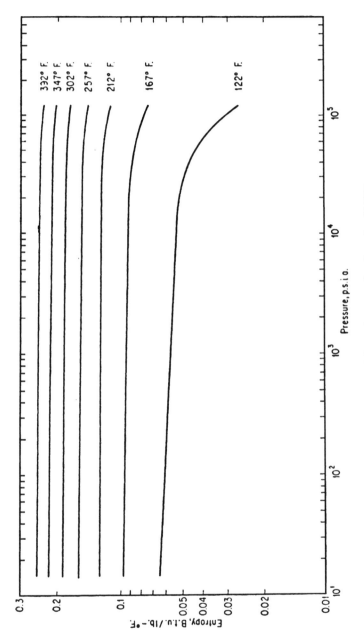

Fig. 2-29 Enthalpy and entropy data for nylon 66 [93].

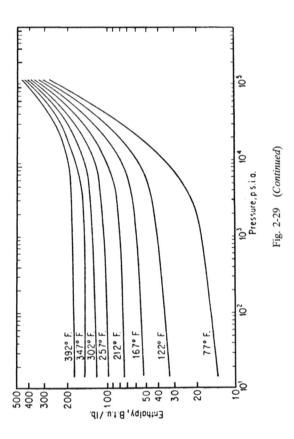

Fig. 2-29 *(Continued)*

REFERENCES

1. MATSUOKA, S., and MAXWELL, B., *J. Polymer Sci.* **32**, 131 (1958).
2. MATSUOKA, S., *Ibid.* **42**, 511 (1950).
3. *Ibid.* **57**, 567 (1962).
4. SPENCER, R.S., and GILMORE, G.D., *J. Appl Physics* **21**, 523 (1950).
5. HELLWEGE, K.H., KNAPPE, W., and LEHMANN, P., *Kolloid, Z.* **183**, 110 (1962).
6. WEIR, C.E., *J. Res. Nat. Bur. Std.* **53**, 245 (1954).
7. PARKS, W., and RICHARDS, R.B., *Trans. Faraday Soc.* **45**, 206 (1949).
8. GRISKEY, R.G., FOSTER, G.N., and WALDMAN, N., *J. Appl. Polymer Sci.* **10**, 201 (1966).
9. GRISKEY, R.G., FOSTER, G.N., and WALDMAN, N., *Modern Plastics* **43**, 245 (1966).
10. GRISKEY, R.G., FOSTER, G.N., and WALDMAN, N., *Polymer Eng. Sci.* **2**, 131 (1966).
11. GRISKEY, R.G., and GRINDSTAFF, D.A., *J. Appl. Polymer Sci.* **12**, 1986 (1968).
12. HEYDEMANN, P., and GUICKING, H.D., *Kolloid, Z.* **193**, 16 (1963).
13. WARFIELD, R.W., "The Compressibility of Bulk Polymers," NOLTR 64-84, U.S. Naval Ordnance Laboratory, White Oak, MD (1964).
14. WARFIELD, R.W., *Polymer Eng. Sci.* **2**, 176 (1966).
15. GRISKEY, R.G., and HAUG, W.A., *J. Appl. Polymer Sci.* **10**, 1475 (1966).
16. SPENCER, R.S., and GILMORE, G.D., *J. Appl. Physics* **20**, 502 (1949).
17. JENCKEL, E., and RINKENS, H., *Z. Elektrochem.* **60**, 971 (1956).
18. WARFIELD, R.W., "The Compressibility of Crosslinked Polymers," NOLTR 64-243, U.S. Naval Ordnance Laboratory, White Oak, MD (1965).
19. BRIDGMAN, P.W., *Proc. Am. Acad. Arts Sci.* **76**, 3, 55 (1948).
20. KOVACS, A.J., *Industrie des Plastiques Modernes* (Paris) **7**, 1, 30 (1955); **7**, 2, 39 (1955); **7**, 3, 36 (1955); **7**, 6, 41 (1955); **7**, 7, 44 (1955); **7**, 8, 41 (1955); **7**, 9, 37 (1955); **8**, 1, 37 (1956); **8**, 2, 38, 41 (1956).
21. VAN KREVELEN, D.W., and HOFTYZER, P.J., *Properties of Polymers*, Elsevier, New York (1972), p. 114.
22. BEAMAN, R.G., *J. Polymer Sci.* **9**, 472 (1953).
23. LEE, W.A., and KNIGHT, G.J., *Brit. Polymer J.* **2**, 73 (1970).
24. FLORY, P.J., ORWALL, R.A., and VRIJO, A., *J. Am. Chem. Soc.* **86**, 3507 (1963).
25. DiBENEDETTO, A.T., *J. Polymer Sci. A-1*, 3459 (1963).
26. NANDA, P., *J. Chem. Phys.* **11**, 3870 (1964).
27. McDONALD, J., *Rev. Mod. Phys.* **33**, 669 (1966).
28. ANDERSON, J., *J. Phys. Chem. Solids* **27**, 547 (1966).
29. KU, P.S., *Equations of State of Organic High Polymers*, AD 678 887, Jan. 1968, Washington, DC.
30. WEIR, C.E., *J. Res. Nat. Bur. Std.* **50**, 153 (1953).
31. GRISKEY, R.G., and WHITAKER, H.L., *J. Appl. Polymer Sci.* **11**, 1001 (1967).
32. SMITH, R.P., *J. Polymer Sci. A-2*, **8**, 1337 (1970).
33. McGOWAN, J.C., *Polymer* **10**, 841 (1969).
34. McGOWAN, J.C., *Polymer* **11**, 436 (1970).
35. KAMAL, M., personal communication.
36. GRISKEY, R.G., and RAO, K., *J. Appl. Polymer Sci.* **17**, 3293 (1973).

37. OLABISI, O. and SIMHA, R., *J. Appl. Polymer Sci.* (in press).
38. RAINE, H.C., RICHARDS, R.B., and RYDER, H., *Trans. Faraday Soc.* **41**, 56 (1945).
39. DOLE, M., HETTINGER, W.P. JR., LARSON, N.R., and WETHINGTON, J.A. JR., *J. Chem. Phys.* **20**, 718 (1952).
40. WUNDERLICH, B., and DOLE, M., *J. Polymer Sci.* **24**, 201 (1957).
41. WARFIELD, R.W., PETREE, M.C., and DONOVAN, P., *SPE Journal* **15**, 1055 (1959).
42. HELLWEGE, K.H., KNAPPE, W., and WETZEL, W., *Kolloid, Z.* **180**, 126 (1962).
43. PASSAGLIA, E., and KEVORKIAN, H.K., *J. Appl. Polymer Sci.* **7**, 119 (1963).
44. TAUTZ, H., GLAUCK, M., HARTMANN, G., and LEUTERITZ, R., *Plastic Kautschuk* **10**, 648 (1963).
45. GRISKEY, R.G., and FOSTER, G.N., *Modern Plastics* **43**, 121 (1966).
46. GRISKEY, R.G., and FOSTER, G.N., *Proceedings 26th Annual Technical Conference SPE* **XIV** (1968).
47. WUNDERLICH, B., *J. Phys. Chem.* **69**, 2078 (1965).
48. GRAY, A.P., and BRENNER, N., *Am. Chem. Soc. Div. Polymer Sic. Preprints* **6**(2), 956 (1965).
49. RICHARDSON, M.J., *Trans. Faraday Soc.* **61**, 1876 (1965).
50. WILKINSON, R.W., and DOLE, M., *J. Polymer Sci.* **58**, 1089 (1962).
51. PASSAGLIA, E., and KERVORKIAN, H.K., *J. Appl. Phys.* **34**, 90 (1963).
52. GRISKEY, R.G., and FOSTER, G.N., *Modern Plastics* **43**, 160 (1966).
53. WILSKI, H., and GREWER, T., *J. Polymer Sci.* **C6**, 33 (1964).
54. KARASZ, F.E., BAIR, H.E., and O'REILLY, J.M., *Polymer* **8**, 547 (1967).
55. FURUKAWA, G.T., and REILLY, M.L., *J. Res. Nat. Bur. Stds.* **56**, 285 (1956).
56. BRICKWEDDE, F.G., Data published in *Styrene*, R.H. BOUNDY and R.F. BOYER, eds., Reinhold, New York (1952).
57. KARASZ, F.E., BAIR, H.E., and O'REILLY, J.M., *J. Phys. Chem.* **69**, 2657 (1965).
58. ABU-TSA, I., and DOLE, M., *J. Phys. Chem.* **69**, 2668 (1965).
59. DAINTON, F.S., EVANS, D.M., HOARE, F.E., and MELIA, T.P., *Polymer* **3**, 297 (1962).
60. WOOD, L.A., and BEKKEDAHL, *Polymer Letters* **5**, 169 (1967).
61. LINTON, W.H., and GOODMAN, H.H., *J. Appl. Polymer Sci.* **1**, 179 (1959).
62. BEAUMONT, R.H., CLEGG, B., GEE, G., HERBERT, J.B.M., MARKS, D.J., ROBERT, R.C., and SIMS, D., *Polymer* **7**, 401 (1966).
63. BRAUN, W., HELLWEGE, K.H., and KNAPPE, W., *Kolloid, Z.* **215**, 10 (1967).
64. CLEGG, C.A., GEE, G., MELIA, T.P., and TYSON, A., *Polymer* **9**, 501 (1968).
65. KARASZ, F.E., BAIR, H.E., and O'REILLY, J.M., *J. Polymer Sci.* A2 **6**, 1141 (1968).
66. DAINTON, F.S., EVANS, D.M., HOARE, F.E., and MELIA, T.P., *Polymer* **3**, 271 (1962).
67. O'REILLY, J.M., KARASZ, F.E., and BAIR, H.E., *J. Polymer Sci.* **C14**, 49 (1966).
68. PARLINOV, L.I., RABINOVICH, I.B., AKLADNOV, N.K., and ARZHAKOV, S.A., *C. Vysokomolekul, Soedin A9*, 483 (1967).
69. GRISKEY, R.G., and HUBBELL, D.O., *J. Appl. Polymer Sci.* **12**, 853 (1968).
70. MARX, P., and DOLE, M., *J. Am. Chem. Soc.* **77**, 4771 (1955).
71. DOUGLAS, T.B., and HARMAN, A.W., *J. Res. Nat. Bur. Stds.* **69A**, 149 (1965).
72. GREIVER, T., and WILSKI, H., *Kolloid, Z.* **226**, 45 (1968).
73. MARX, P., SMITH, C.W., WORTHINGTON, A.E., and DOLE, M., *J. Phys. Chem.* **59**, 1015 (1955).

74. WILHOIT, R.C., and DOLE, M., *J. Phys. Chem.* **57**, 14 (1953).
75. HAUG, W.A., M.S. Thesis, Virginia Polytechnic Inst. and State Univ., Blacksburg, VA (1965).
76. KOLESON, V.P., PAUKOV, I.E., and SKURATON, S.M., *Zh. Fig. Khim.* **36**, 770 (1962).
77. GODOVSKII, Y.K., and LIPATOV, Y.S., *Vysokolmolekul. Soedin, Aro*, 32 (1968).
78. DAINTON, F.S., EVANS, D.M., HOARE, F.E., and MELIA, T.P., *Polymer* **3**, 310 (1962).
79. O'REILLY, J.M., KARASZ, F.E., and BAIR, H.E., *J. Polymer Sci.* **C6**, 109 (1963).
80. SMITH, C.W., and DOLE, M., *J. Polymer Sci.* **20**, 37 (1956).
81. WUNDERLICH, B., and DOLE, M., *J. Polymer Sci.* **32**, 125 (1958).
82. HOFFMAN, J.D., *J. Am. Chem. Soc.* **74**, 1696 (1952).
83. LEBEDEV, B.V., RABINOVICH, I.B., and BUDARINCE, *Vysokomolekul. Soedin, Ser. A9*, 1640 (1967).
84. SHAW, R.J., *Ch. Eng. Data* **14**, 461 (1969).
85. SATOH, S., *J. Sci. Res. Inst.* (Tokyo) **43**, 79 (1948).
84. BERNHARDT, E.C., *Processing of Thermoplastic Materials*, Reinhold, New York (1959).
85. STARKWEATHER, H.W., *J. Polymer Sci.* **45**, 525 (1960).
86. GRISKEY, R.G., and WALDMAN, N., *Modern Plastics* **43**, 9, 245 (1966).
87. GRISKEY, R.G., WALDMAN, N., DIN, M., and GELLNER, C.A., *Modern Plastics* **43**, 10, 103 (1966).
88. GRISKEY, R.G., DIN, M., and GELLNER, C.A., *Modern Plastics* **43**, 11, 119 (1966).
89. GRISKEY, R.G., DIN, M., and GELLNER, C.A., *Modern Plastics* **44**, 1, 165 (1966).
90. *Ibid.*, 3, 129.
91. GRISKEY, R.G., and WANGER, W.H., *Modern Plastics* **44**, 8, 134 (1967).
92. GRISKEY, R.G., and COEL, J., *Proc. Ann. Tech. Conf. Soc. Plastics Engr.* **XXV** (1979).
93. GRISKEY, R.G., and SHOU, J.K.P., *Modern Plastics* **45**, 10, 138 (1968).

PROBLEMS

2-1 Find the coefficient of thermal expansion, $1/V(\partial V/\partial T)_p$, for polystyrene at a pressure of 1.013×10^7 N/m².

2-2 What is the behavior of compressibility, $1/V(\partial V/\partial P)_T$, for polyisobutylene at a temperature of 165°C?

2-3 Determine C_v for polypropylene at a temperature of 45°C.

2-4 Find the change in free energy for polytetrafluoroethylene between an initial state of 75°F and 10,000 psia and a final state of 100°F and 10^5 psia.

2-5 What is the Helmholtz free energy of rigid polyvinyl chloride at 123.8°F and 20,000 psia?

2-6 Polyhexamethylene adipamide (nylon 66) has a melting point of 266°C and a density (at 25°C) of 1.14 g/cm³. What is its specific volume at a temperature of 340°C and a pressure of 2.53×10^7 N/m²?

2-7 The Spencer–Gilmore equation [16] is

$$(P + \pi)(V - W) = R^1 T$$

where P is pressure, V the specific volume, and T the absolute temperature. π and W are empirical constants for a particular polymer, and R^1 is the gas constant divided by the mer weight. Fit this equation to molten polypropylene.

2-8 According to Spencer and Gilmore, the values of π, W, and R^1 for solid poly-ethylene are 3.28×10^8 N/m^2, 875 kg/m^3, and 299.5 m^3 N/m^2/kg K. Calculate the specific volume at 125°C and 3.04×10^7 N/m^2, compare it to the corresponding value from Fig. 2-1.

2-9 The ethylene–propylene copolymer of Fig. 2-3 is isothermally (at 210°C) and reversibly compressed from atmospheric pressure to 3.04×10^7 N/m^2. How much work is done?

2-10 What is the final temperature of molten nylon 6-10 that is compressed from at-mospheric pressure to 1.52×10^7 N/m^2.

2-11 A manufacturer of acrylic copolymer reports the following density data:

Density, kg/m^3	Temperature, °C	Pressure (N/m$^2 \times 10^{-5}$)
1189	23	1.01
1175	149	1317
1122	121	17.58
1154	149	879

From these data, find the copolymer's glass temperature.

2-12 Griskey [94] has shown that there is a significant difference between the com-pressibilities of low-density (0.919–0.920-g/cm^3) and high-density (0.958–0.973-g/cm^3) polyethylenes in the molten state over the range from 130 to 250°C. Does this behavior hold for the following polyethylenes?

Polyethylene	Density g/cm^3 (25°C, 1 atm)	Melting Temperature °C
A	0.923	105
B	0.945	130

2-13 The melting temperature at atmospheric pressure of the polypropylene shown in Fig. 2-2 is 170°C. Based on this and the material in the chapter, delineate the liquid-solid region from 170°C to 180°C.

2-14 An inventor has submitted a patent covering a new device that, when operated adiabatically, can change polychlorotrifluoroethylene from a pressure of 6.96×10^5 N/m^2 and a temperature of 10°C to a pressure of 6.69×10^6 N/m^2 and a temperature of 150°C. Comment on the invention.

2-15 It has been found that the C_p behavior of molten polyolefins (polyethylenes, poly-propylene, etc). with temperature is such that a single correlation line can be used (Fig. 2-15). Is such a correlation possible for molten nylons?

2-16 Molten polyethylene flowing in a heavily insulated tube is at a pressure of 3.48×10^7 N/m^2 prior to entering a series of in-line static mixing units. After leaving the mixers, the polymer is at a pressure of 6.96×10^6 N/m^2 and a temperature of 170°C. What was the temperature of the polymer before it entered the static mixers?

2-17 Estimate the degree of crystallinity in the polyethylene of Fig. 2-1.

2-18 Two engineers are discussing the flow of molten polypropylene through an ex-truder die (entrance conditions of 1000 psia and 380°F). One feels that the poly-mer's exit temperature will stay the same, while the other thinks that it changes. Which is correct? Prove your answer.

2-19 At what pressure will the ethylene–propylene copolymer of Fig. 2-3 crystallize if the temperature is 115°C?

2-20 Isotactic polymethyl methacrylate has a density of 1.22 g/cm^3 at room temperature and pressure. Its glass temperature is 100°C. Calculate three to four isotherms for the molten polymer (i.e., behavior of pressure volume at constant temperature).

2-21 Estimate the volume of an ABS (20% acrylonitrile, 30% butadiene, 50% styrene) copolymer at 1.01×10^8 N/m^2 and 200°C.

MINI PROJECT A

For a semicrystalline polymer of your choice (other than those in Figs. 2-19–2-29), develop charts for enthalpy and entropy as functions of temperature and pressure. Use a datum of 25°C and atmospheric pressure.

Where possible, use experimental data. As needed, use the various equations and correlations in Chapter 2 to obtain data.

MINI PROJECT B

Determine the enthalpy and entropy charts as functions of temperature and pressure for a butadiene-styrene copolymer with 30% butadiene (T_g of 18°C). Use a datum of 25°C and atmospheric pressure.

In butadiene–styrene copolymers, the glass temperature is given by

$$T_g = x_1(T_g \text{ of polystyrene}) + x_2(T_g \text{ of butadiene})$$

the x_1 and x_2 values are mole fractions of styrene and butadiene.

MINI PROJECT C

A widely used *polymer blend* is composed of one part butadiene–styrene copolymer (75/25 by weight) and three parts polystyrene. The copolymer's glass temperature is $-57°C$.

Calculate the enthalpy of the blend over a range of $50–150°C$ and pressures from atmospheric to 1.01 to 10^8 N/m^2.

CHAPTER 3

APPLIED POLYMER RHEOLOGY

3.1 INTRODUCTION

The overall science that considers flow and deformation of all matter, solids as well as fluids, is termed *rheology*. The concept emanates from the Greek philosopher Heraclitus, who wrote, "Panta rhei," translated as "Everything flows." Rheology gives us the key to understanding in the processing of polymers.

Why is this so? There are a number of reasons. First of all, rheology, as we have seen, can be a powerful characterization tool. With it, for example, we can relate structure and properties. Further, all polymer processing operations involve flow and deformation. Indeed, the same can be said for polymer synthesis processes as well.

In essence, then, polymer processing cannot be properly understood, designed, or operated without a knowledge of rheology.

3.2 CLASSIFICATION OF FLUID BEHAVIOR

In order to develop an understanding of the particular concepts of flow needed for polymer processing, let us begin with a simple experiment. Suppose we place a liquid, such as water, at rest between two plates (Fig. 3-1). At a given time, we move the bottom plate with a velocity V.

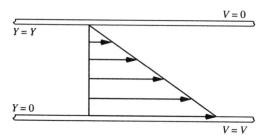

Fig. 3-1   Steady-state velocity profile with lower plate in motion (velocity = *V*).

The result of the motion of the plate is to cause the fluid in the immediate vicinity of the plate to move also. After a period of time, a linear velocity profile is established (i.e., steady-state flow). At this steady state, a constant force *F* is required, which means that

$$\frac{F}{A} = -\mu \frac{0 - V}{y - 0} \tag{3-1}$$

where μ is the fluid's viscosity, that is, a transport coefficient.

In this type of fluid, known as Newtonian, the viscosity is a property of the fluid and is not altered or changed by the shear rate (dV_x/dy). For Newtonian fluids, only temperature and pressure have an effect on viscosity and

$$\tau_{yx} = -\mu \frac{dVx}{dy} \tag{3-1A}$$

In nature, there are many other fluid systems in which this is not the case. Further, there are other cases in which other factors influence the flow behavior. This means that there are a number of categories of fluid behavior. These are:

1. Newtonian
2. "Simple" non-Newtonian (the viscosity is a function of shear rate)
3. "Complex" non-Newtonian (the behavior is a function of both rate of shear and the time parameter)
4. Fluids influenced by external force fields
5. Noncontinuous fluids
6. Relativistic fluids

The *"simple" non-Newtonian fluids* are those in which the rate of shear influences the flow behavior (see Fig. 3-1A). As can be seen, the straight line

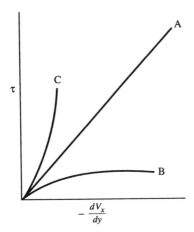

Fig. 3-1A   Comparison shear stress vs shear rate for (A) Newtonian, (B) shear-thinning, (C) shear-thickening fluids.

(i.e., constant-viscosity) behavior of the Newtonian is not matched by two categories of fluids, shear thinning and shear thickening.

In order to get a better feel for these fluids, let us rewrite Newton's law of viscosity so that we have apparent viscosity μ_{APP}.

$$\tau_{yx} = -\mu_{APP} \frac{dV_x}{dy} \tag{3-1B}$$

$$\mu_{APP} = \frac{\left(\dfrac{-dV_x}{dy}\right)}{\tau_{yx}} \tag{3-2}$$

Hence, if we plot the behavior of μ_{APP} on a log-log plot, we obtain Fig. 3-2.

As can be seen, the μ_{APP} for a Newtonian is a constant. However, the other fluids show a decreasing apparent viscosity with increasing shear rate (shear-thinning fluid) or an increasing apparent viscosity with an increasing shear rate (shear-thickening fluid). Both these fluid types have other common names. Shear-thinning fluids are also called *pseudoplastic*, whereas shear-thickening fluids are terms *dilatant*. This latter name is somewhat unfortunate since it confuses the flow behavior of shear-thickening fluids with the concept of volumetric dilation, a different phenomenon that can or cannot occur in shear-thickening fluids.

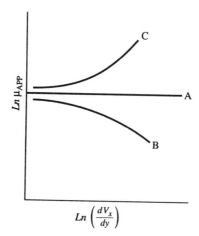

Fig. 3-2   Logarithm of apparent viscosity vs logarithm of shear rate for (A) Newtonian, (B) shear-thinning, and (C) shear-thickening fluids.

It is obvious that the "simple" non-Newtonian cannot be treated by Newton's law of viscosity. As such, other approaches must be taken that lead to often quite complicated rheological constitutive equations. The simplest of these is Eq. (3-3), the apparent viscosity expression, or the Ostwald–De Waele power law:

$$\tau_{yx} = -K \left| \frac{dV_x}{dy} \right|^{n-1} \frac{dV_x}{dy} \qquad (3\text{-}3)$$

where $|dV_x/dy|$ is the absolute value of the rate of shear. The K term is the consistency index, and n is the flow behavior index.

It is also possible to derive the expression

$$\tau_{yx} = K \left(\frac{d\gamma}{dt} \right)^n \qquad (3\text{-}4)$$

where $d\gamma/dt$ is the rate of strain and $(d\gamma/dt = -dV_x/dy)$. The power law is not applicable over the entire range of shear rate behavior but applies only where log (τ_{yx}) vs log ($d\gamma/dt$) is a straight line. Also, note that the flow behavior index n is the slope of such a plot. For the various systems, values for n are presented in Table 3-1.

The behavior of "simple" non-Newtonians will be covered in greater detail later in this chapter. One additional point of interest, however, is the velocity profile when a pseudoplastic fluid flowing in a circular tube is not a parabola

TABLE 3-1 System values for *n*

Fluid Type	n	Examples
Shear-thinning; pseudoplastic	< 1.0	Polymer solutions, polymer melts, foods
Newtonian	1.0	Water, organic fluids
Shear-thickening; dilatant	> 1.0	Continuous and dispersed phases (water–sand, cement, water–cornstarch, etc.)

(not even in laminar flow) but rather a blunted profile, as with Newtonian turbulent flow.

Complex non-Newtonians are those in which the time parameter becomes a factor. In essence,

$$\tau = \phi \left(\frac{d\gamma}{dt}, t \right) \tag{3-5}$$

A way of considering the behavior of these fluids is first to reflect that, in a Newtonian fluid,

$$\tau = \mu \left(\frac{d\gamma}{dt} \right) \tag{3-6}$$

with μ independent of rate of shear. Next, if the behavior of an ideal elastic solid is considered (see Fig. 3-3), we see that Hooke's law applies:

$$\tau = G\gamma \tag{3-7}$$

Further, it is possible to use mechanical analogs to represent the behavior of both these systems. In the case of Hooke's law, we can use a spring (Fig. 3-4) which, when distended by a stress, returns to its original shape and, in so doing, releases the work previously done on it.

The Newtonian fluid can be represented by a dashpot (Fig. 3-4) in which a piston placed in a liquid is attached to move over a pulley. When a stress is exerted, the piston moves to a new location. If the stress in removed, the piston stays in place and does not return to its original position. Further, the work used to move the piston is not regained.

In nature, there are fluids that have the characteristics of both the Newtonian fluid and the elastic solid. A simple representation of such a *viscoelastic* fluid is shown in Fig. 3-5. Here we see that the application and release of stress do not allow the piston to stay in place but rather cause it to recoil.

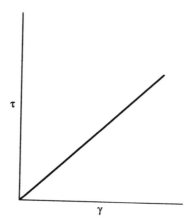

Fig. 3-3   Ideal elastic solid (Hooke's law).

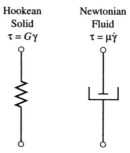

Fig. 3-4   Mechanical models for ideal elastic solid (spring) and Newtonian liquid (dashpot).

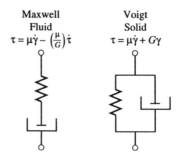

Fig. 3-5   Viscoelastic models [5].

A more intensive and detailed treatment of viscoelasticity gives us the concept of relaxation time, which means, of course, that time is a parameter. Note that the viscoelastic fluid will behave as a simple non-Newtonian after the elastic effects have taken place. Typical viscoelastic fluids are certain polymer melts or polymer solutions.

Two other fluid types in which time is a parameter are those that have been categorized as time-dependent or, more specifically, as thixotropic or rheopectic. These fluids can be thought of as fluids in which

$$\tau = \phi \left(\frac{d\gamma}{dt}, t \right) \tag{3-8}$$

as shown in Figs. 3-6 and 3-7.

Here, it can be seen that the apparent viscosity increases with time for the rheopectic fluid and decreases with time for the thixotropic. Examples of the latter are paints and inks, whereas rheopectic fluids include various lubricants.

Fluids influenced by external force fields are best exemplified by systems in which electrically conducting fields play a role (for example, magnetohydrodynamics). In such cases,

$$\tau = \phi \left(\frac{d\gamma}{dt}, \text{field} \right) \tag{3-9}$$

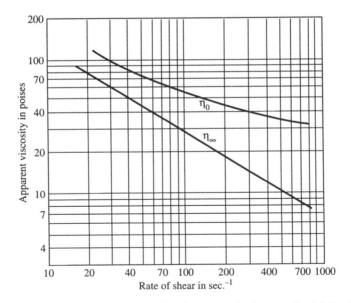

Fig. 3-6   Thixotropic fluid behavior (η_0 is initial viscosity; η_∞ is viscosity after infinite time [11].

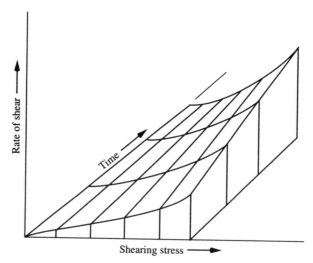

Fig. 3-7   Rheopectic fluid behavior [11].

In essence, in order to handle such systems, we must splice the fluid mechanical behavior with Maxwell's equations for electromagnetic fields.

Originally used for such cases as plasmas and ionized gases, this behavior now has possibilities elsewhere. An interesting aspect of such fluids is that their velocity profile behavior is similar to that of "simple" non-Newtonian in that their profiles are blunted.

In treating the fluid systems described previously, we have assumed that the fluid is a continuum. There are cases in which that is no longer correct. Consider specifically such cases as fogs, aerosols, sprays, smoke, and rarefied gases (i.e., such as might be encountered extraterrestrially). All these are *noncontinuum fluids* and, as such, mandate that special techniques be used.

Finally, we have the case in which a fluid could approach the speed of light. This would invalidate Newtonian mechanics and, as such, mean that a form of *relativistic fluid mechanics* would have be used.

3.3   RHEOLOGICAL EQUATIONS

The best analogy for rheological equations is the equations of state used to describe the *P-V-T* behavior of gases, liquids, and solids. Basically, the equation of state represents an attempt to fit existing data. This is often obscured by derivations of equations of state from thermodynamic micro- and macroscopic approaches, such as virial coefficients and corresponding states. In essentially every instance, however, the resultant equation of state with a number of constants, peculiar to the molecule being studied, must be fit to experimental data.

In a similar manner, the rheological or constitutive equations, some with many constants, are fit to experimental data. Again, however, as in the case of the equations of state, the constitutive equations are derived from theoretical approaches. Basically, the theory itself is not sufficient to enable their use. Instead, a fit to experimental data is required.

The flow behavior of a polymer melt (logarithm apparent viscosity vs logarithm shear rate) is illustrated in Fig. 3-8. Basically, the melt has flow behavior very similar to a Newtonian (relatively constant apparent viscosity) at low shear rates. Similarly, at very high shear rates, a second Newtonian type of behavior is encountered. As such, therefore, the constitutive equation should represent, as closely as possible, the behavior shown in Fig. 3-8. The goodness of this representation is directly proportional to the number of constants in the constitutive equation: The more constants, the truer the representation.

The most direct and least complex of all the constitutive equations is the previously described Ostwald–DeWaele power law. Although the equation does not describe the entire range of behavior shown in Fig. 3-8, it gives an excellent fit over useful ranges that are encountered in processing. As such, it can be widely utilized, as long as its limitations are borne in mind.

A number of more complex equations are shown in Table 3-2.

In Table 3-2, μ_0 is the zero shear rate viscosity (i.e., the Newtonian viscosity at low shear rate; $\tau_{1/2}$ is the shear stress value of $\mu_0/2$; $(\alpha - 1)$ is the slope of a plot of the logarithm of $(\mu_0/\mu_{APP} - 1)$ vs the logarithm of $(\tau/\tau_{1/2})$; μ_∞ is the Newtonian viscosity at high shear rate; and A and B are constants.

There are also more complicated constitutive equations, many of which at-

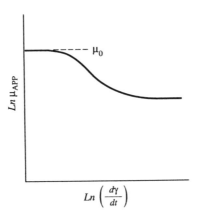

Fig. 3-8   Polymer melt flow behavior.

TABLE 3-2   Constitutive Equations

Equation	Apparent Viscosity Term
Ellis	$\dfrac{\mu_0}{1 + \left(\dfrac{\tau}{\tau_{1/2}}\right)^{a-1}}$
Eyring	$\mu_0 \left[\dfrac{(\tau/B)}{\sinh\,(\tau/B)}\right]$
Reiner–Philippoff	$\mu_\infty + \left[\dfrac{\mu_0 - \mu_\infty}{1 + (\tau/A)^2}\right]$

tempt to deal with viscoelastic behavior as well. Summaries and extensive discussions of these are given elsewhere [12].

3.4   MORE COMPLEX ASPECTS OF POLYMER RHEOLOGY

As previously mentioned, the subject of rheology is an extremely broad and complicated one. Indeed, any facet of polymer rheology is in itself *proper* material for an entire textbook. In this book, consideration of the important area of polymer rheology has necessarily been reduced to the bare essentials.

Even so, it is worthwhile at least to mention some aspects of polymer behavior that, although complex, should be made known to practitioners of polymer processing.

One quantity that may be encountered is the normal stress. An instance of the effect of this type of stress occurs when certain polymeric fluids are placed between a cone and plate or two parallel disks and then sheared. Such fluids will exert a normal force such that the cone and plate or parallel disks tend to separate. Measurements of such cases show that the normal stresses are functions of the shear rate (see Figs. 3-9 and 3-10) but increase at a much more rapid rate than shear stresses do.

In rheology, the normal stress behavior is described by the normal stress differences:

$$N_1 = \tau_{11} - \tau_{22} \tag{3-10}$$

$$N_2 = \tau_{22} - \tau_{33} \tag{3-11}$$

The τ's are the normal stresses in the coordinate directions indicated. Figure 3-11 illustrates both shear (two different letter subscripts) and normal (same letter subscripts). The shear stresses act along a plane, whereas the normal

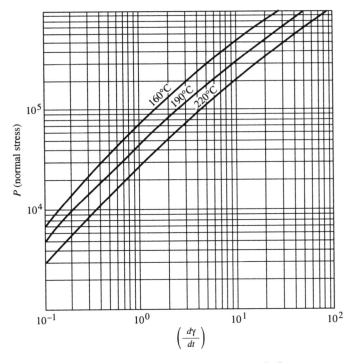

Fig. 3-9   Normal stress as a function of shear rate for polyethylene [10].

stresses act perpendicular to a plane. As shown, subscripts identify the direction and action of each stress.

The normal stress functions corresponding to the apparent viscosity for shear stress are called the first (ψ_1) and second (ψ_2) normal stress coefficients. As shown,

$$\psi_1 = \frac{N_1}{\gamma^2} \tag{3-12}$$

and

$$\psi_2 = \frac{N_2}{\gamma^2} \tag{3-13}$$

are the appropriate normal stress difference divided by the shear rate (or strain rate) squared.

Knowledge of the behavior of apparent viscosity and the normal stress coefficients is inversely proportional to the case of the experimental measurement needed. The most is known about the apparent viscosity and the least about the second normal stress coefficient (ψ_2).

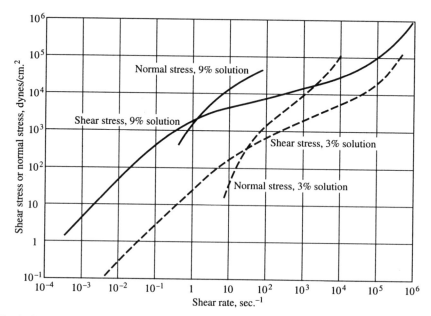

Fig. 3-10   Normal and shear stress as a function of shear rate for polyisobutylene solutions [13].

An additional area is that of extensional flows. In this case, the stress is the difference between the normal stresses:

$$\tau_E = \tau_{11} - \tau_{22} = \tau_{11} - \tau_{33} \tag{3-14}$$

where τ_E is the net tensile stress.

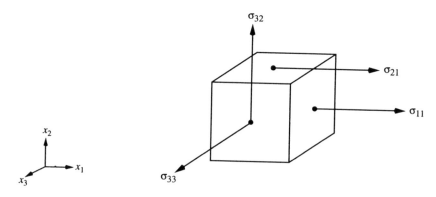

Fig. 3-11   Shear and normal stresses ($\sigma = \tau$) [14].

Although extensional flows are quite complex, there are some important limiting cases. One such instance is for small or slow deformations, where

$$\mu_{ext} = 3\mu_{APP} \tag{3-15}$$

Another is the case of the take-up end of a filament (see Fig. 3-12), in which

$$\mu'_{ext} = \frac{\tau_E}{\epsilon_L} \tag{3-16}$$

where μ'_{ext} is the apparent extensional viscosity and ϵ_L the tensile strain rate at take-up.

The foregoing discussion, while limited, should provide additional needed insight for polymer processing.

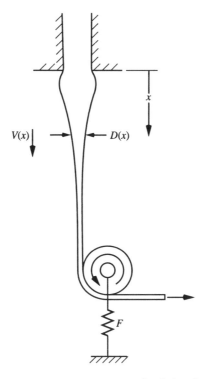

Fig. 3-12   Melt spinning measurement for apparent extensional viscosity [14].

3.5   APPLIED RHEOLOGY

The most basic and fundamental approach to scientific and technological problems is always the best. However, in a variety of circumstances, engineers and scientists cannot always take this route. Instead, a less sophisticated, more empirical, and very applied approach must be used.

Polymer rheology is no different in this respect. Although the use of theory would be most commendable, a variety of reasons, not the least of which is the inherent complexity of the polymer's rheological behavior, necessitate the use of what is essentially applied rheology.

As an example, consider the data aspect of the situation. The first requirement in applying rheology is, of course, the need for rheological data. Such data can be determined from sophisticated rheological devices, such as capillary viscometers and rotational devices. Data, however, can also be determined from operating process data or by making certain in-line measurements.

Next, methods have to be used that make use of these varied sources in such a way as to produce meaningful and correct designs, proper understanding of operations, and real insight into polymer processing. The sections that follow will, it is hoped, accomplish these goals.

3.6   ISOTHERMAL FLOW OF MOLTEN OR THERMALLY SOFTENED POLYMERS IN CIRCULAR CONDUITS

Polymers are generally processed in the molten or thermally softened states. The former represent semicrystalline polymers above the melting point, whereas the latter are amorphous polymers well above their glass temperature. The difference between them can be seen in Fig. 3-13, which represents specific heat (C_p) data for both types. As can be seen, the semicrystalline materials peak at a given temperature (i.e., have a heat of fusion), but the thermally softened types show no such effect.

Generally, this means that both types can be treated as "fluids" as long as the temperature is high enough. For semicrystalline polymers, we add a qualification; pressure should also be specified. The reason for this will be discussed in more detail later.

Let us consider now the steady isothermal flow of a molten or thermally softened polymer through a tube.

It has been shown that, for steady, isothermal, laminar flow, the relationship between τ_{wall} and the term $8V/D$ (where V is the average fluid velocity) must be independent of tube diameter as long as:

1.   Shear stress–shear rate relations are not time-dependent.
2.   Slippage does not occur at the tube wall.

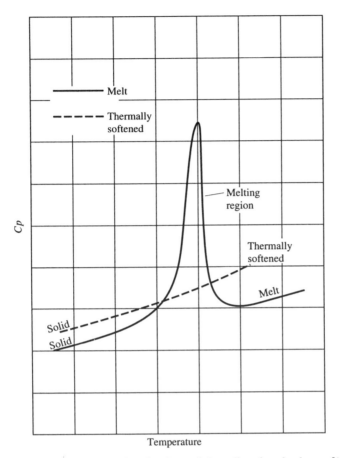

Fig. 3-13 Comparison melting behavior of molten and thermally softened polymers [1].

In general, both effects are not found to any extent in the flow of polymer melts through a capillary.

For the case of a polymer melt or solution flow through a tube (assuming power law behavior), the pertinent equation is

$$\left(-\frac{dV_z}{dr}\right)_{wall} = 3\left(\frac{8Q}{\pi D^2}\right) + \frac{D\Delta P}{4L}\frac{d\left(\frac{8Q}{\pi D^2}\right)}{d\left(\frac{D\Delta P}{4L}\right)} \tag{3-17}$$

where Q = volumetric flow rate.

Since $V = 4Q/\pi D^2$,

$$\left(-\frac{dV_z}{dr}\right)_{\text{wall}} = \frac{3}{4}\left(\frac{8V}{D}\right) + \frac{1}{4}\left(\frac{8V}{D}\right)\frac{d\ln(8V/D)}{d\ln(D\Delta P/4L)} \tag{3-18}$$

Also, a relation can exist between $D\Delta P/4L$ and $8V/D$ such that

$$\left(\frac{D\Delta P}{4L}\right) = \tau_{\text{wall}} = K'\left(\frac{8V}{D}\right)^{n'} \tag{3-19}$$

In this equation, n' and K' are *not* the power law parameters n and K. The relationships between n, n' and K, K' are given in Eqs. (3-20) and (3-21),

$$n = \frac{n'}{1 - \frac{1}{3n' + 1}(dn'/d\ln\tau)} \tag{3-20}$$

but since $(dn'/d\ln\tau) \cong 0$,

$$n \cong n' \tag{3-21}$$

and

$$K' = K\left(\frac{3n + 1}{4n}\right)^n. \tag{3-22}$$

The combination of Eqs. (3-17 through 3-22) can be used to compute the flow of molten polymer or polymer solution through the tube or to determine velocity profiles.

It is possible to develop a friction factor–modified Reynolds number correlation for power law fluids. The correlation is given in Fig. 3-14, where

$$Re' = \frac{D^n V^{2-n'}\rho}{g_c K' 8^{n'-1}} \tag{3-23}$$

The friction factor f in Fig. 3-14 is a function of Re'. Furthermore,

$$\frac{\Delta P}{L} = \frac{2f\rho V^2}{g_c D} \tag{3-24}$$

Hence, pressure drops (or flow rates) can be computed using Fig. 3-14.

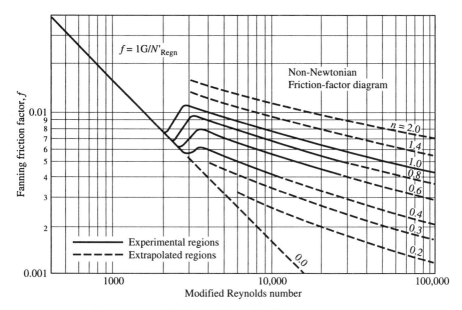

Fig. 3-14   Friction factor vs modified Reynolds number [2].

A word about Fig. 3-14: In the laminar flow region (usually the ycase for molten or thermally softened polymers), the friction factor f is related to Re' as shown:

$$f = \frac{16}{Re'} \qquad (3\text{-}25)$$

This holds for $Re' < 2100$.

At this point, it is worthwhile to comment on the shape of the velocity profiles found in the tube flow as a function of n, the power law index. This relation is given by

$$\frac{V_z}{V} = \frac{1 + 3n}{1 + n}\left[1 - \left(\frac{r}{R}\right)^{(n+1)/n}\right] \qquad (3\text{-}36)$$

Figure 3-15 shows the relation between the velocity and the position in the tube cross section.

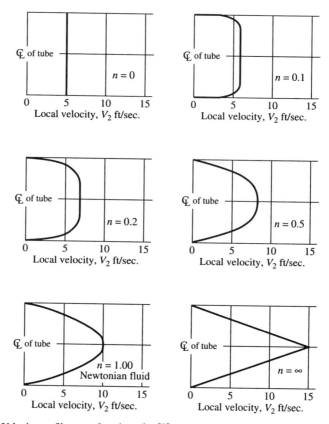

Fig. 3-15   Velocity profiles as a function of *n* [3].

3.7   COMPUTATION OF FLOW RATES AND PRESSURE DROPS FOR MOLTEN AND THERMALLY SOFTENED POLYMERS IN ISOTHERMAL TUBE FLOW

In the preceding section, we considered ways and means of describing isothermal flow in conduits. We next direct our attention to the methodologies used to calculate pressure drops or volumetric flow rates. Three approaches will be presented: calculation from flow curves, use of friction factors, and computation using a flow equation.

If rheological flow curve data are available (as in Fig. 3-16), we can directly compute the pressure drop in tube flow as well as determine the velocity profile and flow rate.

To illustrate this, consider Fig. 3-17, which represents a force balance on an

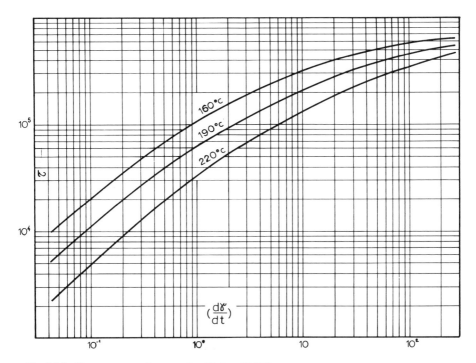

Fig. 3-16   Shear stress vs shear rate for Alathon 10 [10].

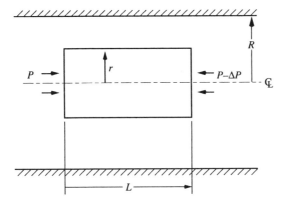

Fig. 3-17   Force balance on fluid element [3].

element of fluid flowing inside a tube. For such a system, the pressure and shearing forces must be in equilibrium:

$$\Delta P \,(\pi r^2) = \tau \,(2\pi r L) \tag{3-27}$$

and

$$\tau = \frac{r\Delta P}{2L} \tag{3-28}$$

Equation (3-28) shows that τ varies from a value of zero at the center to a maximum of $R\Delta P/2L$ at the wall. Since, from the flow curve (Fig. 3-16), there must be some relationship between shear rate and shear stress (at a given temperature), we can write

$$\tau = \frac{r\Delta P}{2L} = \phi \left(\frac{-dV_z}{dr} \right) \tag{3-29}$$

or

$$\frac{dV_z}{dr} = \psi \left(\frac{r\Delta P}{2L} \right) \tag{3-30}$$

The ϕ and ψ merely indicate some unknown function.

Then we can compute a given V_z at an r from

$$\int_0^{V_z} dV_z = \int_{r=R}^{r=r} \psi \left(\frac{r\Delta P}{2L} \right) dr \tag{3-31}$$

$$V_z = \int_{r=R}^{r=r} \psi \left(\frac{r\Delta P}{2L} \right) dr \tag{3-32}$$

The procedure for doing this is as follows. Suppose $\Delta P/L$ is fixed and we want to calculate Q, the volumetric flow rate. We then first calculate values of $r\Delta P/2L$ for various r values. From the flow curve (Figure 3-16), we then determine values of dV_z/dr for each r. This gives us a table of dV_z/dr vs r data. Next, we plot dV_z/dr vs r and integrate graphically from $r = R$ ($V_z = 0$) to $r = r$ ($V_z = V_z$). This yields V_z vs r data. Finally, Q can be calculated by

$$Q = \int_{r=R}^{r=0} V_z \,(2 \,\pi \,r) \,dr \tag{3-33}$$

This last equation can be solved by plotting $2 \pi r V_z$ vs r and integrating graphically from $r = R$ to $r = 0$.

If the Q value is fixed and we wish to find $\Delta P/L$, the technique is slightly more complicated. In order to do this, we *assume a value of $\Delta P/L$* and then carry out the Q calculation to see if it matches the actual value. If not, we assume a different $\Delta P/L$ until calculated and actual Q values match. When this is done, we have assumed the correct $\Delta P/L$ value.

Another method of accomplishing the same result is by the use of the friction factor–modified Reynolds number relation (see Fig. 3-14). Here, we must have a value of n and K; notice that we do not need an entire flow curve but only an n and K value. With these values, we can obtain n' and K' [Eqs. (3-21) and (3-22)].

Now, if flow rate Q is given, we can calculate V, the average velocity since

$$Q = V \cdot \pi R^2 \tag{3-34}$$

Next, we use this V value together with n' and K' to get the modified Reynolds number [Eq. (3-23)]. This, in turn, together with Fig. 3-14, gives us a value of f. Then, since

$$\frac{\Delta P}{L} = \frac{2 f \rho V^2}{g_c D} \tag{3-35}$$

we can calculate $\Delta P/L$.

If Q is to be calculated, we first assume a value for V and calculate $\Delta P/L$. When the calculated and actual $\Delta P/L$ values match, we have assumed the correct V. The Q value can be computed from Eq. 3-34.

Finally, the pressure drop or flow rate can be calculated by using an appropriate equation. Consider, for example, the situation if the fluid obeys the power law. Then, integration and appropriate mathematical manipulation will yield

$$V_z = \left(\frac{\Delta PR}{2KL}\right)^{1/n} \frac{R}{\frac{1}{n} + 1} \left[1 - \frac{r^{(n+1)/n}}{R}\right] \tag{3-36}$$

and

$$Q = \pi \left[\frac{\Delta PR}{2KL}\right]^{1/n} \frac{nR^3}{1 + 3n} \tag{3-37}$$

Hence, if we know n and K, we can calculate either Q or $\Delta P/L$ for a flow case. The velocity profile is given by Eq. 3-36.

3.8   ISOTHERMAL FLOWS OF MOLTEN AND THERMALLY SOFTENED POLYMERS IN NONCIRCULAR CONDUITS

In certain instances, we wish to treat flows in noncircular conduits such as slits, square ducts, and rectangular ducts.

Flow problems in such geometries can be calculated directly from the flow curve data if proper force balances are set up between shearing and pressure forces. This, in essence, would repeat the first method covered in section 3-7.

The friction factor–modified Reynolds number approach cannot be used directly for noncircular geometries. In order to use techniques such as this, we must resort to the hydraulic radius concept, in which

$$D_{\text{equivalent}} = 4R_{\text{hydraulic}} = 4 \left(\frac{\text{cross section/available for flow}}{\text{wetted perimeter}} \right) \qquad (3\text{-}38)$$

For example, for a square duct (each side equal to e),

$$D_{\text{equivalent}} = 4R_H = 4 \left(\frac{e^2}{4e} \right) = e \qquad (3\text{-}39)$$

Then, the $D_{\text{equivalent}}$ can be used in place of D in Eq. 3-23 for modified Reynolds number. Once the modified Reynolds number is found, the remainder of the calculation can proceed as before. Care must be exercised, however, in using the hydraulic radius approach, since it is more appropriate for turbulent flows. Nevertheless, it will give at least a reasonable estimate for noncircular cross sections.

It is, of course, also possible to solve directly for V, using an appropriate flow equation. For example, for flow in a very wide slit (i.e., broad parallel plates separated by a small distance (2B), where power law holds,

$$V_2 = \left(\frac{\Delta P}{KL} \right)^{1/n} \left(\frac{n}{n+1} \right) \left[1 - \left(\frac{b}{B} \right)^{(n+1)/n} \right] \qquad (3\text{-}40)$$

Here, b is the distance measured from the centerline. Also,

$$\frac{B\Delta P}{L} = K \left(\frac{2n+1}{2n} \frac{Q^1}{B^2} \right)^n \qquad (3\text{-}41)$$

where Q^1 = volumetric flow rate per foot of width.

A number of solutions for various cross sections were obtained by Kozicki, et al. [4] in the form

$$\frac{2V}{R_H} = \frac{n}{a + bn}\left(\frac{\tau_{\text{wall}}}{K}\right)^{1/n}$$
(3-42)

where a and b are dependent on system geometry. Values of a and b are summarized in Ref. 4 for annuli, rectangular ducts, elliptical ducts, and isoceles triangular ducts. Application of Eq. 3-42 permits the computation of flow rates and pressure drops.

3.9 END EFFECTS, KINETIC ENERGIES, AND OTHER FACTORS AFFECTING FLOW RESULTS

A velocity profile in a fluid is not immediately established at the entrance to a tube. Instead, it takes a specified axial distance for this profile to be established. This distance in tube diameters is usually in the range of 0.05 times the modified Reynolds number. Hence, for a case in which Re^1 is 2100, about 105 tube diameters would be required whereas, for an Re^1 of 100, only 5 diameters would be needed.

As can be seen, problems arise with short tubes or, more specifically, capillaries. For such cases, pressure drop must be corrected since the entrance length acts as an additional resistance. To better understand this, consider Fig. 3-18 which plots flow data taken with various capillaries. These data appear to be parameters of L/D. If we assume that the effect of entrance length can be attributed to an additional ''length'' (see Fig. 3-19), then the data of Fig. 3-18 can be brought to one line (Fig. 3-20). Hence, for cases where entrance length

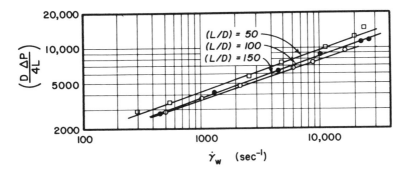

Fig. 3-18 Melt rheology data showing L/D effects [5].

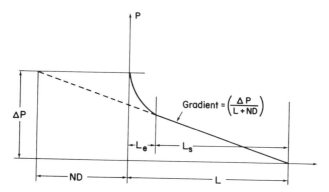

Fig. 3-19   Entrance effect correction [5].

is significant, the pressure gradient should be $\Delta P/(L + ND)$ not $\Delta P/L$. Usually N is determined by experiment. However, a value of about 5.0 appears to be reasonable for molten polyethylene.

Still another effect that must be considered is correction for kinetic energy changes. This occurs if we have a high-speed discharge from a capillary or short tube. For this case,

$$\Delta P = \Delta P_{observed} - \frac{\rho Q^2}{\alpha \pi^2 R^4} \tag{3-43}$$

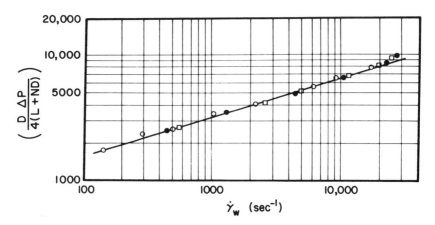

Fig. 3-20   Corrected melt rheology [10].

where α is related to n, the power law index as

$$\alpha = \frac{(4n + 2)(5n + 3)}{[3(3n + 1)^2]} \tag{3-44}$$

Elastic effects can and do occur in molten or thermally softened polymer systems. This is a case of the fluid having both the characteristics of a Newtonian liquid and a Hookean elastic solid. The elastic effects will generally manifest themselves in entrance or exit situations. Hence, they can become important in capillaries. However, it should be pointed out that normally elastic effects are compensated for in the entrance length correction.

3.10 NONISOTHERMAL FLOWS OF MOLTEN OR THERMALLY SOFTENED POLYMERS

In many processing operations, the fluid temperature changes. This means that the effect of temperature must be taken into account in describing flow.

Generally, increasing temperature increases the fluidity of a material. Non-Newtonians are no exception to this pattern. The situation is complicated, however, by the complex behavior of the fluids.

If power law behavior is assured, we can separately consider the effect of temperature on n, the flow behavior index. Table 3-3 shows the effect of temperature on n.

As can be seen, the effect of temperature on n is slight. In practical processing, n can be assumed to be relatively constant, provided that temperature does

TABLE 3-3 **Effect of Temperature on** n

Material	Shear Rate, s^{-1}	Flow-Behavior Index and Corresponding Temp., °C	Increase in Flow-Behavior Index per 100°C Temp. Increase
Polyethylene (melt index 2.1)	0.01	0.84 at 108° and 1.00 at 230°	0.13
	100	0.32 at 108° and 0.49 at 230°	0.14
Polyethylene (melt index 2.0)	0.1	0.59 at 112° and 0.88 at 250°	0.21
	10	0.33 at 125° and 0.59 at 250°	0.21
Plasticized polyvinyl butyral	10	0.24 at 125° and 0.365 at 155°	0.42
"Vistanex" LM-S polyisobutylene	1,000	0.30 at 38° and 0.49 at 149°	0.17
X-672 GR-S rubber	100	0.17 at 38° and 0.25 at 93°	0.14

Source: Ref. 6.

not change by more than 30°C. Incidentally, the apparent large effect on n for polyvinylbutyral is masked by the plasticizer effect.

Whereas, the effect of temperature on n is not great, the same cannot be said for K, the consistency index. Here it is found that temperature has an exponential effect as in an Arrhenius relation:

$$K = A_{exp} \; (E/RT) \tag{3-45}$$

Values of E at low shear rates are given in Table 3-4.

At higher shear rates, the behavior is somewhat more complicated as shown in Table 3-5.

In nonisothermal situations, it is necessary to compensate for changes in flow behavior as just discussed.

Before nonisothermal conditions can be dealt with, it is necessary to be able to compute temperature changes such as heat transfer. The very nature of polymer melts so complicates the situation that normal approaches do not necessarily suffice.

In order to get better insight into this behavior, consider the transfer of heat to a flowing molten polymer. The fluid itself is not only non-Newtonian but also, to a certain extent, compressible. Furthermore, the possibility of elastic effects exists. Finally, molten polymer flow gives rise to viscous dissipation or internal heat generation. Basically, this means that heat is generated internally in the fluid according to the relation

$$\phi_v = \frac{\text{Viscous}}{\text{heating factor}} = \mu_{APP} \left(\frac{dV_z}{dr} \right)^2 \tag{3-46}$$

A rough estimate of the importance of viscous dissipation can be made using the relation

$$\text{Temperature rise} \cong \frac{\Delta P}{Jc_p} \tag{3-47}$$

where ΔP is the pressure drop, J the mechanical equivalent of heat, and C_p, the polymer specific heat. An estimated value in the order of magnitude of 1°C, as in Eq. 3-47 would mean that viscous dissipation was not significant.

Higher estimated temperatures would mean that the heat transfer itself must be calculated. Methods for doing this will be presented in the next chapter.

TABLE 3-4 *E* **Values at Low Shear Rates**

Polymer	E Value
Polyethylene	11-12.8 kcal/g mole
Polyisobutylene	15.7-16.4 kcal/g mole
GR-S rubber	20.8 kcal/g mole
Polystyrene	22.0-23.0 kcal/g mole
Polyvinyl butyral	25.9 kcal/g mole
Polyvinyl chloride acetate	35.0 and 60.0 kcal/g mole
Cellulose acetate	70.0 kcal/g mole

Source: Ref. 6.

TABLE 3-5 *E* **Values at High Shear Rates**

Material	Average Temp. °C	Shear Rate s^{-1}	Activation Energy, kcal/g mole
Polyethylene (melt index 2.1)	150	1,000	6.9
Polyethylene (melt index 2.1)	200	1,000	5.7
Polyethylene (melt index 2.1)	150	100	8.2
Polyethylene (melt index 2.1)	200	100	7.0

3.11 APPLIED RHEOLOGY EXAMPLES

Example 3-1

Consider an extruder barrel and screw as shown in Fig. 3-22. The pertinent dimensions are:

D = 2.50 in. $\quad t$ = 2.50 in.
L = 10.0 in. $\quad n$ = 1
h = 0.100 in. $\quad g$ = 0.007 in.
e = 0.250 in.

It can be shown that the volumetric flow rate is made of drag flow and pressure flow:

$$q = q_d + q_p \tag{3-48}$$

Furthermore, using the proper geometrical considerations:

$$q = 0.775\, N - 3.13 \times 10^{-4}\, \frac{\Delta P}{\mu} \tag{3-49}$$

Units of q are in.^{3}/min, N is in rpm, ΔP is in psi, and μ is in lbf s/in.^{2}

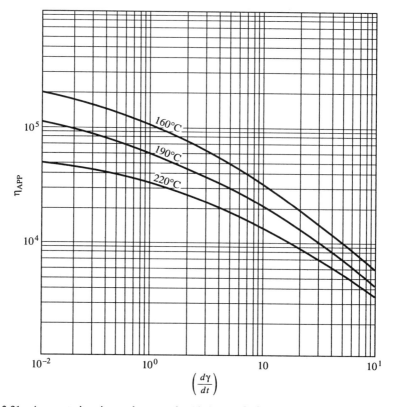

Fig. 3-21 Apparent viscosity vs shear rate for Alathon 10 [10].

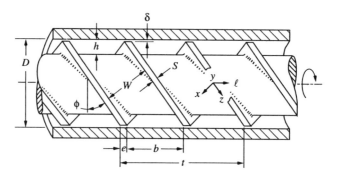

Fig. 3-22 Single-screw extruder (double-flighted) [3].

Calculate the discharge rate at 50 rpm if the polyethylene whose flow curve is shown in Fig. 3-21 is pumped at 374°F against an overall ΔP of 1000 psi.

Since we have a non-Newtonian fluid, we must have a shear rate from which to select a value of μ_{APP}. That is,

$$\tau = \mu_{APP} \frac{dV_x}{dr}$$

In order to arrive at the shear rate, let us assume that a mean value will suffice and that the value is

$$\gamma = \frac{dV_x}{dr} = \frac{\pi DN}{h}$$

$$\gamma = 65.5 \text{ s}^{-1}$$

which is an average or mean shear rate in the channel.

For this shear rate, the μ_{APP} value (from Fig. 3-21) is 0.098 lbf s/in.^{2} Substituting this and the values of N and ΔP in Eq. 3-49 yields

$$q = (38.7 - 3.6) \text{ in.}^3/\text{min}$$

$$q = 35.1 \text{ in.}^3/\text{min}$$

or, on a lbm basis,

$$w = \frac{(47.4 \text{ lbm})}{\text{ft}^3} \frac{(35.1 \text{ in.}^3)}{\text{min}} \frac{(\text{ft}^3)}{1728 \text{ in.}^3} \frac{(60 \text{ min})}{h} \quad w = 57.6 \frac{\text{lbm}}{h}$$

Example 3.2:   Design Analysis of a Composite Die

Suppose we have a composite die (Fig. 3-23), where R is 1.414 in., h is 0.08 in. (height of slit), w is 24 in., and L (land length) is 0.20 in.

Analyze the situation if 2 in.^{3}/s are pumped from an extruder to this die. The polymer is the polyethylene whose flow behavior is given in Fig. 3-21. The temperature is 374°F.

First, estimate the shear rates in the cylindrical and slit portions of the die. For the cylindrical section,

$$\gamma_{APP} = \text{apparent shear rate} = \frac{1}{2} \frac{4q}{\pi R^2}$$

$$\gamma_{APP} = \frac{4(2 \text{ in.}^3/\text{s})}{2 \pi (1.414 \text{ in.})^2} = 0.64 \text{ s}^{-1}$$

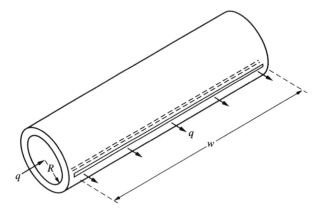

Fig. 3-23   Diagram of an end-fed manifold sheeting die. Land length of lips L is measured in direction of small arrows emerging from lips [3].

The 1/2 is used because all the fluid has gone out of the slit before the cylinder end is reached.

For the slit,

$$\gamma_{APP} = \frac{6q}{wh^2} = \frac{6(2 \text{ in.}^3/\text{s})}{(24 \text{ in.}) (0.08 \text{ in.})^2} = 78 \text{ s}^{-1}$$

Then, the μ_{APP} values can be obtained from Fig. 3-21.

Section	γ_{APP}, s^{-1}	μ_{APP}, lbf s/in.^{2}
Cylinder	0.64	0.54
Slit	78	0.0769

and

$$\Delta P_{\text{cylinder}} = \frac{8q\mu_{APP}W}{\pi\gamma^4}$$

using half the volumetric flow.

$$\Delta P_{\text{cylinder}} = \frac{8(1 \text{ in.}^3/\text{s}) (0.54 \frac{\text{lbf}}{\text{in.}^2}) (24 \text{ in.})}{\pi (1.414 \text{ in.})^4} = 8.26 \text{ psi}$$

also

$$\Delta P_{slit} = \frac{12q\mu_{APP}L}{wh^3}$$

$$\Delta P_{slit} = \frac{12 \ (2 \ \text{in.}^3/\text{s}) \ (0.0769 \ \text{lbf/in.}^2) \ (0.20 \ \text{in.})}{(24 \ \text{in.}) \ (0.08 \ \text{in.})^3} = 29.98 \ \text{psi}$$

Thus, the pressure drop in the slit is almost four times that in the cylinder. This would mean that the film would be thicker at the feed end than at the far end.

Example 3-3

For the extruder of Example 1, what length die should be used if a 0.5-in. diameter is specified. The polymer has a n of 0.46 and a K of 3.0 lbf s^n/in.^{2}
The extruder output is

$$q = 35 \ \text{in.}^3/\text{min}$$

then the extruder q must be the q die and

$$V \text{ for die} = \frac{35 \ \text{in.}^3/\text{min} \ (\text{min}/60 \ \text{s})}{\pi \ (0.5)^2 \ \text{in.}^2} = 2.98 \ \text{in./s}$$

Next, compute a modified Reynolds number for the die:

$$Re' \ \frac{D^n V_\gamma^{2-n}}{g_c K' 8^{n-1}}$$

$$n = n' = 0.46$$

$$K' = (3.0 \ \text{lbf s}^n/\text{in.}^2) \left(\frac{3n+1}{4n}\right)^n$$

$$K' = 3.376 \ \text{lbf s}^n/\text{in.}$$

$$Re' = \frac{(0.5 \ \text{in.})^{0.46} \ (2.98 \ \text{in./sec})^{1.54} \ (47.4 \ \text{lbm ft}^3) \ (\text{ft}^3/1728 \ \text{in.}^3)}{(32.2 \ \text{lbm ft/lbf s}^2 \ (12 \ \text{in.})/\text{ft} \ (3.376 \ \text{lbf s}^n/\text{in.}^2) \ (8)^{-0.54}}$$

$$Re' = 2.7 \times 10^{-4}$$

From Fig. 3-14 for $Re' = 2.7 \times 10^{-4}$,

$$f = 59,300$$

and

$$L = \frac{\Delta P g c D}{2 f \rho V^2}$$

$$L = \frac{(1000 \text{ lbf/in.}) \ (32.2 \text{ lbm ft/lbf s}) \ (12 \text{ in./ft}) \ (0.5 \text{ in.})}{(59{,}300) \ (47.4 \text{ lbm/ft}^3) \ (\text{ft}^3/1728 \text{ in.}) \ (2.98 \text{ in./s})^2}$$

$$L = 6.67 \text{ in.}$$

REFERENCES

1. GRISKEY, R.G., CHOI, M.H. and SISKOVIC, N., *Polymer Eng. Sci.* **13**, 287 (1973).
2. METZNER, A.B., and REED, J.C., *AIChE J.* **1**, 434 (1955).
3. METZNER, A.B., *Processing of Thermoplastic Materials*, E.C. Bernhardt, ed., Reinhold, New York (1965), Chap. 1.
4. KOZICKI, W., CHOU, C.H., and TIU, C., *Chem. Eng. Sci.* **21**, 665 (1966).
5. MCKELVEY, J.M., *Polymer Processing*, Wiley, New York (1962), p. 91.
6. PHILIPPOFF, W., and GASKINS, F.H., *J. Polymer Sci.* **21**, 205 (1956).
7. BESTUL, A.P., and BELCHER, H.V., *J. Appl. Phys.* **24**, 696 (1953).
8. GRISKEY, R.G., SISKOVIC, N., and SALTUK, I., *Polymer Eng. Sci.* **12**, 397 (1971).
9. *Ibid.* **12**, 402 (1972).
10. GREGORY, D.R., GRISKEY, R.G., and SISKOVIC, N., *AIChE J.* **17**, 281 (1971).
11. VAN WAZER, J.R., LYONS, J.W., KIM, K.Y., COLWELL, R.E., *Viscosity and Flow Measurement*, Interscience, New York (1963).
12. BIRD, R.B., ARMSTRONG, R.C., and Hassager, O., *Dynamics of Polymeric Liquids*, Vols. I and II, Wiley, New York (1977).
13. BRODNYAN, J.G., GASKINS, F.H., and PHILIPOFF, W., *Trans. Soc. Rheol.* **1**, 109 (1957).
14. DEALY, J.M. and WISSBRUN, K.F., *Melt Rheology and Its Role in Plastics Processing*, Van Nostrand Reinhold, New York (1990).
15. GRISKEY, R.G., and GREGORY, D.R., *AIChE J.* **13**, 122 (1967).
16. BIRD, R.B., and SADOWSKI, T., *Trans. Soc. Rheol.* **9**, 243 (1965).

PROBLEMS

3-1   Various materials are placed between two parallel plates. A series of stresses are applied to the upper plate as listed:

Time Interval	Stress Values
$0-t_1$	0
t_1-t_2	S
t_2-t_3	0
t_3-t_4	2S
t_4-t_5	0

All the time intervals are equal.

Plot the strain–time behavior for the following systems: Newtonian fluid, Hookean solid, non-Newtonian fluid, a Maxwell fluid (strain is made up of an elastic viscous component), $\gamma = \gamma_e + \gamma_v$.

3-2 A polyethylene melt is flowing through a 3-m-long pipe (diameter 0.05 m) with a volumetric flow rate of $9.4 \times 10^{-4} \text{m}^3/\text{s}$. What is the maximum point velocity in the tube? Indicate all assumptions.

3-3 Non-Newtonian fluid behavior in pastes and fine suspensions can be described by the Bingham model. The fluid remains rigid below a yield stress τ_0 and then flows when τ exceeds τ_0.

$$\tau = -\eta_0 \gamma \pm \tau_0 \qquad |\tau_{yx}| > \tau_0$$

$$\gamma = 0 \qquad |\tau_{yx}| < \tau_0$$

A vertical tube filled with such a fluid has a flat plate held over its bottom end. If the plate is removed, what is the criterion for flow to occur?

3-4 The apparent viscosity of a non-Newtonian fluid can be written as one of two functions:

$$\eta_{APP} = \eta_{APP} (\tau, T)$$

or

$$\eta_{APP} = \eta_{APP} (\gamma, T)$$

Find the relation between $(\partial\eta_{APP}/\partial T)_\tau$ and $(\partial\eta_{APP}/\partial T)_\gamma$ for both a dilatant fluid and a pseudoplastic fluid.

3-5 Estimate the apparent viscosity of the polyethylene of Tables 3-4–3-6 at a temperature of 260°C and a shear rate of 600 s^{-1} if the value at 174°C is 0.315 kg/m-s. Justify any assumptions.

3-6 Pressure drop data for polyethylene flowing in various capillaries are given below as a function of shear rate.

	Pressure Drop N/m$^2 \times 10^{-5}$		
Shear Rate, s^{-1}	(L/D = 60)	(L/D = 30)	(L/D = 10)
40	28.96	16.89	6.90
60	38.62	21.38	11.73
90	48.28	27.25	13.80
120	55.18	31.38	15.86
250	75.87	45.33	22.07

Use the end correction method to bring all the data on a common line.

3-7 What is the value, or what are the values, of N in Fig. 3-20?

3-8 A small-scale piping system (diameter, 0.025 m; length, 0.25 m) gives the following results for a non-Newtonian fluid:

Mass Flow Rate kg/s	Pressure Drop N/m^2
1.92×10^{-4}	17,243
4.08×10^{-4}	34,486
8.8×10^{-4}	68,975
1.95×10^{-3}	1.38×10^5
4.58×10^{-3}	2.76×10^5

If the pressure drop in a 0.03-m-diam pipe is limited to 13.8×10^5 N/m² at a volumetric flow rate of 6.26×10^{-5} m³/s, what length of pipe is needed?

3-9 A polymer solution has the following Ellis equation constants ($\eta_0 = 26.1$ dyne-s/cm²; $\alpha = 1.412$; $\tau_{1/2} = 6.172$ dynes/cm²). What is the pressure drop for a 3-m-long round pipe (diameter, 0.05 m) with a flow rate of 6.3×10^{-4} m³/s?

3-10 The rheological data for polymer melt are as follows:

τ, dynes/cm²	γ, s^{-1}
14.1	0.9
58.5	10
228	100
344	200
435	300
502	400
780	800
820	1000

What diameter of circular pipe will give a pressure drop of 109.1×10^5 N/m² for a volumetric flow of 1.9×10^{-3} m³/s?

3-11 A power law fluid (n' of 0.29; k' of 2.6 N-s$^{n'}$/m²) flows in a 0.04-m tube with an average velocity of 6.4 m/s. What would the pressure drop be if the tube were 28 m long?

3-12 In Fig. 3-14, the transition Reynolds number for turbulent flow increases in value as n' decreases. The line for n' at 0.0 does not show a transition. Explain why this occurs.

3-13 A molten polymer is to be extruded through a heptagonal die with an average velocity of 0.04 m/s. The available pressure drop through the 0.1-m-long die is 55.18×10^5 N/m². The power law parameters for the polymer are $n = 0.53$ and $k = 3 \times 10^4$ N-s^n/m². What is the characteristic dimension for the die (i.e., side of heptagon)?

3-14 An engineer is designing a capillary rheometer to be used with a polymer whose n value is 0.3. The polymer's density is 695 kg/m³. Capillary radius is 0.005 m.

What are the limits on volumetric flow rate if the actual and observed pressures must be within 10% of each other?

3-15   Fit the power law, the Ellis equation, the Eyring equation, and the Reiner–Phillipoff equation to the data of Figs. 3-19 and 3-21.

3-16   Compare the pressure drop per unit length value found using Eq. (3-41) in the text with that determined from Fig. 3-14 (using hydraulic radius) for a wide but narrow rectangular channel (height, 0.02 m; width, 0.3 m). The rheological parameters for the fluid are an n of 0.7 and a K of 6×10^4 N-s^n/m$_2$.

3-17   Develop a correlation between the critical Reynolds number (point at which laminar flow ends) and the n' value (i.e., see Fig. 3-14) from 1.0 to 0.0.

3-18   The residence time ratio (RTR), which is an important scale-up parameter for flow reactors, can be determined either by dividing the maximum velocity by the average velocity or by using the ratio of average residence time by minimum residence time. Typical values in Newtonian flow systems are 1.0 (plug flow), 1.25 (turbulent flow), and 2.0 (laminar flow). Can these cases be matched by flow situations for pseudoplastic fluids? Prove your answers.

3-19   A molten polymer (n' of 0.71 and a K' of 2100 N-s^n/m^2) flows through a tube (diameter, 0.01 m; length, 0.5 m). If the pressure drop is 87.7×10^5 N/m^2, what is the fluid's average velocity? Also, what is its maximum velocity?

3-20   The data of Fig. 3-14 do not show any critical Reynolds number data for shear-thickening (i.e., $n' > 1.0$) fluids. Speculate as to how the transition to turbulent flow takes place in these fluids. Clearly explain and provide support for your position.

MINI PROJECT A

Develop a single rheological equation that describes both shear stress and normal stress behavior as function of shear rate (use data from Figs. 3-9, 3-10, 3-17, as well as other published data).

MINI PROJECT B

Bird and Sadowski [16] developed equations to treat the flow of polymer solutions through porous media using the Ellis equation. Later work found pressure drop and flow data for molten polymers through porous media [10,15]. This later work used the power law as its rheological base. Check to determine whether the Ellis equation approach can describe these data.

MINI PROJECT C

The flow behavior of polymer melts and solutions (Figs. 3-8 and 3-10) shows at low and high shear rates what appear to be Newtonian viscosities with a transition regime in between. Develop a model based on the molecular structure of a polymer melt or solution to describe these phenomena.

CHAPTER 4

HEAT TRANSFER IN POLYMER SYSTEMS

4.1 INTRODUCTION

The transfer of heat is an important aspect of every polymer processing operation. In fact, without heat transfer, the shaping and forming of polymers would not be possible.

Although considerable progress has been made in understanding heat transfer in polymeric systems, much still remains to be done. There are a number of reasons why this is so, including lack of good physical data such as (thermal conductivities, specific heats, and rheological behavior), and failure to cope with such complexities of polymer behavior as elastic effects, compressibility, and viscous dissipation.

The present effort will be directed to providing a fundamental and useful treatment of heat transfer in polymeric systems with the goals of affording both understanding and practical techniques.

For a cylindrical coordinate system (as shown in Fig. 4-1), the equation of energy is:

$$
\begin{aligned}
\rho \hat{C}_v \left(\frac{\partial T}{\partial t} + v_r \frac{\partial T}{\partial t} + \frac{v_\theta}{r} \frac{\partial T}{\partial \theta} + v_z \frac{\partial T}{\partial z} \right) &= - \left[\frac{1}{r} \frac{\partial}{\partial r} (r q_r) + \frac{1}{r} \frac{\partial q_\theta}{\partial \theta} + \frac{\partial q_z}{\partial z} \right] \\
&\quad - T \left(\frac{\partial p}{\partial T} \right)_\rho \left(\frac{1}{r} \frac{\partial}{\partial r} (r v_r) + \frac{1}{r} \frac{\partial v_\theta}{\partial \theta} + \frac{\partial v_z}{\partial z} \right)
\end{aligned}
\tag{4-1}
$$

$$-\left\{\tau_{rr}\frac{\partial v_r}{\partial r}+\tau_{\theta\theta}\frac{1}{r}\left(\frac{\partial v_\theta}{\partial\theta}+v_r\right)\right.$$

$$+\tau_{zz}\frac{\partial v_z}{\partial z}\right\}-\left\{\tau_{r\theta}\left[r\frac{\partial}{\partial r}\left(\frac{v_\theta}{r}\right)+\frac{1}{r}\frac{\partial v_r}{\partial\theta}\right]\right.$$

$$+\tau_{rz}\left(\frac{\partial v_z}{\partial r}+\frac{\partial v_r}{\partial z}\right)+\tau_{\theta z}\left(\frac{1}{r}\frac{\partial v_z}{\partial\theta}+\frac{\partial v_\theta}{\partial z}\right)\right\}+A_0$$

where:

r, θ,z = coordinate directions
v_r, v_θ, v_z = velocity components in the r, θ, and z coordinates
$\tau_{r\theta}$ τ_{rz}, $\tau_{\theta z}$ = shear stresses
τ_{rr}, $\tau_{\theta\theta}$, τ_{zz} = normal stresses
ρ = density
p = pressure
T = point temperature
q_r, q_z, q_θ = components of energy flux
$\hat{C}_v$ = constant-volume specific heat
A_0 = a heat-generation term

In verbal form, Eq. (4-1) shows that the temperature of a fluid element in motion is affected by heat conduction, the q terms in the first bracket; expansion effects, the second term on the right multiplied by T $(\partial p/\partial t)\rho$; and viscous heating or viscous dissipation, the remainder of the terms on the right-hand side of the equation excepting A_0. A_0 is for all other types of heat generation, such as phase changes, chemical sources, and electrical sources.

The viscous dissipation effects take place in all fluids and results because the

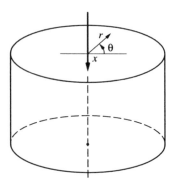

Fig. 4-1   Cylindrical coordinates.

energy used to move the fluid becomes dissipated. The size of this effect is related both to the velocity gradient and the fluid apparent viscosity. Hence, for fluids with large apparent viscosities, such as molten polymers, this effect can be quite sizable.

If the density ρ and thermal conductivity can be considered constant, Eq. (4-1) becomes

$$
\begin{aligned}
\rho \, \hat{C}_p & \left(\frac{\partial T}{\partial t} + v_r \frac{\partial T}{\partial r} + \frac{v_\theta}{r} \frac{\partial T}{\partial \theta} + v_z \frac{\partial T}{\partial z} \right) \\
& = k \left[\frac{1}{r} \frac{\partial}{\partial r} \left(r \frac{\partial T}{\partial r} \right) + \frac{1}{r^2} \frac{\partial^2 T}{\partial \theta^2} + \frac{\partial^2 T}{\partial z^2} \right] \\
& + 2\mu \left\{ \left(\frac{\partial v_r}{\partial r} \right)^2 + \left[\frac{1}{r} \left(\frac{\partial v_\theta}{\partial \theta} + v_r \right) \right]^2 \right. \\
& + \left. \left(\frac{\partial v_z}{\partial z} \right)^2 \right\} + \mu \left\{ \left(\frac{\partial v_\theta}{\partial z} + \frac{1}{r} \frac{\partial v_z}{\partial \theta} \right)^2 \right. \\
& + \left. \left(\frac{\partial v_z}{\partial r} + \frac{\partial v_r}{\partial z} \right)^2 + \left[\frac{1}{r} \frac{\partial v_r}{\partial \theta} + r \frac{\partial}{\partial r} \left(\frac{v_\theta}{r} \right) \right]^2 \right\} + A_0
\end{aligned}
\tag{4-2}
$$

Forms of the energy equation in rectangular and spherical coordinates are given elsewhere [1].

4.2 FLUID THERMAL PROPERTIES

Proper thermal physical data are among the necessary ingredients for the treatment of heat transfer and the use of the energy equation. The case of polymers is no exception. At times, however, there are difficulties in this area because both the data and correlation are not as extensive as those for solids, liquids, and gases.

The situation with respect to specific heat or heat capacity data and correlation has already been discussed. The other important physical thermal parameter, thermal conductivity, represents an even more poorly defined case.

4.3 THERMAL CONDUCTIVITY

Anderson [2] and Knappe [3] have summarized most of the available polymer thermal conductivity data. Where possible, of course, experimental data should be used. In many instances, however, such data are not available. It therefore becomes necessary to use correlations for estimating the needed information.

Separate correlations exist for (1) amorphous solid, (2) semicrystalline solid, and (3) molten polymers. For example, a generalized correlation [4] for solid amorphous polymers is given in Fig. 4-2. This requires the T_g for the polymer and the thermal conductivity at T_g (this can be bypassed if one thermal conductivity is known at a given temperature). Hence, Fig. 4-2 requires specific data before it can be used.

For cases in which T_g and the needed data are not available, the thermal conductivity can be estimated in other ways. In the region above the T_g ($T/T_g >$ 1.1) Eq. 4-3,

$$k = 6.3 \times 10^{-3} \frac{[1 - 0.00015 \ (T - T_g)]}{(T_g^{0.216})} M^{-0.3} \qquad (4\text{-}3)$$

where k is thermal conductivity in cal/cm s °C and M is the mer weight, gives reasonable estimates (error < 2%).

For amorphous solid polymers below T_g, it is found that k correlates well with the refractive index [5]. Such correlation (see Fig. 4-3) can be used to get the value of k at a given T needed to use Fig. 4-2.

Thermal conductivity values for solid semicrystalline polymers can be estimated using the method of Eiermann [6]. This involves the relation

$$\frac{kc}{ka} - 1 = 5.8 \left(\frac{\rho c}{\rho a} - 1 \right) \qquad (4\text{-}4)$$

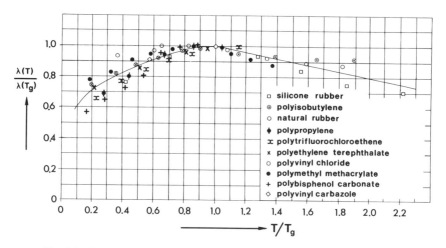

Fig. 4-2   Generalized curve for amorphous polymer thermal conductivities [4].

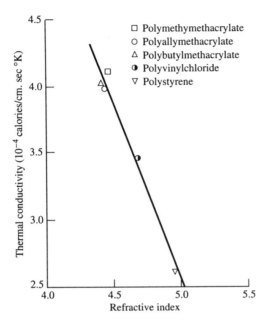

Fig. 4-3   Amorphous polymer thermal conductivities vs refractive index [5].

where *ka, kc* are the thermal conductivities for the pure amorphous and crystalline states and ρa, ρc the corresponding densities. This relation can be used in conjunction with Figs. 4-4 and 4-5 to estimate the thermal conductivity for a polymer of known crystallinity.

If values of both ρa or ρc and *ka* or *kc* are not known, a correlation of the type shown in Fig. 4-6 can be used to estimate *k* at a given temperature. All that is needed is the specific heat and density at room temperature.

Molten polymer thermal conductivities represent a somewhat different case than that of the solid phase. In essence, the thermal conductivities of the melt vary little with temperature. Consequently, an acceptable value of *k* can be estimated from the relation

$$k = \frac{1.2 \times 10^{-2} \, (Cp) \, (\rho)^{1.33}(M)^{0.3}}{(T_m)^{0.216}} \tag{4-5}$$

where T_m is the melting temperatures in °K, and C_p and ρ are the values for the specific heat and density of the melt. Equation (4-5) was found to represent experimental data for a number of polymers to within 2%.

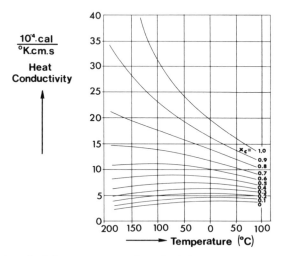

Fig. 4-4   Thermal conductivity vs temperature for polyethylene as a function of crystallinity [4].

The effect of orientation of a solid and its resultant anistropy is given in the review papers by Anderson [2] and Knappe [3]. Pressure effects generally can be handled [3] as in the equation

$$\frac{dk}{dp} = (k_{HP})\ (5.25)\ (\beta) \tag{4-6}$$

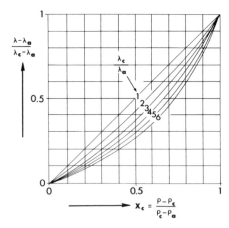

Fig. 4-5   Correlation polymer thermal conductivity with temperature and crystallinity [4].

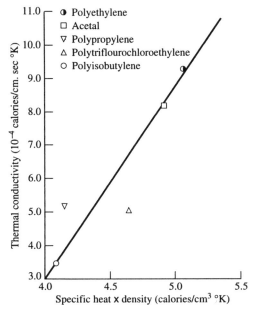

Fig. 4-6   Thermal conductivity for crystalline polymers vs product of specific heat and density [5].

where β is the isothermal compressibility and k_{HP} is the high-pressure value of thermal conductivity.

Example 4-1:   Estimating Thermal Conductivity of an Amorphous Polymer Using a Generalized Correlation

Given that the thermal conductivity for polymethyl methacrylate is 5.2 × 10^{-4} cal/cm K s at 25°C, what is the value at 200°C?

At 25°C,

$$T/T_g = 0.77$$

and, from Fig. 4-2,

$$\frac{k \, (298 \, K)}{k \, (T_g)} = 0.96 = \frac{5.2 \times 10^{-4} \, \dfrac{cal}{cm \, K \, s}}{k \, (T_g)}$$

so that

$$k\ (T_g) = 5.4 \times 10^{-4} \ \frac{\text{cal}}{\text{cm K s}}$$

and, at 200°C,

$$\frac{T}{T_g} = 1.22$$

Then, from Figure 4-2,

$$\frac{k\ (473 \text{ K})}{k\ (T_g)} = \frac{k\ (473 \text{ K})}{5.4 \times 10^{-4} \ \dfrac{\text{cal}}{\text{cm K s}}} = 0.95$$

Hence,

$$k\ (473 \text{ K}) = 5.1 \times 10^{-4} \ \frac{\text{cal}}{\text{cm K s}}$$

Example 4-2: Estimating Thermal Conductivity of a Thermally Softened Amorphous Polymer Using Eq. (4-3)

What is the thermal conductivity of polystyrene at 496 K?

Mer weight M = 104.5
T_g = 373 K.

Then,

$$k = \frac{6.3 \times 10^{-3} \ [1 - 0.00015 \ (496\text{-}373)]}{(373)^{0.216} \ (104.5)^{0.300}}$$

$$k = 4.2 \times 10^{-4} \ \frac{\text{cal}}{\text{cm K s}}$$

compared to an experimental value of 4.19×10^{-4} cal/cm K s.

Example 4-3:   Estimating Thermal Conductivity of a Molten Semicrystalline Polymer Using Eq. (4-5),

What is the thermal conductivity of a molten polyethylene (density of 0.976 g/cm^3 at 25°C)?

$$M = 28.1$$

$$T_m = 409 \ K$$

$$C_p = 0.54 \ \frac{\text{cal}}{\text{g K}}$$

$$k = \frac{1.2 \times 10^{-2} \ (0.54) \ (0.976)^{1.33}}{(409)^{0.216} \ (28.1)^{0.300}} = 6.29 \times 10^{-4} \ \frac{\text{cal}}{\text{s cm K}}$$

compared to an experimental value of $6.15 \times 10^{-4} \ \dfrac{\text{cal}}{\text{cm s K}}$.

4.4 CONDUCTION HEAT TRANSFER IN POLYMER SYSTEMS

If a system is not in motion, the general equation of energy can be greatly simplified to

$$\rho C_v \frac{\partial T}{\partial t} = - \left[\frac{1}{r} \frac{\partial}{\partial r} (rq_r) + \frac{1}{r} \frac{\partial q_\theta}{\partial \theta} + \frac{\partial q_z}{\partial z} \right] + A_0 \tag{4-7}$$

or, if density and thermal conductivity are constant, to

$$\rho \ C_p \frac{\partial T}{\partial t} = k \left[\frac{1}{r} \frac{\partial}{\partial r} \left(r \frac{\partial T}{\partial r} \right) + \frac{1}{r^2} \frac{\partial^2 T}{\partial \theta^2} + \frac{\partial^2 T}{\partial Z^2} \right] + A_0 \tag{4-8}$$

These equations are then applicable to many aspects of polymer processing, such as cooling material in a mold, heating a sheet, and thermoforming. Fundamentally, they represent cases of unsteady-state heat transfer, and the solutions of Eq. (4-7) and (4-8) can be quite complex. However, generalized solutions for various shapes do exist. The basis for these is to neglect A_0 and then to solve the remaining equation:

$$\rho C_p \left(\frac{\partial T}{\partial t} \right) = k \left[\frac{1}{r} \frac{\partial}{\partial r} \left(r \frac{\partial T}{\partial r} \right) + \frac{1}{r^2} \frac{\partial^2 T}{\partial \theta^2} + \frac{\partial^2 T}{\partial Z^2} \right] \tag{4-9}$$

for a given shape.

Graphical versions [8] of these solutions are given in Figs. 4-7–4-10. In Fig. 4-7–4-9, the solutions are summarized, respectively, for large slabs, spheres, and long cylinders. Figure 4-10 gives the midplane temperatures for various solid shapes. The nomenclature for the charts is as follows: t_a, t_m, t_x, t, and t_b, respectively, are the ambient, midplane, surface, point, and original temperatures; k is the polymer thermal conductivity; θ is the time; ρ and C_p are the polymer density and specific heat; and r and r_m are the position and radius or half-thickness.

In some cases, average temperatures are also of interest. These can be found from Fig. 4-11 [9]. Another situation that is of interest from the viewpoint of polymer processing is the unsteady-state transfer of heat to polymer granular particles. This is especially important, for example, in the solids conveying a section of an extruder. Figure 4-12 [10] gives such data for polymer particles resembling chips (average size 0.084 to 0.134 in.) and powders.

4.5 CONDUCTION EXAMPLES

Example 4-4: Bonding Together Two Plastic Sheets

Two sheets of glass-reinforced polyester are to be bonded together with an adhesive that fuses at 110°C. The press used to heat the system has platens capable of attaining 200°C. How long should it take to bond the sheets if each is 2 cm thick?

In this case, we assume that the interface (adhesive) must reach 110°C to bond. Then,

$$y = \frac{t_a - t_m}{t_a - t_b} = \frac{200 - 110}{200 - 25} = 0.515$$

This assumes that resistance between platens and sheet surfaces is negligible. The x value corresponding to the y value is 1.32. Hence,

$$x = 1.32 = \frac{k\theta}{\rho\, C_p\, r^2\, m}$$

$$\theta = \frac{1.32\, r^2 m}{(k/\rho\, C_p)}$$

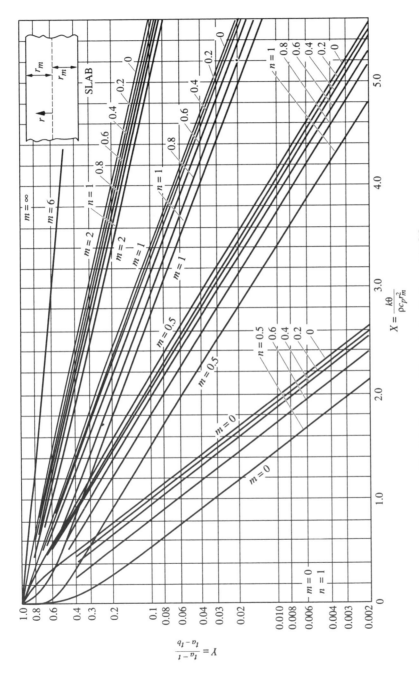

Fig. 4-7   Unsteady-state conduction for large slabs [8].

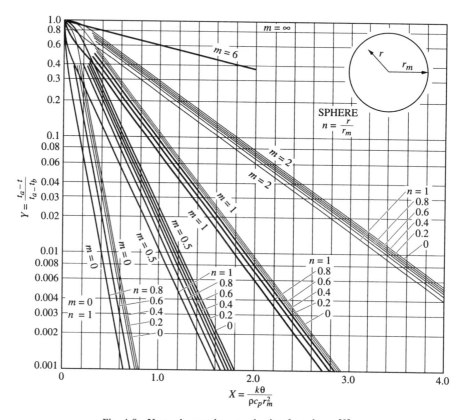

Fig. 4-8   Unsteady-state heat conduction for spheres [8].

The quantity $(k\ \rho C_p)$ is α, the thermal diffusivity. For the glass-reinforced polyester, α is about 2.6×10^{-3} cm²/s. Hence,

$$\theta = \frac{(1.32)\ (2\ \text{cm})^2}{(2.6 \times 10^{-3}\ \text{cm}^2/\text{s})} = 2.034 \times 10^3\ \text{s}$$

or

$$\theta = \frac{2034}{3600}\ \text{h} = 0.565\ \text{h}$$

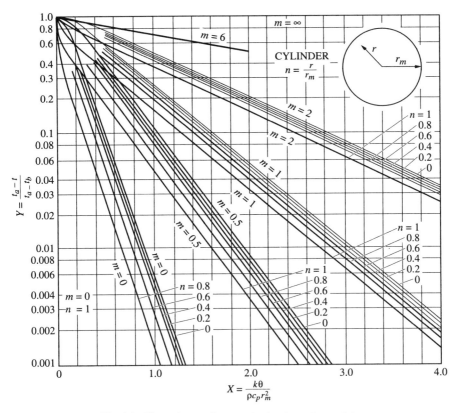

Fig. 4-9   Unsteady-state heat conduction for cylinders [8].

Example 4-5:   Cooling an Injection-Molded Object

Suppose a polystyrene object whose shape can be approximated by a slab (2 cm thick) is to be cooled from 200 to 38°C in a mold. What is the cooling time required if the surroundings are at 25°C.

Since average temperature is required, we use Fig. 4-11. For this case,

$$y = \frac{25 - 38}{25 - 200} = \frac{13}{175} = 0.0742$$

Assume that $h >>> k$ and, consequently, $m = k/h \, rm \cong 0$. Then, from Fig. 4-11, the x value is

$$x = 0.97$$

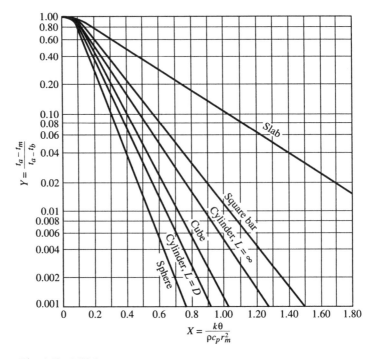

Fig. 4-10   Midplane temperatures for unsteady-state heat conduction [9].

and,

$$x = \frac{\alpha\theta}{r^2 m} = 0.97$$

Then,

$$\theta = \frac{(0.97)\ r^2 m}{\chi}$$

For polystyrene at an average temperature of $\left(\dfrac{200\ +\ 38}{2}\right)$°C or 119°C, the α value is 1.235×10^{-3} cm^2/s.

Thus,

$$\theta = \frac{(0.97)\ (1\ \text{cm})^2}{1.235 \times 10^{-3}\ \text{cm}^2/\text{s}} = 785\ \text{s}$$

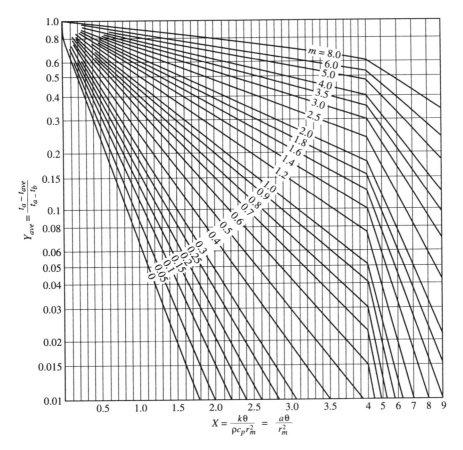

Fig. 4-11   Average slab temperatures for unsteady-state heat conduction [8].

or

$$\theta = \frac{785}{3600} = 0.218 \ h$$

Example 4-6:   Heating Resin in the Solids-Conveying Section of an Extruder

A nylon 610 material is charged to a screw extruder whose barrel and screw walls are heated to 200°C. If the depth of the helical flow channel is 1.2 cm, what is the maximum time required for the midplane temperature to reach 175°C?

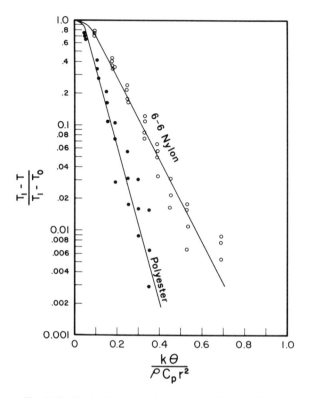

Fig. 4-12   Unsteady-state heat-transfer to polymer chips.

In this case, the granular solids can be represented by a bed of the material. The maximum required time will occur when heat is supplied only by the extruder barrel and screw walls (i.e., no frictional heating assumed).

Figure 4-12 is therefore applicable. For this case,

$$y = \frac{T_1 - T}{T_1 - T_0} = \frac{25 - 175}{25 - 200} = \frac{150}{175} = 0.855$$

The x value corresponding to this, for 610 nylon, is

$$x = 0.09$$

and

$$\frac{k\theta}{\rho \, C_p r^2} = 0.09$$

$$\theta = \frac{0.09 \, r^2}{k/\rho \, C_p}$$

The $k/\rho \, C_p$ or α value for 610 nylon at $(175 + 25)/2$, or $100,°C$ is 9.55×10^{-5} cm^2/s. Hence,

$$\theta = \frac{(0.09) \, (0.6 \text{ cm})^2}{9.55 \times 10^{-5} \, \dfrac{\text{cm}^2}{\text{s}}} = 340 \text{ s}$$

If the velocity of the granular solids along the flow channel were known, then, the distance to reach the desired temperature could be calculated.

4.6 CONVECTION HEAT TRANSFER IN POLYMERIC SYSTEMS; SOLUTIONS OF THE ENERGY EQUATION IN CIRCULAR CONDUITS

A fairly sizable technical literature has accumulated over the years in the area of convective heat transfer in polymeric systems. This literature can be roughly divided into those efforts that are mainly experimental and those that are essentially solutions to the equation of energy.

Many of these solutions to the equation of energy for tube flow used the form

$$V_z \frac{\partial T}{\partial Z} = \frac{k}{\rho C_p} \frac{1}{r} \left[\frac{\partial}{\partial r} \left(r \frac{\partial t}{\partial r} \right) \right] \tag{4-10}$$

which assumed only a V_z; no viscous dissipation; no compressibility effects; no internal heat sources; constant ρ, C_p, and k; and that Z direction convection ($\rho \, C_p \, V_z \, \partial T/\partial z$) far exceeded Z direction conduction $[k(\partial^2 T/\partial Z^2)]$.

This equation was then solved with both a form of the equation of motion,

$$\frac{-\partial P}{\partial Z} = \frac{1}{r} \frac{\partial}{\partial r} (r \, T_{rz}) \tag{4-11}$$

and an appropriate rheological equation.

A tabulation of these solutions is given in Table 4-1.

In Fig. 4-13, the Nusselt–Graetz solutions for plug flow and Newtonian fluids are compared to power law fluids solutions ($n = 1/2$, $n = 1/3$) as derived by Lyche and Bird [10]. In Fig. 4-14, calculated results for a power law fluid of $n = 1/3$ are shown [15]. The $\psi(E)$ is the measure of temperature dependence for the apparent viscosity.

These types of solutions, while interesting, usually do not provide the nec-

TABLE 4-1 Analytical Solutions of Energy Equation for Polymeric Systems

Authors and References	Rheological Equation Used	Remarks
Lyche and Bird (10)	Power law	—
Grigull (11)	Power law	—
Schenk and Van Laar (12)	Prandtl–Eyring	—
Whiteman and Drake (13)	Power law	—
Bird (14)	Power law	—
Christiansen and Craig (15)	Form of power law	Temperature-dependent "viscosity"
Christiansen and Jensen (16)	Powell–Eyring	Temperature-dependent "viscosity"
Kwant et al. (93)	Power law	
Mahalingham et al. (94)	Power law	
Joshi and Bergeles (103)	Power law	Constant heat flux, temperature-dependent properties

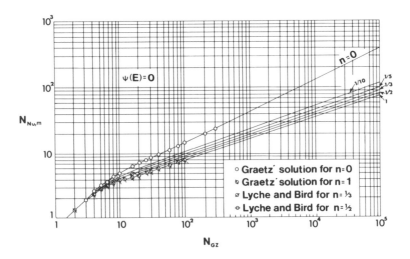

Fig. 4-13   Nusselt–Graetz solution for heat transfer in non-Newtonian systems [10].

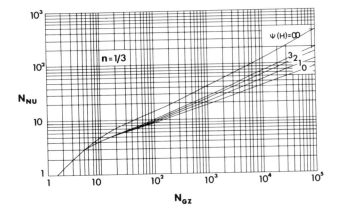

Fig. 4-14 Nusselt–Graetz solution for non-Newtonian systems (temperature-dependent rheology) [15].

essary means for dealing with polymer processing that involves molten or thermally softened polymers. For such cases, the effects of fluid compressibility, viscous dissipation, and even elasticity can be important.

There are also solutions that deal specifically with the thermal entrance region [96–98]. One of these will be considered later in this chapter since it involves experimental data.

In this respect, more rigorous solutions of the equation of energy were developed by a number of authors [17–21]. These authors assumed that convective Z direction $[\rho\,C_p\,V_z\,(\partial T/\partial z)]$ heat transfer exceeded conductive direction $[k\,(\partial^2 T/\partial Z^2)]$ heat transfer, that normal stresses could be neglected, and that there was only a V_z (and hence only a τ_{rz}). The energy equation then becomes

$$\rho\,C_v\left(V_z\,\frac{\partial T}{\partial Z}\right) = -\left[\frac{1}{r}\frac{\partial}{\partial r}\,(r\,q_r)\right] - \tau_{rz}\left(\frac{\partial V_z}{\partial r}\right) \tag{4-12}$$

Notice that Eq. (4-12) does not neglect viscous dissipation, compressibility effects, and temperature and pressure effects on ρ, C_v, and k.

Equation (4-12) can be transformed into a more amenable form by several substitutions. The first of these is that (from Fourier's law),

$$r\,q_r = -kr\,\frac{\partial T}{\partial r} \tag{4-13}$$

Also, since the system is now a compressible one $(C_p \neq C_v)$ and, consequently,

$$C_p - C_v = \frac{T\epsilon^2}{\rho\beta} \tag{4-14}$$

where $\epsilon = -1/\rho\ [(\partial\rho/\partial T)_p]$, and coefficient of thermal expansion, and $\beta = 1/\rho$ $[(\partial\rho/\partial p)_T]$, the compressibility.

If Eqs. (4-13) and (4-14) are substituted into Eq. (12), then the following form results:

$$\rho\ C_p\ V_z \left(\frac{\partial T}{\partial Z}\right) = \frac{1}{r}\frac{\partial}{\partial r}\left(kr\ \frac{\partial T}{\partial r}\right) + T\ \epsilon V_z \left(\frac{\partial P}{\partial z}\right) - \tau_{rz}\left(\frac{\partial V_z}{\partial r}\right) \tag{4-15}$$

This equation was then solved, together with the equation of motion,

$$\frac{\partial P}{\partial Z} = \frac{1}{r}\frac{\partial}{\partial r}\ (r\ \tau_{rz}) \tag{4-16}$$

and appropriate relations for the system's rheology and its physical property behavior with temperature and pressure. A summary of these solutions is given in Table 4-2.

There are other solutions for internal heat generation. These, however, treat the generation as being uniform [22–24], as linearly dependent [25], or as some function of radial position [26]. As such, they are not appropriate for viscous dissipation. None of the foregoing [22–26] considered the temperature dependence of physical properties or the effect of thermal expansion.

Mention should also be made of two other pieces of work, Refs. 101 and 102, which essentially represent review articles, together with some order-of-magnitude estimates of viscous dissipation effects.

Figures 4-15–4-20 show some of the results obtained in the cases of Table 4-2. Figure 4-15 shows the effect of thermal expansion [17] for a non-Newtonian fluid ($n = 0.25$) with negligible viscous dissipation and constant fluid properties. The various ϵ' values represent the average value of $T\epsilon$ across the tube. As can be seen, the effect of thermal expansion or compressibility cooling is to depress the point temperatures in the center of the tube.

In Fig. 4-16 [19], this effect is coupled with viscous dissipation and temperature-dependent physical properties. Here the centerline temperature depression occurs, together with a temperature peak near the wall. The Z values represent reduced lengths L/L_∞.

TABLE 4-2   Solutions to Eq. (4-15)

Authors and References	Rheological Equation	Remarks
Toor (17)	Power law	Neglects viscous dissipation; assumes constant physical properties; analytical solution
Toor (18)	Power law	Constant properties; treated inlet regions; analytical solution
Gee and Lyon (19)	Empirical equation $\frac{1}{\mu} = \left(\frac{1}{\mu_0}\right)(1 + k\tau^n)$	Temperature-dependent physical properties; computer solution
Toor (20)	Power law	Constant physical properties; forced convection; analytical solution
Forsyth and Murphy (21)	Power law	Temperature-dependent physical properties; computer solution
Popovska and Wilkinson (99)	Power law	No effect of compressibility cooling
Dang (100)	Power law	No effect of compressibility cooling

Let us examine these effects more closely, especially in light of their geometrical locations. The overall thermal expansion effect is $T\,\epsilon V_z\,(\partial p/\partial z)$ where

$$\frac{\partial P}{\partial Z} = -\frac{1}{r}\frac{\partial}{\partial r}(r\,\tau_{rz})$$

Hence, the thermal expansion effect is a function of

$$(T\epsilon V_z/r)\,(\partial/\partial r)\,(r\,\tau_{rz})$$

In moving from the center to the wall, the portion of this term that undergoes the greatest change is $1/r$ (for $r/R = 0$, the $1/r \rightarrow \infty/R$; at the wall, $r/R = 1$, the $1/r = 1/R$).

In contrast, the viscous dissipation term $\tau_{rz}\,(\partial V_z/\partial r)$ depends directly on τ_{rz} and the velocity gradient.

If we consider separately the effect of the tube center and the tube wall regions, we see in Tables 4-3 and 4-4 that

From the foregoing, it can be seen that the effect of expansion cooling reaches a maximum where viscous dissipation reaches a minimum, and vice versa. This, together with the fact that a molten or thermally softened polymer is a poor heat

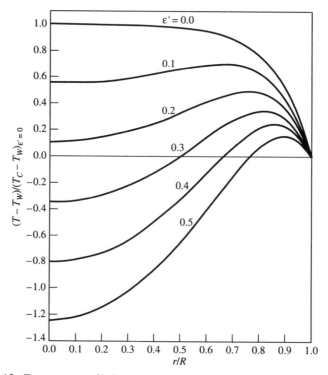

Fig. 4-15   Temperature profile in non-Newtonian system with expansion effects [17].

conductor, accounts for the characteristic shape of the curves in Fig. 4-16. In essence, large heat generation near the wall cannot be dissipated to the central part of the tube because of poor heat conduction. Additionally, there is little cooling effect of expansion in this region. Finally if, as in Fig. 4-16, the higher wall temperature prevents the wall from acting as a sink, the net result is the peak temperature found near the wall.

Likewise, the characteristic centerline depression of temperature (Fig. 4-16; see also Figs. 4-17 and 4-18) is due to expansion cooling being at a maximum with little viscous dissipation and the possibility of radial conduction heat transfer being muted by poor conductivity.

In Figs. 4-16–4-18, the families of curves show the effect of increasing axial length. Note that, even at reduced length values of 1.0 or ∞, the characteristic profile curve holds. Figure 4-18 illustrates the effect of increased compressibility cooling.

Figure 4-19 [21] presents still another solution, which shows the shape of calculated temperature profiles.

One of the solutions listed in Table 4-2 also presented data in the form of the

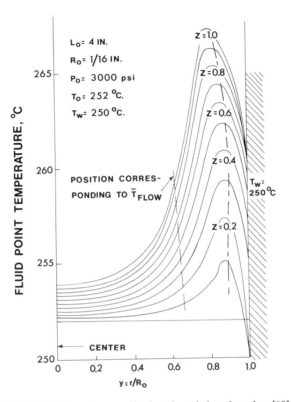

Fig. 4-16   Temperature profiles in polymethyl methacrylate [19].

Nusselt–Graetz solution. These are shown in Fig. 4-20, which uses the term Br' as a parameter (Br' is a measure of internal heat generation). If Br' is small, heat generation can be neglected. Negative values indicate situations in which the fluid is being heated ($T_w > T_1$), and positive Br' values indicate a cooled fluid ($T_w < T_1$).

The Nusselt (Nu) and Graetz (Gz) numbers are defined, respectively, as hD/k and $W C_p/kL$, where h is a heat-transfer coefficient (Btu/h ft^2 °F), k the thermal conductivity (Btu/h ft °F), D the tube diameter (ft), w the mass flow rate (lbm/h), C_p the specific heat (Btu/lbm °F), and L the axial length (ft).

4.7   EXPERIMENTAL AND EMPIRICAL STUDIES OF CONVECTIVE HEAT TRANSFER FOR LAMINAR FLOW OF POLYMER SOLUTIONS

Experimental studies of heat transfer in flowing polymeric systems can be broken down into two categories. The first of the involved experimental measure-

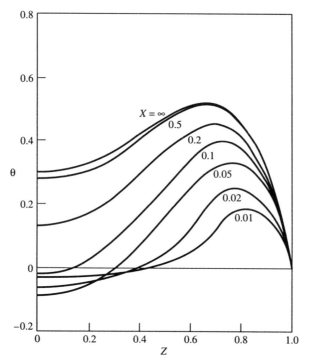

Fig. 4-17   Temperature profile development for a non-Newtonian fluid ($\epsilon' = 0.2$; $n = 1/2$) [18].

ments of heat transfer in flowing polymer solution systems combined with the Leveque approach to yield a Nusselt–Graetz correlation. The second category considered measurement of temperature profile data for flowing molten or thermally softened polymer systems. Leveque-type correlations were also developed in the second group.

The studies of Metzner and co-workers [27,28] and Oliver and Jensen [29] typify the first category mentioned above (i.e., polymer solutions). Their findings can be summarized as follows: If $Gz' > 10$ and $n' > 0.10$, then,

$$Nu = 1.75 \left(\frac{3n' + 1}{4n'}\right)^{1/3} (Gz)^{1/3} \left(\frac{K_b'}{K_w'}\right)^{0.14} \tag{4-17}$$

If $Gz < 20$ and/or $n < 0.10$,

$$Nu = 1.75\, \Delta^{1/3}\, (Gz)^{1/3} \left(\frac{K_b'}{K_w'}\right)^{0.14} \tag{4-18}$$

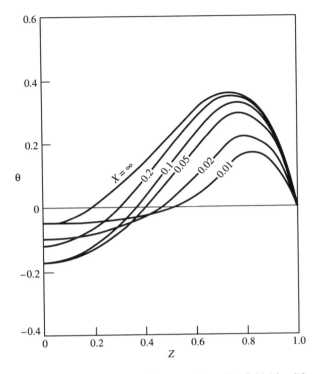

Fig. 4-18 Temperature profile development for a non-Newtonian fluid ($\epsilon' = 0.3$; $n = 1/2$) [18].

where $\Delta^{1/3}$ is obtained from Figs. 4-21 and 4-22, and K_b and K_w are the power law consistency indexes evaluated at bulk fluid and wall conditions.

Other experimental studies for polymer solutions involved the work of Bassett and Welty [97], Popovska and Wilkinson [99], and Joshi and Bergles [103]. The first of these dealt with the thermal entry region and a constant wall flux. Popovska and Wilkinson used a numerical solution of the energy equation to test experimental data over Graetz numbers that ranged from 80 to 1600. Joshi and Bergles used a constant wall flux and correlated for power law fluids the effect of fluid property variation with temperature.

Certain investigators [28, 29] found, in some cases, that free or natural convection occurred in certain systems. These were generally confined to the low Graetz number ($Gz = 20$) region. In addition, the ratio of Grashof number to Reynolds number, $\dfrac{Gr}{(Re)^{\frac{2}{2-n'}}}$ should be 5.0 [30] or larger for free convection to

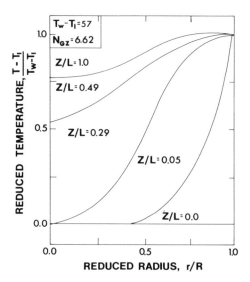

Fig. 4-19   Temperature profile development [21].

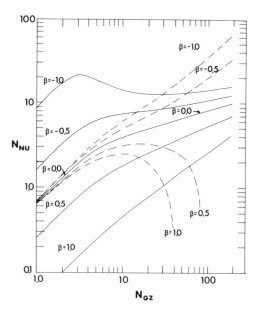

Fig. 4-20   Nusselt–Graetz relation with viscous heating [20]. $\beta = Br' =$ modified Brinkman number.

TABLE 4-3   Effect of Location on Compressibility Cooling

Location	$1/r$ Value	Effect on $T\epsilon V_z/r \dfrac{\partial}{\partial r}(r\tau_{rz})$ Term	Result
Tube center region	$\dfrac{1}{r} \rightarrow \dfrac{\infty}{R}$	Term largest	Large cooling expansion effect
Tube wall region	$\dfrac{1}{r} = \dfrac{1}{R}$	Term small	Small cooling expansion effect

TABLE 4-4   Effect of Location on Viscous Dissipation

Location	$\left(\dfrac{\partial V_{rz}}{\partial r}\right)$	τ_{rz}	Effect on $\tau_{rz}\left(\dfrac{\partial V_{rz}}{\partial r}\right)$	Result
Tube center region	$\left(\dfrac{\partial V_z}{\partial r}\right)$ is small (zero at Center)	τ_{rz} is small (zero at center)	$\tau_{rz}\left(\dfrac{\partial V_z}{\partial r}\right)$ is small (zero at center)	Viscous dissipation is small
Tube wall region	$\left(\dfrac{\partial V_z}{\partial r}\right)$ is large (maximum at wall)	τ_{rz} is large (maximum at wall)	$\tau_{rz}\left(\dfrac{\partial V_z}{\partial n}\right)$ is large (maximum at center)	Viscous dissipation is large

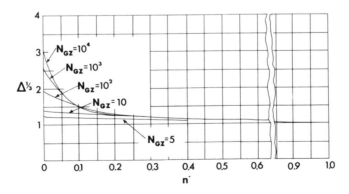

Fig. 4-21   Factor $\Delta^{1/3}$ for laminar flow [27].

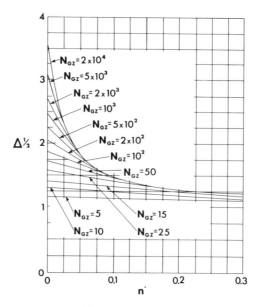

Fig. 4-22   Factor $\Delta^{1/3}$ for laminar flow and low n' values [27].

be sizable. In such cases, the expression to be used is

$$Nu = 1.75 \left[Gz + 0.0083 \left(Pr_w \, Gr_w \right)^{0.75} \right]^{1/3} \left(\frac{K'_b}{K'_w} \right)^{0.14} \tag{4-19}$$

where Gr is the Grashof number defined as $(\epsilon \, \Delta T \, D^3 \, \rho^2)/(\mu_{APP})_w$, and $(\mu_{APP})_w$ the apparent viscosity at wall conditions.

4.8 EXPERIMENTAL AND EMPIRICAL STUDIES OF CONVECTIVE HEAT TRANSFER FOR LAMINAR FLOW OF MOLTEN OR THERMALLY SOFTENED POLYMERS

The earliest experimental studies of heat transfer to flowing molten or thermally softened polymers were the papers of Beyer and Dahl [31] and Schott and Kaghan [32]. These dealt mainly with determining the radial point at which temperature approximated the mass average fluid temperature. Later Bird [33] measured a few experimental points that did not check well with theory. Gee and Lyon [19] also indirectly checked their theory by showing that calculated and experimental average flow rates of a thermally softened polymer undergoing heat transfer compared favorably.

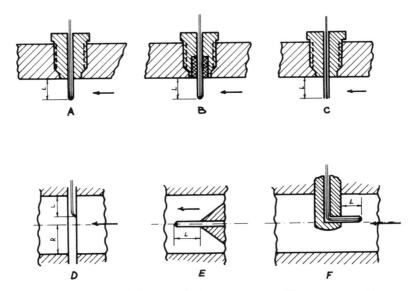

Fig. 4-23   Placement of thermocouples for temperature profile measurement [37].

The independent studies of Van Leeuwen [34, 35] and Griskey and Wiehe [36] set the stage for a considerable number of experimental measurements. Basically, those studies developed the concept of using a thermocouple parallel to axial flow and pointing upstream (as in Fig. 4-23F). Comparison of this arrangement to the others of Fig. 4-23 is given in Table 4-5. In addition to the points covered in Table 4-5, it should also be recognized that configuration *F*, being parallel to flow and pointed upstream, will produce the least flow disturbance. This is quite important since large distortions of flow pattern would make

TABLE 4-5   Comparison Various Thermocouple Arrangements Shown in Fig. 4-23

Thermocouple Configuration	Dynamic Response	Thermal Errors	Mechanical Stability
A	Poor	Poor	Poor
B	Poor	Improved over A but still not good	Poor
C	Poor	Poor	Poor
D	Poor	Poor	Improved over A,B,C
E	Poor	Poor	Adequate
F	Good	Good	Adequate

temperature measurements questionable. Finally, configuration *F* would mini-mize shear heating at the thermocouple.

A number of investigators [36–45] made temperature profile measurements of flowing molten or thermally softened polymer systems using configuration *F*. Caution must be taken, however, in interpreting these results. For example, a number of these studies [34, 35, 38–40] were directed to measuring temper-atures in the exit of a screw extruder or within plunger or injection-molding machines. These placements resulted in large temperature fluctuations (anywhere from 10 to 30°C). There are a number of possible reasons for this occurrence. First, both temperature and velocity profiles would undergo rearrangement and hence influence readings. Next, although configuration *F* was used, the ther-mocouple holder [34, 35, 38–40] was quite sizable. This could have caused flow disturbances and considerable viscous heating, either of which would have influenced the readings.

The remaining investigations [36, 37, 41–45] made measurements in tube flow under more controlled conditions. The earliest of these efforts [36] found that temperatures could be measured quite precisely at a given circumference (deviation of only 0.5°C or less). This result for (*Gz* < 5) and subsequent cor-roboration in later studies [37, 41–45] showed that natural or free convection was not a problem in heat transfer to flowing molten or thermally softened polymers. In addition, evidence was found for both cooling by expansion and viscous dissipation.

The next chronological study [41] further verified the effect of viscous heat-ing. Although, some apparent anomalies were found, these appear to be results of limitations in experimental capacity (for example, low Graetz numbers and limited temperature ranges).

The earlier studies [34–41] led to more sophisticated investigations, which delineated in detail the temperature profiles in flowing molten or thermally soft-ened polymers. These measurements were made in a large-scale unit consisting of 15 ft of 0.957-in. i.d. smooth stainless steel tubing. The molten and thermally softened polymers were pumped into the test unit by means of a 2-in. extruder equipped with a gear pump.

The device used to measure the temperature profiles is represented in Fig. 4-24. The thermocouples were iron–constantan, enclosed first in hypodermic tubing and then in a stainless steel sheath. Teflon® tubing was used as an in-sulation between the tubing and the sheath. This construction insured against heat-conduction errors (Fig. 4-25).

The outside diameter of each probe was 0.0625 in. This meant that the given thermocouple surface seen by the flowing polymer was only 0.4% of the total tube cross-sectional area.

Additionally, all probes were set at different reduced radius (*r/R*) values (0.00, 0.186, 0.373, 0.560, 0.746, and 0.931). This gave a spiral-like pattern. Hence,

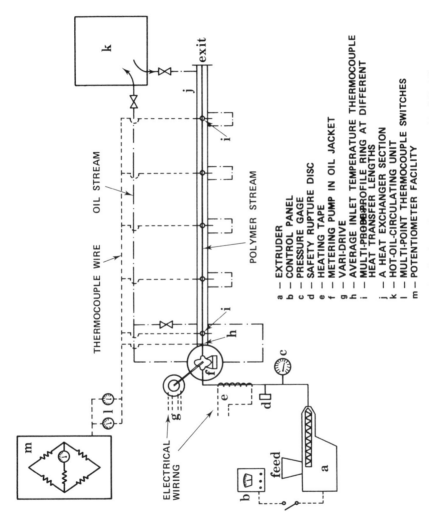

a — EXTRUDER
b — CONTROL PANEL
c — PRESSURE GAGE
d — SAFETY RUPTURE DISC
e — HEATING TAPE
f — METERING PUMP IN OIL JACKET
g — VARI-DRIVE
h — AVERAGE INLET TEMPERATURE THERMOCOUPLE
i — MULTI-PROBE PROFILE RING AT DIFFERENT
    HEAT TRANSFER LENGTHS
j — A HEAT EXCHANGER SECTION
k — HOT-OIL-CIRCULATING UNIT
l — MULTI-POINT THERMOCOUPLE SWITCHES
m — POTENTIOMETER FACILITY

Fig. 4-24 Device for measuring flowing polymer temperature profiles [42, 43].

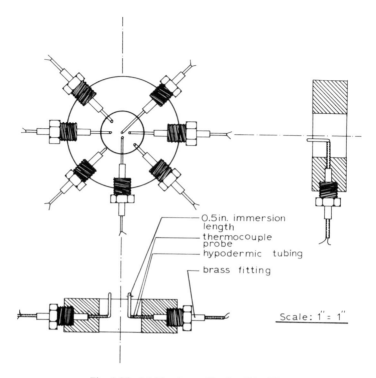

Fig. 4-25   Multiprobe profile ring [42, 43].

any probe had minimal interference from its neighbors. Furthermore, the spiral pattern was changed at each axial position (distances measured from entrance were 0.525, 3.525, 6.525, 9.525, and 12.525 ft, respectively.

4.9   HEAT-TRANSFER DATA FOR MOLTEN POLYMER SYSTEMS

Results from these studies for polyethylene (Dart Industries Inc. type 107; density 0.920 g/cm^3) are given in Figs. 4-26–4.31. In each figure, the weight flow rate is constant. This means that the changing Graetz number is due to axial length (i.e., low axial length, high Graetz number). Consequently, Figs. 4-26–4.31 show the effects not only of axial length for a given weight flow rate but also of changing weight flow rate for an axial length.

The curves in Figs. 4-26–4.31 share certain characteristics. First of all, there is a definite depression in temperature from $r/R = 0$ to $r/R = 0.3$ (expansion cooling) and a peak temperature that occurs near the wall (viscous dissipation).

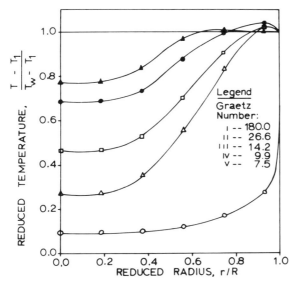

Fig. 4-26 Experimental dimensionless temperature profiles at different heat-transfer lengths with a flow rate = 155.5 g/min.

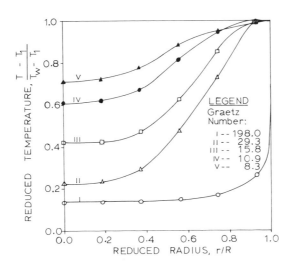

Fig. 4-27 Experimental dimensionless temperature profiles at different heat-transfer lengths with a flow rate = 167.5 g/min.

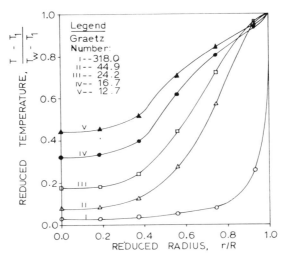

Fig. 4-28   Experimental dimensionless temperature profiles at different heat-transfer lengths with a flow rate = 262.2 g/min.

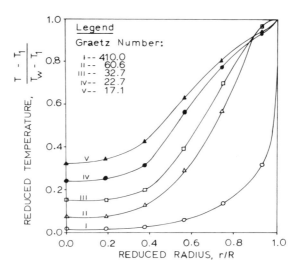

Fig. 4-29   Experimental dimensionless temperature profiles at different heat-transfer lengths with a flow rate = 354.0 g/min.

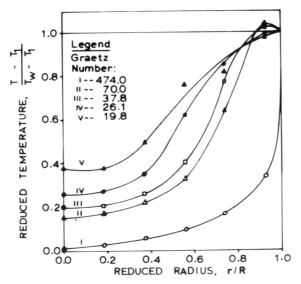

Fig. 4-30 Experimental dimensionless temperature profiles at different heat-transfer lengths with a flow rate = 410.0 g/min.

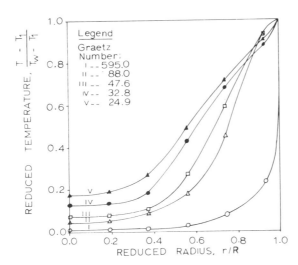

Fig. 4-31 Experimental dimensionless temperature profiles at different heat-transfer lengths with a flow rate = 515.4 g/min.

In essence, the shapes of the curves resemble those predicted in earlier theoretical studies [17–20].

Furthermore, as weight flow increases, the reduced point temperatures decrease, and the curves for the various Graetz numbers seem to come together. This result is explained by the concept of residence time. As mass flow rate increases, the time available for heat transfer decreases. Hence, the difference in the positioning of the temperature profiles between Fig. 4-26 (at 155.5 g/min) and Fig. 4-31 (at 515.4 g/min) is due to changes in residence time.

As mentioned, Figs. 4-26–4.31 show definite effects of both compressibility cooling and viscous dissipation. The latter effect can be further illustrated by Figs. 4-32 and 4-33. Here, a reduced temperature $(T - T_w)/(T_1 - T_w)$, is plotted against r/R for given values of B^* (i.e., Br'). As can be seen, the actual experimental behavior exceeds that predicted by Toor's theoretical approach [20].

If the molten polymer is cooled, the resultant data take the form shown in Figs. 4-34–4.37 for polyethylene (Phillips Marlex 6001). Once again, the curves come together as mass flow rate is increased because of decreased residence time (compare Fig. 4-34 at 191.0 g/min to Fig. 4-37 at 338.9 g/min although it is not obvious that there is a definite effect of viscous dissipation in cooling.

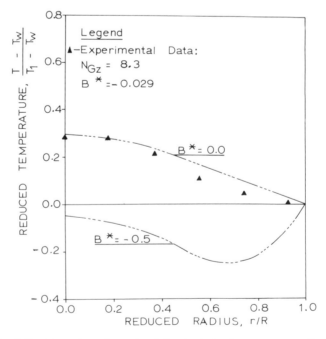

Fig. 4-32 Comparison of experimental and calculated temperature profiles [42, 43].

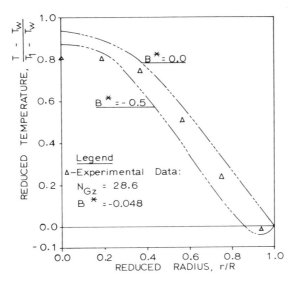

Fig. 4-33 Comparison of experimental and calculated temperature profiles [42, 43].

This will be demonstrated in a following section, which will consider the heat transferred from the standpoint of Nusselt number data.

4.10 HEAT-TRANSFER DATA FOR THERMALLY SOFTENED POLYMER SYSTEMS

Thermally softened polymers are those that are amorphous in the solid state. Figure 2-8 illustrates the difference between these polymers and those that form melts (i.e., semicrystalline in the solid state); as can be seen, the thermally softened polymers do not have a melting point as do the melts.

Figures 4-38–4-42 [44] show temperature profile data (heating for a flowing polymethyl methacrylate, Rohm and Haas VM100). The behavior is similar to that found earlier for polyethylene, namely, depression at the tube center (compressibility cooling) and peaking near the wall (viscous dissipation). Also, the effect of residence time is clearly shown by the bunching of the profiles at higher mass flow rates.

The data of Figs. 4-38–4-42 compare favorably with the calculated profiles of Gee and Lyon [19] for a similar material. A direct comparison is shown in Fig. 4-43. As can be seen, the curves of point temperature are quite similar for both heating and cooling.

As a point of interest, the extent of the compressibility cooling and viscous

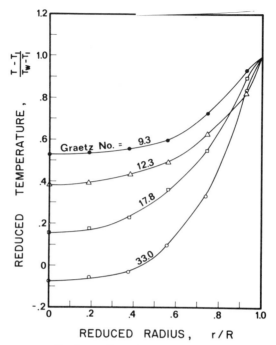

Fig. 4-34   Temperature profiles for cooling a flowing molten polymer (flow rate of 191 g/min) [48].

dissipation effects (as in Figs. 4-44 and 4-45) appears more marked for the thermally softened polymer than for polyethylene. These occurrences can be explained as follows: First, compressibility cooling is a function of ϵT, the product of the coefficient of thermal expansion and the temperature. A comparison of the thermal expansion coefficients of polyethylene and polymethyl methacrylate shows that the latter has a larger ϵ value for a given temperature and pressure. This means that the expansion cooling effect would then be greater for polymethyl methacrylate (as the data show). Likewise, in terms of viscous dissipation, the polymethyl methacrylate apparent viscosity (μ_{APP}) was, in some cases, an order of magnitude higher than that of polyethylene for a given shear rate and temperature. Thus, according to the equation for viscous dissipation factor,

$$\phi_v = \mu_{APP} \left(\frac{dV_z}{dr} \right)^2 \tag{4-19}$$

in terms of apparent viscosity. As can be seen, a considerably larger μ_{APP} will mean a much greater viscous heating effect for polymethyl methacrylate.

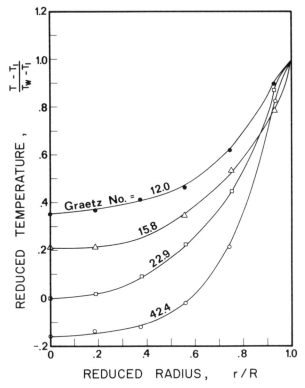

Fig. 4-35   Temperature profiles for cooling a flowing molten polymer (flow rate of 245.1 g/min) [48].

A set of cooling profile curves for the thermally softened polymer are given in Figs. 4-46–4-48. Once again, the effect of viscous dissipation is considerable although not apparent. This will be considered later under a discussion of Nusselt number behavior.

4.11   THE NUSSELT–GRAETZ RELATIONSHIP FOR MOLTEN AND THERMALLY SOFTENED POLYMERS

The temperature profile data for heat transfer to and from molten and thermally softened polymers give considerable insight into the intimate behavior of such systems. However, it is equally important to be able to treat the overall heat-transfer situation as well. These data can then be used for engineering evaluations and design purposes.

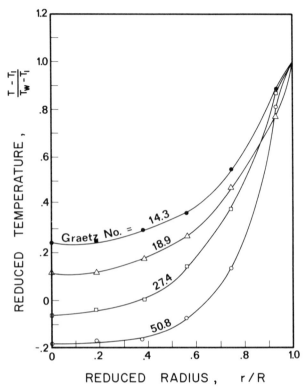

Fig. 4-36   Temperature profiles for cooling a flowing molten polymer (flow rate of 293.5 g/min) [48].

Such heat-transfer data are based on heat-transfer coefficients (h_a) and the Nusselt number ($h_a D/k$). This particular coefficient is based on an average fluid temperature. Figure 4-49 summarizes the relation between Nusselt and Graetz numbers for both molten and thermally softened polymers. The $B^* = 0.0$ curve is the case for heat transfer without viscous dissipation (both heating and cooling). The negative B^* curves represent Toor's [20] theoretical calculation for heating with viscous dissipation, whereas the positive B^* curves represent Toor's theoretical data for cooling with viscous dissipation.

The experimental data of Fig. 4-49 show some interesting trends. In heating, for example, the experimental results for B^* of -0.043 and -0.141 (from Gz of $4-100$) trend upward much more rapidly than the theoretically derived curves predict (see B^* of -0.5 and -1.0). Hence, it can be concluded that the actual effect of viscous dissipation on the Nusselt number is greater than that derived from theory.

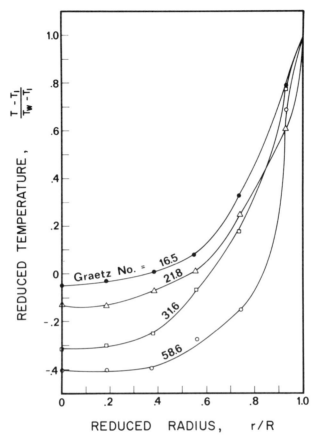

Fig. 4-37   Temperature profiles for cooling a flowing molten polymer (flow rate of 338.9 g/min) [48].

Two trend lines (dotted lines labeled A–A and B–B) are shown in Fig. 4-49 for the B^* data of -0.043 and -0.141. As can be seen, the former ranges from Gz of 6 to 45, whereas the latter goes from 15 to 100. It is recommended that these trend lines be used to estimate the actual effect of viscous dissipation on Nusselt number for B^* values higher than -0.141.

The suggested method is to read the Nusselt value for either of the trend lines at a given Graetz number and then read the value from the B^* used (i.e., -0.141 or -0.043) from the theoretical data (i.e., interpolate between 0.0 and -0.5). The ratio of these values should be used with the given theoretical value for the B^* desired.

As an example, let Graetz be 40, and estimate the Nusselt number for a B^*

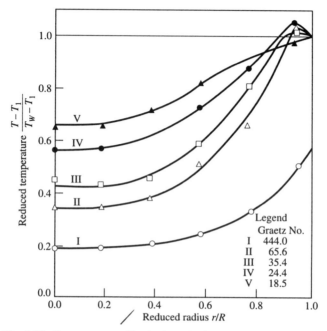

Fig. 4-38   Temperature profiles for heated polymethyl methacrylate [44].

of -1.0. For $Gz = 40$, the actual value of Nusselt for $B^* = -0.141$ is 12.0, whereas the theoretical value is 6.6. At B^* of -1.0, the value from theory is 13.0. This gives an estimated value of an actual Nusselt number value for B^* of -1.0 (at $Gz = 40$) of $(12.0/6.6)$ 13.0 or 23.7.

If the $A-A$ line ($B^* = -0.043$) is used, the estimated actual Nusselt number at $B^* = -1.0$ would be 22.2. Averaging both would give a 23.0 Nusselt value.

Estimated Nusselt values should be determined separately from both trend lines (i.e., if they both exist for the given Graetz number) and then averaged. If only one trend line exists for the Graetz number, then, of course, only that line can be used.

It is further suggested that the estimated actual value be used in tandem with the theoretical Nusselt number for process decisions. Both Nusselt numbers can be used to calculate the upper and lower limits of the expected temperature rise. Such calculations could then be used in making the most conservative engineering decisions with respect to a process.

The cooling data (positive B^* values) also show large deviations from the theoretical predictions. These data can also be used to estimate actual Nusselt numbers for B^* values higher than the experimental data. In the case of the cooling, the short segments for B^* of 0.078 and 0.108 can be utilized for the

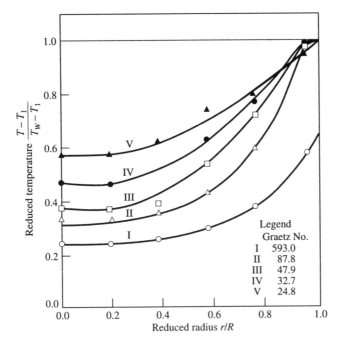

Fig. 4-39   Temperature profiles for heated polymethyl methacrylate [44].

Graetz number range of 15–35, whereas the 0.150 and 0.186 data would apply from 30 to ~50. The methodology would be the same as that described for the heating case.

In any case, the use of the $B^* = 0.0$ curve of Fig. 4-49 should not be relied on for engineering calculations since it will obviously result in large errors in Nusselt numbers for situations of even modest viscous dissipation.

4.12   THERMAL ENTRANCE LENGTHS IN FLOWING MOLTEN AND THERMALLY SOFTENED POLYMER SYSTEMS

The thermal entrance length is the axial distance required for the fluid's temperature profile to develop completely. This distance is well defined for Newtonian fluids [45] but not so well delineated for non-Newtonians. The papers of Toor [18], Forsyth and Murphy [21], and Charm [46] all dealt with the axial temperature profile development. These, however, were mainly theoretical and analytical efforts that did not compare results to experimental data.

Such a comparison is made in Fig. 4-50 for the data of Fig. 4-27 with Forsyth

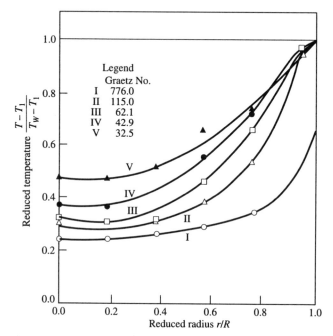

Fig. 4-40   Temperature profiles for heating polymethyl methacrylate [44].

and Murphy's solution [21]. As can be seen, the solution curves for distances less than the thermal entrance length ($z/L < 1$) do not precisely describe the situation at short lengths. However, as the thermal entrance length is approached, the fit improves. For the case of Fig. 4-27, the thermal entrance was found to be established at the 12.6-ft length.

Other predicted thermal entrance lengths from other studies [18, 45, 46] gave considerably different results. For example, a modification of Kay's relation for Newtonians [45] yielded a thermal entrance length several orders of magnitude smaller than experimental data, whereas the studies of Charm [46] and Toor [18] gave predicted values that were several orders of magnitude higher than the 12.6 ft indicated by Forsyth's and Murphy's work in Fig. 4-50.

It also becomes apparent that, as mass flow rate increases, the thermal entrance length increases. Consider, for example, Fig. 4-31 where, for a mass flow rate of 515.4 g/min, the temperature profile at 12.6 ft seems far below the fully established value. In this case, a thermal entrance length was estimated to be over 30 ft.

It is recommended that the method of Forsyth and Murphy [21] be used to establish the thermal entrance length for a given polymer system.

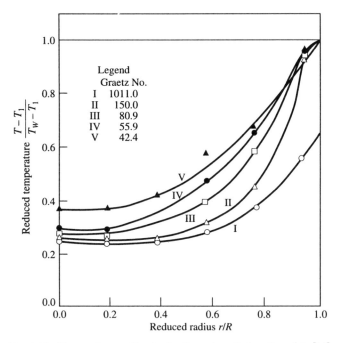

Fig. 4-41   Temperature profiles for heating polymethyl methacrylate [44].

4.13   THE EFFECT OF ELASTICITY ON HEAT TRANSFER IN FLOWING POLYMER SYSTEMS

As has been pointed out, viscoelastic behavior is frequently associated with flowing polymeric systems. Usually, such behavior is especially pronounced in molten or thermally softened polymers. Elastic effects should be contrasted with the viscous effects (i.e., viscous heating). In the latter case, the energy necessary to move the fluid is dissipated as heat. In the former instance, however, energy put into an elastic system can be recovered. This suggests the possibility of a form of energy storage in a viscoelastic system.

The effect of a polymer fluid's elastic behavior on heat transfer has been generally neglected although its possible importance has been cited [47].

An experimental study was made [48] that compared the heat-transfer behavior of a polyethylene having marked elastic behavior to one whose rheological data showed only slight elasticity (i.e., much lower normal stress values). Some typical temperature profile curves for the polymer are shown in Figs. 4-51–4-56.

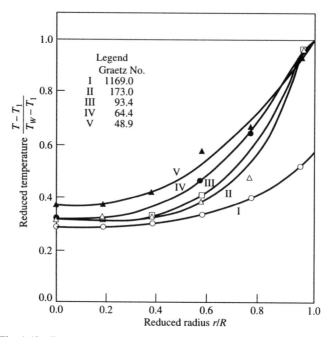

Fig. 4-42   Temperature profiles for heating polymethyl methacrylate [44].

Generally, the temperature profiles for Figs. 4-51–4-56 were higher than those for corresponding mass flow rates in Figs. 4-27–4-31 (less elastic polyethylene). This is further illustrated in Figs. 4-57 and 4-58, where the reduced temperature profiles for the polymer of Fig. 4-53 are higher than those for the less elastic material. From this behavior, it would appear that any effect of elasticity on energy storage seems to be overwhelmed by either lessened compressibility cooling or increased viscous dissipation (or both).

One important effect apparently related to elasticity was found, however. This was the inability to measure temperature profiles for the elastic system at high Graetz number (short axial lengths of 0.525 ft). Unlike the data of Figs. 4-27–4-31, the temperature profiles at 0.525 ft scattered so badly that no effective representation of them could be made in Figs. 4-51–4-56.

It would therefore appear that the principal effect of elasticity on heat transfer to flowing molten polymer systems is at short axial lengths. Hence, systems involving such elastic materials can be described reasonably well by the procedures previously discussed as long as short axial lengths are avoided.

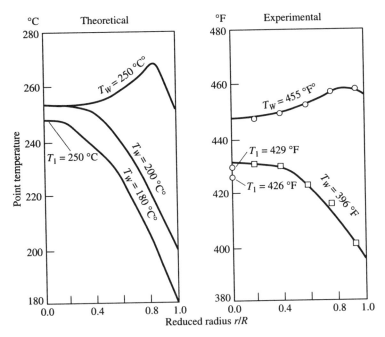

Fig. 4-43   Comparison experimental data [44] to calculated values [19].

4.14   CONVECTIVE HEAT TRANSFER IN NONCIRCULAR CONDUITS

Studies on noncircular conduit cases of heat transfer to and from polymer systems have been quite limited. What little work that has been done has concentrated on the parallel-plate case.

The earliest studies were those of Tien [49], Suckow et al. [50], and Crozier et al. [51]. The work of Tien involved taking the approximate velocity profile found by Schecter [52] and then using it to solve the energy equation (assuming constant physical properties and no viscous dissipation) to yield average temperature and the Nusselt number. Suckow et al. used the exact velocity distribution instead of the approximate profile to solve the energy equation. Crozier and co-workers used the Leveque technique to find appropriate equations for heat transfer.

The equations developed by Crozier and co-workers [51] were

$$\frac{h_a De}{k} = \frac{1}{2} \frac{D_e^2 \, CpG}{kL} \left(\frac{\gamma_B}{\gamma_w}\right)^{0.14} \tag{4-20}$$

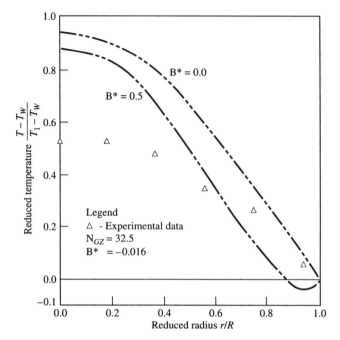

Fig. 4-44   Effect of viscous dissipation on heat transfer to a thermally softened polymer [44]. B* is modified Brinkman number.

when the Graetz number is less than 3.0, and

$$\frac{h_a De}{k} = 1.86 \left(\frac{D_e^2\, G\, C_p}{kL}\right)^{1/3} \left(\frac{2n\,+\,1}{3n}\right)^{1/3} \left(\frac{\gamma_B}{\gamma_w}\right)^{0.14} \tag{4-21}$$

for Graetz number >20. The De is simply twice the plate separation and γ_B and γ_w are $g_c K'\,8^{n'-1}$ at the fluid's bulk and wall temperatures, respectively.

For the range $20 > Gz > 3$, either the work of Tien [49] or Suckow et al. [50] can be used. In the latter case, point temperature is given as a function of reduced length (for a power law fluid of $n = 1/2$):

$$\theta = 1.222\,(1 - \lambda^3)\, e^{-2.787\psi} + (0.226)\,(1\,-\,14\lambda^3 + 13\lambda^6)e^{-31.934\psi} \tag{4-22}$$

where λ is the reduced distance y/b (y is the vertical distance measured from the center line, and b is the half-plate separation) and ψ is the reduced length $kx/[\rho C_p b^2 (V_x)_{\max}]$, with x the axial distance. Also, the average temperature is

$$\theta_{AV} = 1.047 e^{-2.787\psi} - 0.058 e^{-31.934\psi} \tag{4-23}$$

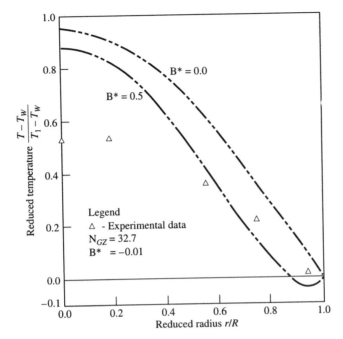

Fig. 4-45   Effect of viscous dissipation on heat transfer to a thermally softened polymer [44].

Figures 4-59 and 4-60 show the behavior of θ and θ_{AV} as functions of ψ and λ. Heat-transfer coefficients can be determined from the average temperature behavior.

A more recent study was the work of Vlachopoulos and Keung (89), who solved the problem numerically using the exact velocity profile for situations with and without viscous dissipation. Their results for a power law fluid ($n = 1/2$) are shown in Fig. 4-61 as a function of ψ.

The β in Fig. 4-61 is defined as

$$\beta = \left(\frac{n+1}{n}\right)^{n+1} \frac{KV_{max}^{n+1} \, b^{1-n}}{k(T_0 - T_w)} \tag{4-24}$$

Other work that considered parallel plates or slits were the papers of Lin and Hsu [105], Ybarra and Eckert [106], and Dinh and Armstrong [107]. All three studies involved theoretical solutions of the energy equation that included vis-

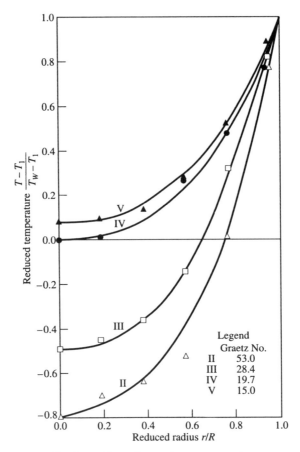

Fig. 4-46   Temperature profiles for cooling polymethyl methacrylate (flow rate of 158.6 g/min) [44].

cous dissipation but neither temperature-dependent properties nor compressibility cooling. The power law was used for flow behavior in each case.

   Lin and Hsieh [108] also reported a theoretical solution for heat transfer with Couette flow in, an annulus with viscous dissipation, and a moving inner cylinder.

   There appear to be no other heat-transfer data available for flowing polymeric systems in noncircular cross sections. However, the situation can be handled by using the geometric parameter technique developed by Kozicki et al. [53]. Ac-

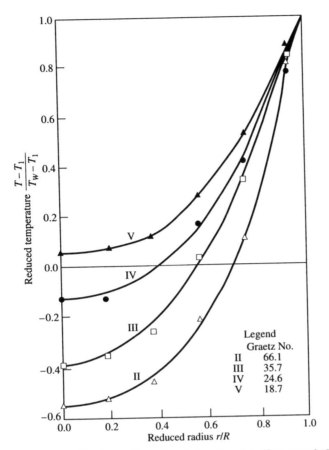

Fig. 4-47   Temperature profiles for cooling polymethyl methacrylate (flow rate of 197.8 g/min) [44].

cording to this work, a modified Reynolds number is given by

$$Re'' = \frac{4R_H V\rho}{N_{\text{eff}}^{(a+b)}} \tag{4-25}$$

where R_H is the hydraulic radius for the conduit (cross-sectional area available for flow-wetted perimeter), N_{eff} an effective viscosity, $\tau_{\text{wall}}/(2V/R_H)$; V the average velocity, and ρ the density; a and b are the empirical constants [53]. Equa-

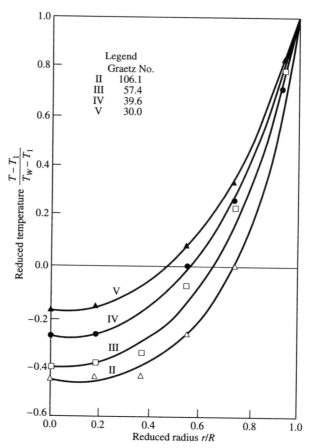

Fig. 4-48   Temperature profiles for cooling polymethyl methacrylate (flow rate of 317.9 g/min) [44].

tion (4-25) can then be used either with the expression

$$Re = \frac{D_{eq} V \rho}{\eta_{APP}} \qquad (4\text{-}26)$$

or

$$Re' = \frac{D_{eq}^{n'} V^{2-n'} \rho}{g_c K' 8^{n-1}} \qquad (4\text{-}27)$$

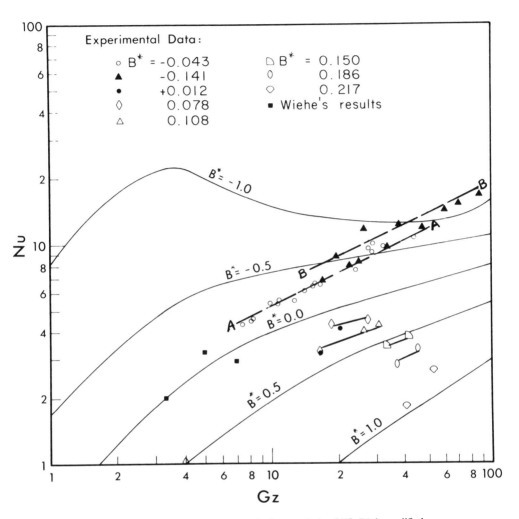

Fig. 4-49   Effect of viscous dissipation on Nusselt–Graetz relation [47]. B* is modified
Brinkman number.

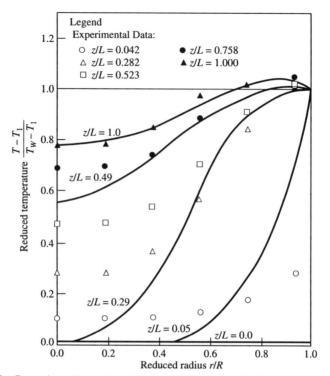

Fig. 4-50   Comparison of experimental [42, 43] and calculated [21] temperature profiles.

to give a relation for D_{eq}, the equivalent diameter. In one case,

$$D_{eq} = \frac{4 R_H \eta_{APP}}{\left(\dfrac{\tau_{wall}}{2V/R_H}\right)^{a+b}}$$

(4-28)

and, in the other,

$$D_{eq} = \left(\frac{4R_H V_{gc}^{n'-1} K' 8^{n'-1}}{\left(\dfrac{\tau_{wall}}{2V/R_H}\right)^{a+b}}\right)^{1/n}$$

(4-29)

The resultant D_{eq} can then be used with Eqs. (4-17), (4-18), or Fig. 4-49 to give the appropriate heat-transfer coefficient.

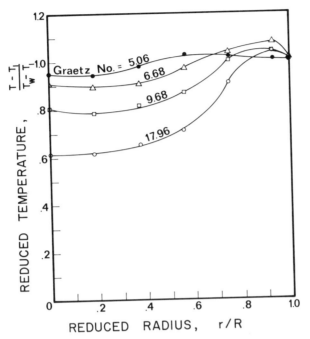

Fig. 4-51 Temperature profiles for heating a polymer with elastic characteristics (flow rate of 103.8 g/min) [48].

4.15 HEAT TRANSFER IN TURBULENT FLOW

In general, situations in which polymer turbulent flows are encountered are limited. Furthermore, when they are found, it is usually only for polymer solutions. As such, then, heat transfer to turbulent flowing polymer systems is not too important. This is reflected in the limited literature on this subject.

Available work includes some data on slurries [61–63], experimental studies on power law fluids through tubes [64], and a combined theoretical and experimental approach [65].

A slurry equation is [62]

$$\frac{hD}{k_{\text{slurry}}} = 0.027 \left(\frac{DV\rho}{\mu_s}\right)^{0.8} \left(\frac{C_p\,\mu_s}{k_{\text{slurry}}}\right)^{1/3} \left(\frac{\mu_L}{\mu_{\text{LW}}}\right)^{0.14} \tag{4-30}$$

where k_{slurry} is the thermal conductivity of the slurry, μ_s is the slurry viscosity, μ_L and μ_{LW} are viscosities of the suspending liquid in the bulk and at the wall.

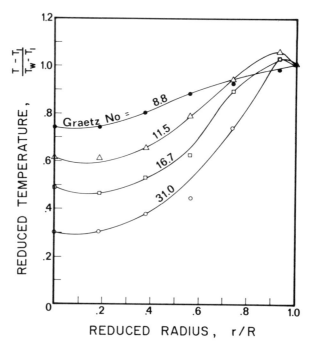

Fig. 4-52   Temperature profiles for heating a polymer with elastic characteristics (flow rate of 179.3 g/min) [48].

The μ_s is obtained from the relation

$$\mu_s = \frac{\mu_L}{[1 - (X_v/X_{vB})]^{1.8}} \tag{4-31}$$

where X_v is the volume fraction of suspended particles and X_{vB} the volume fraction of solid in a suspended bed (after prolonged settling).

Another slurry relationship is given by [63].

$$\left(\frac{h}{C_p V}\right)\left(\frac{C_p \eta_{APP}}{k_{slurry}}\right)^{2/3}\left(\frac{\eta_{APP\ wall}}{\eta_{APP}}\right)^{0.14} = 0.027 \left(\frac{DV\rho}{\eta_{APP}}\right)^{-0.2} \tag{4-32}$$

The general relation recommended for turbulent heat transfer to flowing non-Newtonians is that developed by Metzner and Friend [65]. This was done by analogy with momentum transport, assuming several points. These included: constant heat flux, equality of eddy thermal and momentum diffusivities at all

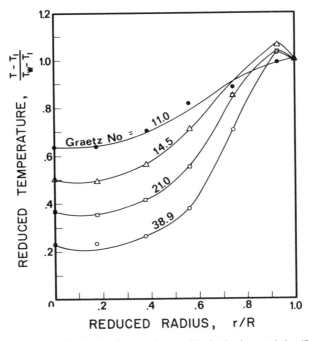

Fig. 4-53 Temperature profiles for heating a polymer with elastic characteristics (flow rate of 225 g/min) [48].

radii, no dependence of fluid on time, no fluid elasticity, large Prandtl numbers and the assumption that heat flux is a function of tube radius.

The result was the relation

$$\frac{h}{C_p \, V_\rho} = \frac{f/2}{1.20 \, + \, 11.8 \, \sqrt{f/2} \, (Pr_{\text{wall}}^{-1}) \, (Pr_{\text{wall}})^{-1/3}} \tag{4-33}$$

with the limitation that

$$(Pr) \, (Re)^2 \, f > 500{,}000 \tag{4-34}$$

Equation (4-33) correlated not only various polymer solutions but also slurry data determined in another study [63].

No published heat-transfer data appear to be available for turbulent flow of polymer systems through other geometries. If such situations are encountered, it is recommended that Eq. (4-33) be used with four times the hydraulic radius substituted for the circular conduit diameter. Based on Newtonian fluid results, this should give at least reasonable estimates of the heat-transfer coefficient.

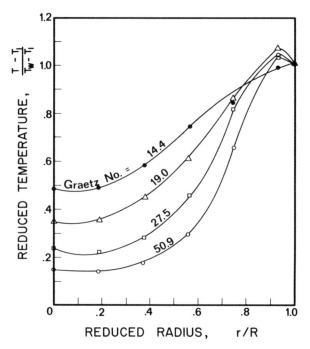

Fig. 4-54   Temperature profiles for heating a polymer with elastic characteristics (flow rate of 294.5 g/min) [48].

4.16   CONVECTION EXAMPLES

Example 4-7

A polymer solution (n of 0.5; K at 90°F of 51 lbm s^{n-2} ft^{-1}; viscosity activation energy of 14,900 Btu/lb mole) is fed into a 1-in. i.d. stainless steel tube (10 ft long) at a mass flow rate of 750 lbm/h and a temperature of 90°F. The velocity profile is fully developed before the solution enters the heated tube. Heat is supplied by steam condensing at 20 psia.

Remaining fluid properties are: density = 58 lbm/ft^3, specific heat = 0.6 Btu/lbm °F, thermal conductivity = 0.5 Btu/ft°F h.

Calculate the exit temperature of the fluid.

The temperature of the condensing steam is 227.96°F. For purposes of calculation, we will take this as 228°F.

Next, we use Eq. (4-17), where

$$Nu = 1.75 \left(\frac{3n + 1}{4n} \right)^{1/3} (Gz)^{1/3} \left(\frac{K_B}{K_W} \right) 0.14$$

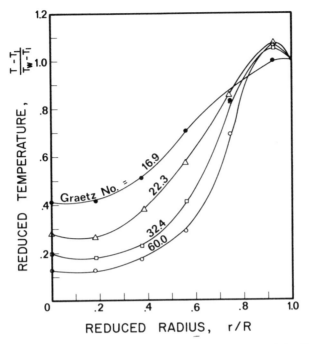

Fig. 4-55   Temperature profiles for heating a polymer with elastic characteristics (flow rate of 346.8 g/min) [48].

From the given data,

$$\left[\frac{3n + 1}{4n}\right] = \left(\frac{2.5}{2.0}\right) = 1.25$$

$$Gz = \frac{WCp}{kL} = \frac{(750 \text{ lbm/h}) (0.6 \text{ Btu/lbm°F})}{(0.5 \text{ Btu/ft°F h}) (10 \text{ ft})} = 90$$

The value of K_B is taken at the average of the entering and wall temperatures: $(90 + 228)/2°F$ or $159°F$ K_W is taken at $228°F$.

Hence,

$$K_B = K_{159°F} = K_{90°F} \exp\left[\frac{E}{R}\left(\frac{1}{619} - \frac{1}{550}\right)\right]$$

$$K_W = K_{228°F} = K_{90°F} \exp\left[\frac{E}{R}\left(\frac{1}{688} - \frac{1}{550}\right)\right]$$

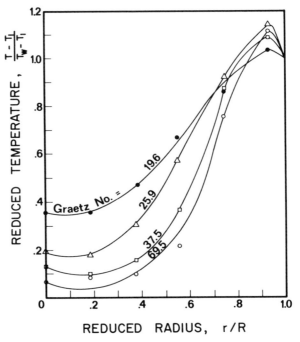

Fig. 4-56 Temperature profiles for heating a polymer with elastic characteristics (flow rate of 401.5 g/min) [48].

Solving these equations yields

$$K_B = 11.182 \text{ lbm s}^{2-n} \text{ ft}^{-1}$$

$$K_W = 3.315 \text{ lbm s}^{2-n} \text{ ft}^{-1}$$

Then,

$$h = \frac{k}{D} \, 1.75 \left(\frac{3n+1}{4n}\right)^{1/3} (Gz)^{1/3} \frac{K^{0.14}}{K_W}$$

$$h = \frac{0.5 \text{ Btu/ft°F h}}{1/12 \text{ ft}} (1.75) (1.25)^{1/3} (90)^{1/3} \left[\frac{11.182}{3.315}\right]^{0.14}$$

$$h = 69.73 \, \frac{\text{Btu}}{\text{ft}^{2}\text{°F h}}$$

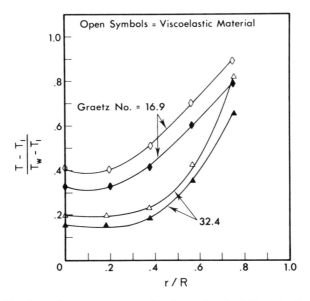

Fig. 4-57   Comparison temperature profiles for varying elasticities [42, 43, 48].

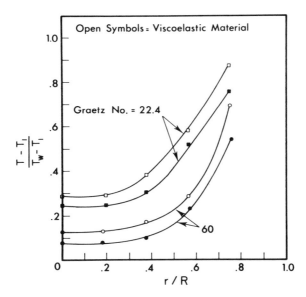

Fig. 4-58   Comparison temperature profiles for varying elasticities [42, 43, 48].

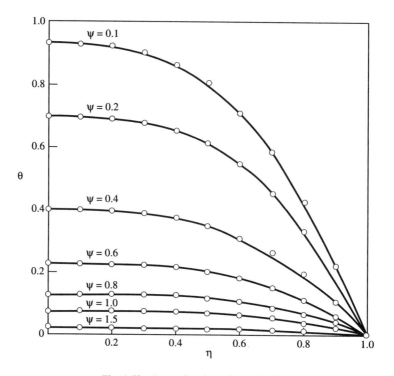

Fig. 4-59 θ as a function of ψ and λ [50].

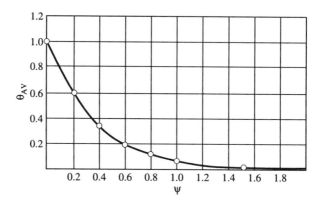

Fig. 4-60 θ_{AV} as a function of ψ [50].

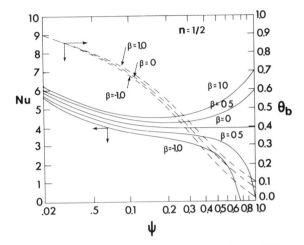

Fig. 4-61 Nusselt–Graetz relation for parallel plates [89].

Next, by enthalpy balance,

$$w\, C_p\, \Delta T_{\text{fluid}} = h_A\, (T_{\text{wall}} - T_{\text{fluid}})_{\text{AV}}$$

$$w\, C_p\, (T_2 - T_1) = h_A \left\{ T_{\text{wall}} - \left[\frac{T_2 + T_1}{2} \right] \right\}$$

$$\left(750\, \frac{\text{lbm}}{\text{h}} \right) \left(\frac{0.6\ \text{Btu}}{\text{lbm}^\circ\text{F}} \right) \left(T_2 - 90^\circ\text{F} \right) = \left(69.73\, \frac{\pi}{12} \times 10 \right) \text{ft}^2 \left\{ 228 - \left[\frac{T_2 + 90}{2} \right] \right\}$$

Solving for T_2 gives a value of 137°C.

Example 4-8

Thermally softened polymethylmethacrylate entering at 426°F is being cooled in a 9.6-ft-long tube (i.d. of 1 in.). The flow rate of the polymer is 317.9 g/min. Wall temperature is 395°F. Compare the exit temperature calculated using the appropriate B^* value to that for $B^* = 0.0$.

The B^* value is given by

$$B^* = \frac{\tau_{\text{wall}}\, w\, n}{(3n + 1)\, k\, \rho \pi R\, (T_1 - T_{\text{wall}})}$$

For polymethyl methacrylate, the T_{wall} and n values can be obtained from the rheological data presented by Westover [86]. Density and thermal conductivity are taken from the work of Heydemann and Guicking [reference 12; Chapter Two] and Griskey et al. [5], respectively. The computation gives a value of 0.150 for B^*.

The Graetz number for this case is 39.7. For this G_z, the corresponding Nusselt numbers are 5.8 (at $B^* = 0.0$) and 3.5 (at $B^* = 0.150$), respectively.

Then, $q = h_A \Delta T$

By enthalpy balance,

$$ w \, C_p \, \Delta T_{fluid} = h_A \, (T_{wall} - T_{fluid}) $$

and

$$ w \, C_p \, (T_2 - T_1) = h_A \left[T_{wall} - \frac{(T_1 + T_2)}{2} \right] $$

Then,

$$ \frac{WC_p}{hA} (T_2 - 426°F) = \left[410°F - \frac{426°F}{2} - \frac{T_2}{2} \right] $$

The w is 317.9 g/min, C_p is 0.9 Btu/lbm°F and A is given by $\pi \, DL$. If the h values for $(B^* - 0.150$ and $B^* = 0.0)$ 4.02 and 6.67 Btu/h ft²°F are substituted, the results are:

$$ T_2 = 425.6 \text{ (for } B^* = 0.150) $$
$$ T_2 = 415.9 \text{ (for } B^* = 0.0) $$

Example 4-9

Molten polyethylene at a flow rate of 3.575 lbm/min is heated in a 1-in. i.d. tube 10 ft long. The polymer enters the tube (wall temperature of 430°F) at an inlet temperature of 284°F.

Conservatively, what would be the effect of neglecting viscous dissipation on the polymer's exit temperature?

Calculating Graetz number:

$$ Gz = \frac{wC_p}{kL} \frac{(3.575 \text{ lbm/min}) (0.606 \text{ Btu/lbm°F})}{(0.002167 \text{ Btu/min-ft-f}) (10 \text{ ft})} $$

$$ Gz = 100 $$

Without viscous dissipation ($B^* = 0.0$ in Fig. 4-49):

$$Nu = 7.8$$

With viscous dissipation (a conservative approach is to use the data points):

$$Nu = 18.5$$

Then, for the polymer:

$$wC_p (T_2 - T_1) = hA \left\{ [T_{wall} - [(T_2 + T_2)/2)] \right\}$$

$$\frac{wC_p}{hA} (T_2 - T_1) = [T_{wall} - T_1/2 - T_2/2]$$

$$T_2 \left(\frac{wC_p}{hA} + 1/2 \right) = T_{wall} + T_1 \left(\frac{wC_p}{hA} - 1/2 \right)$$

$$h \text{ (for no dissipation)} = \frac{Nuk}{D}$$

$$h = \frac{(7.8)\ (0.002167\ \text{Btu/min-°F-ft})}{(1/12\ \text{ft})}$$

$$h = 0.2028\ \text{Btu/min-°F-ft}^2$$

Likewise, with viscous heating:

$$h = \frac{(18.50)\ (0.002167\ \text{Btu/min-°F-ft})}{(1/12\ \text{ft})}$$

$$h = 0.4811\ \text{Btu/min-°F-ft}^2$$

Then solving:

$$T_2 = 285.5°\text{F (no dissipation)}$$

$$T_2 = 287.5°\text{F (conservative calculation)}$$

Example 4-10

A flowing polyethylene melt (300 g/min) is being heated in a rectangular duct (1 in. × 2 in.). Initial polymer temperature is 410°F. The exit temperature is 425°F. Determine the wall temperature needed to accomplish this for a 10-ft length of duct.

The Graetz number for this case is

$$Gr = \frac{wC_p}{kL} = \frac{(500 \text{ g/min}) \text{ (lbm/453.6 g) (60 min/h) (0.68 Btu/lbm°F)}}{(0.148 \text{ Btu/ft h °F) (10 ft)}} = 30.5$$

where data are from Refs. 5 and 88.

For this Graetz number, the Nusselt number (with viscous dissipation) is, from Fig. 4-49,

$$Nu = \frac{h \, De}{k} = 10$$

Now, in order to determine h, a value is needed for De, the equivalent diameter. The values of a and b for Eq. (4-25) are 0.2440 and 0.7276, respectively. This means that De is then given by

$$De = \frac{4R_H \, N_{APP}}{\left(\frac{\tau_w}{2V/R_H}\right)^{a+b}}$$

$$De = \frac{4 \, (1/36 \text{ ft}) \, (0.725 \text{ lbf/in.}^2 \text{ s})}{\left[\frac{2.9 \times 10^{-2} \text{ lbf/in.}^2}{(2) \, (5.27 \times 10^{-4} \text{ ft/s})/(36 \text{ ft})}\right]} \, 0.9716$$

$$De = 0.1045 \text{ ft}$$

Then,

$$h = \frac{k}{De} Nu = \frac{(0.148 \text{ Btu/h ft°F)}}{(0.1045 \text{ ft})} \, 10$$

$$h = 14.18 \text{ Btu/h ft}^2°F$$

Once again,

$$wC_p \, (T_2 - T_1) = h_A \left[T_{wall} - \left(\frac{T_1 + T_2}{2} \right) \right]$$

$$\frac{wC_p}{h_A} \, (T_2 - T_1) + \frac{T_1}{2} + \frac{T_2}{2} = T_{wall}$$

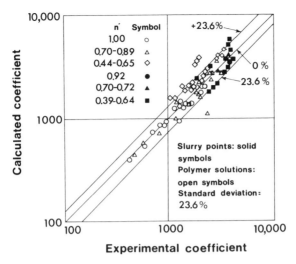

Fig. 4-62 Comparison of calculated [90] and experimental [90, 65] values of heat-transfer coefficients for turbulent flow.

Substituting in the equation we obtain

$$\frac{(300 \text{ g/min}) (60 \text{ min/h}) (\text{lbm}/453.6 \text{ g}) (0.68 \text{ Btu/lbm}^\circ\text{F})}{(14.8 \text{ Btu/h ft}^{2\circ}\text{F}) (2\text{m}^2) (\text{ft}^2/144 \text{ m}^2) (10 \text{ ft})} (15^\circ\text{F}) + 417.5^\circ\text{F} = T_{\text{wall}}$$

Solving

$$T_{\text{wall}} = 623.5^\circ\text{F}$$

4.17 RADIATION HEAT TRANSFER IN POLYMERIC SYSTEMS—GENERAL

Substances can emit various forms of radiant energy such as X-rays. However, the only form produced by virtue of temperature is thermal radiation. At temperatures near or below room temperature this mode of heat transfer is not important. However, at temperatures in the range of 1000°F, the radiant thermal energy can be significant.

If thermal radiation is conceived of as a "photon gas," it is possible from thermodynamics to show that the energy density of the radiation is

$$E_b = \sigma \, T^4 \tag{4-35}$$

where σ is the Stefan–Boltzmann constant (0.1714×10^{-8} Btu/h ft^2 °F or 1.36×10^{-12} cal/cm^2 s°C), T is absolute temperature, and E_b is the energy emitted from a blackbody (one in which absorptivity and emissivity are unity).

The blackbody represents an idealization that can only be closely approximated experimentally. For any other body (termed a gray body), the emissivity ϵ is

$$\epsilon = \frac{E}{E_b} \tag{4-36}$$

or the ratio of intensity emitted by the body relative to a blackbody. For all bodies in radiation equilibrium,

$$\alpha' = \epsilon \tag{4-37}$$

the α' is that fraction of incident radiation absorbed.

A very important aspect of radiative heat transfer is the system geometry. This is accounted for by using radiation shape factors, also called view factors, angle factors, or configuration factors and defined as follows

F_{12} = fraction of energy leaving surface 1 that reaches surface 2

F_{21} = fraction of energy leaving surface 2 that reaches surface 1

Ultimately, it can be shown that

$$A_1 F_{12} = A_2 F_{21} = \int_{A_1} \int_{A_2} \cos\phi_1 \cos\phi_2 \, \frac{dA_1 \, dA_2}{\pi \, r^2} \tag{4-38}$$

where ϕ_1 and ϕ_2 are angles between A_1 and A_2, respectively, and the normal, and r is the length of the ray joining A_1 and A_2.

Radiation shape factors can be computed for different situations and then used with the equations

$$q_{net} = A_1 F_{12} (E_{b1} - E_{b2}) = A_2 F_{21} (E_{b1} - E_{b2}) \tag{4-39}$$

or

$$q_{net} = A_1 F_{12} E_{b1} \left(1 - \frac{E_{b2}}{E_{b1}}\right) = \sigma A_1 F_{12} T_1{}^4 \left[1 - \left(\frac{T_2}{T_1}\right)^4\right] \tag{4-40}$$

If $T_1 \gg T_2$ then,

$$q_{net} = \sigma A_1 F_{12} T_1^4 \tag{4-41}$$

Also, it is possible to write the q_{net} expression in the form

$$q_{net} = h_r A_1 (T_1 - T_2) \tag{4-42}$$

where h_r is a radiative heat-transfer coefficient.

For nonblackbodies connected by a third surface not exchanging heat,

$$q_{net} = \frac{\sigma A_1 (T_1^4 - T_2^4)}{\dfrac{A_1 + A_2 - 2A_2 F_{12}}{A_2 - A_1 (F_{12})^2} + \left(\dfrac{1}{\epsilon_1} - 1\right)} + \frac{A_1}{A_2}\left(\frac{1}{\epsilon_2} - 1\right) \tag{4-43}$$

where ϵ_1 and ϵ_2 are the emissivities of surfaces 1 and 2, respectively.

For the special case of infinite parallel planes,

$$q = \frac{\sigma A_1 (T_1^r - T_2^4)}{\dfrac{1}{\epsilon_1} + \left(\dfrac{A_1}{A_2}\right)\left(\dfrac{1}{\epsilon_2} - 1\right)} \tag{4-44}$$

4.18   RADIATION HEAT TRANSFER IN OPAQUE POLYMER SYSTEMS

In many instances, polymer systems are essentially opaque to thermal radiation. This means that the incident radiation absorbed by the body is converted to heat at the surface. This surface heat either is lost or flows into the polymer by conduction. For such cases, the temperature distribution of such a polymer will depend not only on the radiant energy flux but also on the conduction and surface convection behavior.

McKelvey [91] has considered this case for a slab at constant initial temperature T_i. For such a case, the temperature distribution within the slab is

$$(T - T_i) = \frac{q_s a}{k}\left[\left(\frac{\alpha t}{a^2}\right) + \left(\frac{3\alpha^2 - a^2}{6a^2}\right) - \frac{2}{\pi^2}\sum_{n=1}^{\infty}\frac{(-1)}{n^2} e^{\frac{-n^2\pi^2\alpha\theta}{a^2}}\cos\left(\frac{n\pi\alpha}{a}\right)\right] \tag{4-45}$$

where a is the half thickness of the slab, t is time, x is the distance from the centerline of the slab, and q_s is the heat flux across the surface.

If this equation is rewritten for T_c (centerline temperature) and T_s (the surface temperature), expressions for $(T_c - T_i)$ and $(T_s - T_i)$ are obtained. If the ratio of $(T_s - T_i)$ to $(T_c - T_i)$ is taken as I, a uniformity index, then

$$I = \frac{(T_s - T_i)}{(T_c - T_i)} = \frac{\dfrac{\alpha t}{a_2} + \dfrac{1}{3} - \dfrac{2}{\pi^2}\sum_{n=1}^{\infty}\dfrac{1}{n^2}\exp\dfrac{-n^2\pi^2\alpha\theta}{a^2}}{\dfrac{\alpha t}{a^2} - \dfrac{1}{6} - \dfrac{2}{\pi}\sum_{n=1}^{\infty}\dfrac{(-1)^n}{n^2}\exp\dfrac{-n^2\pi^2\alpha\theta}{a^2}} \tag{4-46}$$

$$I = \phi\left(\frac{\alpha t}{a^2}\right) \tag{4-47}$$

The relation of I to $\alpha t/a^2$ is given by Fig. 4-63. It can be used to set the slab dimensions or required time of heating for a given energy interchange situation.

4.19 RADIATION HEAT TRANSFER IN SEMITRANSPARENT POLYMER SYSTEMS

In certain heating situation, polymers behave as semitransparent or transparent materials. For such cases, incident thermal radiation will pass not only through the surface but also through the entire slab. This means that such polymers will heat at a much lower rate than predicted for opaque materials.

Progelhof et al. [92] have treated this problem, assuming also that convection heat transfer could be an additional factor. In essence, they solved the Fourier equation with a value radiant energy absorption term as a lumped parameter.

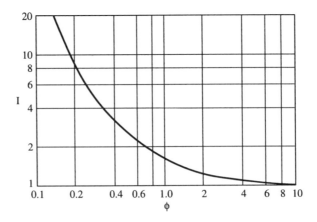

Fig. 4-63   Evenness index I as a function of ϕ [91].

The solution for the semitransparent case with a finite number of constant absorption coefficients J (see Figs. 4-64 and 4-65 and Table 4-6) was

$$q(x) = \sum_{i=1}^{J} A_i \cosh \alpha_i X \qquad (4\text{-}48)$$

where $q(x)$ is the rate of energy generation within the sheet. Also,

$$A_i = \frac{2\alpha_i (1 - \rho_r)}{e^{\alpha i a} - \rho_r e^{-\alpha_i a}} \int_{\lambda_i}^{\lambda_i + \Delta \lambda_i} Q_r (\lambda) \, d\lambda \qquad (4\text{-}49)$$

where Q_R is radiant heat flux, ρr the fraction of radiation reflected, x the absorption coefficient, and λ the wavelength of the radiation; the T_0 of Fig. 4-66 is air ambient temperature.

In dimensionless form, Eq. (4-49) becomes

$$\phi = \sum_{n=1}^{\infty} a_n \, e^{-\gamma_n^2 \tau^1} + b - \sum_{i=1}^{J} \eta_i \, Fi \, \frac{\cosh \alpha_{ix}}{(\alpha_i a)^2} \qquad (4\text{-}50)$$

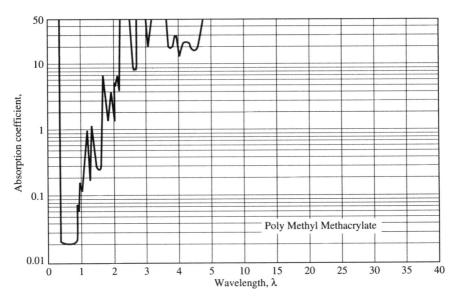

Fig. 4-64   Absorption spectrum for polymethyl methacrylate [92].

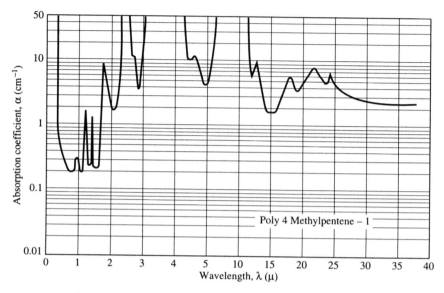

Fig. 4-65   Absorption spectrum for poly-4-methylpentene (TPX) [92].

TABLE 4-6   Step Function Approximation to Absortion Spectrum

Wavelength, λ, μ	Absorption Coefficient, α, cm^{-1}
(Polymethyl methacrylate)	
0 –0.4	∞
0.4 –0.9	0.02
0.9 –1.65	0.45
1.65–2.2	2.0
2.2 –∞	∞
(Poly-4-methylpentene-1)	
0 –0.3	∞
0.3 –1.6	0.25
1.6 –2.2	2.5
2.2 –2.7	∞
2.7 –2.9	5.0
2.9 –4.7	∞
4.7 –5.5	5.0
5.5 –12.0	6.0
12.0–13.5	2.5
13.5–17.0	5.0
17.0–27.5	2.5
27.5–∞	

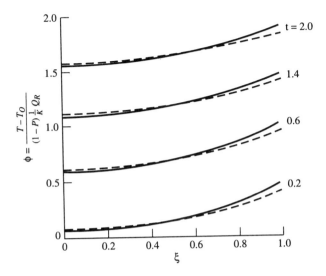

Fig. 4-66   Dimensionless temperature vs dimensionless distance (dimensionless time; solid line PMMA; dotted line TPX) [92].

and

$$b = \frac{k}{ha}\left\{1 + \sum_{i=1}^{J} F_i\left[\eta_i\left(\frac{\sinh\alpha_i a}{\alpha_i a} + \frac{ha}{k}\frac{\cosh\alpha_i a}{(\alpha_i a)^2}\right) - 1\right]\right.$$ (4-51)

$$a_n = \sum_{i=1}^{J} \eta_i F_i\left[\frac{x_i a \sinh\alpha_i a \cos\gamma_n + \gamma_n \cosh\alpha_i a \sin\gamma_n}{(\alpha_i a)^2 (\alpha_i^2 a^2 + \gamma_n^2)}\right.$$

$$- \frac{b\sin\gamma_n}{\gamma_n}$$ (4-52)

The ϕ is a dimensionless temperature $(T - T_0)/[a/k\,(1 - \rho_r)\,Q_r]$, γ_n an eigenvalue, η_i a radiation parameter, $(2\alpha_i a)/(e^{\alpha}ia - \rho_r e^{-\alpha}ia)$, τ' a dimensionless time, kt/a^2, h a convection heat-transfer coefficient, and F_i the fraction of radiant energy in the wavelength band $\lambda_i + \Delta\lambda_i$ that falls on the sheet.

Results for polymethyl methacrylate (PMMA) and poly-4-methylpentene-1 (TPX) are presented in Figs. 4-66 and 4-67. Also, the same data were analyzed with an index *EI* similar to the one used by McKelvey for opaque materials. This index is defined as

$$EI = \frac{T_{\max} - T_{\min}}{T_{\max} - T_0}$$

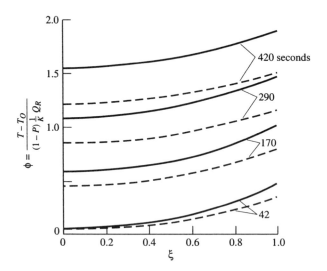

Fig. 4-67 Dimensionless temperature vs dimensionless distance (real time; solid line PMMA; dotted line TPX) [92].

Figure 4-68 shows a general plot of *EI* vs *kt/a* for various Biot numbers (*ha/k*). The actual index for the polymers is shown in Fig. 4-69.

4.20 RADIATION EXAMPLES

Example 4-11

A 3-ft-wide strip of an opaque polymer (0.02 in. thick) is being heated con-

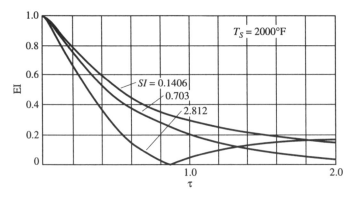

Fig. 4-68 Evenness index as a function of dimensionless time [92].

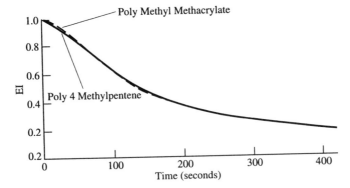

Fig. 4-69   Evenness index as a function of real time [92].

tinuously between banks of infrared heaters at 1500°F; each heater bank has an area of 0.08 ft². Assuming that heat is transferred entirely by radiation, calculate the net gain in enthalpy of the polymer if the rate remains 25 ft/min: Take polymer density as 80 lb/ft³.

Equation (44) for infinite parallel plates can be used with the additional stipulations that radiant energy absorption by the heating elements is negligible and that the temperature of the heating elements is much greater than the polymer's temperature.

For this case,

$$q_{net} = \sigma \, A_{heater} \, \epsilon_{heater} \, T_{heater}^4$$

and

$$q_{net} = \left(0.1714 \times 10^{-8} \, \frac{Btu}{h \, ft^2 \, °R^4}\right) (2)(0.08 \, ft^2)(0.90)(1960°R)^4$$

$$q_{net} = 3650 \, Btu/h$$

Then,

$$\text{Mass rate of polymer} = \left(\frac{25 \, ft}{min}\right)\left(\frac{60 \, min}{h}\right)(3 \, ft)\left(\frac{0.02}{12} \, ft\right)\left(\frac{80 \, lbm}{ft^3}\right)$$

$$\text{Mass rate of polymer} = 600 \, \frac{lbm}{h}$$

and,

$$\text{Enthalpy rise} = \frac{3650 \text{ Btu/h}}{600 \text{ lbm/h}} = 6.25 \frac{\text{Btu}}{\text{lbm}}$$

Example 4-12

A sheet of polymethyl methacrylate 0.20 in. thick is heated radiantly to a surface temperature of 2000°F for 519 s. The convective heat-transfer coefficient is 8.85 Btu/h ft^2 °F. What heat transfer would be needed to yield the same evenness index for the same time of exposure? Explain the difference in the two situations.

For the polymethyl methacrylate,

$$\tau' = \frac{\alpha t}{L^2} = \frac{(1.585 \times 10^{-4} \text{ cm}^2/\text{sec})}{(0.254 \text{ cm})^2} (519 \text{ s}) = 1.27$$

$$\text{Biot number} = \frac{hL}{k} = \frac{\left(8.85 \dfrac{\text{Btu}}{\text{m ft}^2 \text{ °F}}\right)\left(\dfrac{0.1}{12 \text{ ft}}\right)}{\left(0.105 \dfrac{\text{Btu}}{\text{h ft °F}}\right)}$$

Biot number = 0.703

Hence, from Fig. 4-68,

$$EI = \text{evenness index} = 0.12$$

Now, if exposure time is the same, the τ' value will be identical. Then, if evenness index is to be the same, the only change is in Biot number (see Fig. 4-68).

$$\text{Biot number}_2 = 2.812$$

and

$$h_2 = 35.4 \text{ Btu/h ft}^2 \text{ °F.}$$

The explanation for the second case is that minimum temperature lies not at the sheet centerline but somewhere between that point and the surface. This results because the higher heat transfer cools the region near the surface but does not affect the radiant heat transfer to the remainder of the sheet.

REFERENCES

1. BIRD, R.B., STEWART, W.E., and LIGHTFOOT, E.N., *Transport Phenomena*, Wiley (1966).
2. ANDERSON, D.R., *Chem. Rev.* **66**, 677 (1966).
3. KNAPPE, W., *Advances Polymer Sci.* **7**, 477 (1971).
4. VAN KREVELEN, D.W., and HOFTYZER, P.J., *Properties of Polymers*, Elsevier, New York, (1972).
5. GRISKEY, R.G., LUBA, M., and PELT, T., *J. Appl. Polymer Sci.* **23**, 55 (1979).
6. EIERMANN, K., *Kunststoffe* **55**, 335 (1965).
7. CARSLAW, H.C., and JAEGER, J.C., *Conduction of Heat in Solids*, Oxford University Press, (1959).
8. GURNEY, H.P., and LURIE, J., *Ind. Eng. Chem.* **15**, 1170 (1923).
9. WILLIAMS, E.D., and ADAMS, L.H., *Phys. Rev.* **14**, 99 (1919).
10. LYCHE, B.C., and BIRD, R.B., *Chem. Eng. Sci.* **6**, 35 (1957).
11. GRIGULL, V., *Chem. Ing. Tech.* **28**, 553, 665, (1956).
12. SCHENK, J., and VAN LAAR, J., *Appl. Sci. Res.* **A7**, 449 (1958).
13. WHITEMAN, I.R., and DRAKE, D.B., *Trans. ASME* **80**, 728 (1958).
14. BIRD, R.B., *Chem. Ing. Tech.* **31**, 569 (1959).
15. CHRISTIANSEN, E.B., and CRAIG, S.E., *AIChEJ.* **8**, 154 (1962).
16. CHRISTIANSEN, E.B., and JENSEN, G.E., *Progress in International Research on Thermodynamics and Transport Properties*, Academic Press, New York (1962).
17. TOOR, H.L. *Ind. Eng. Chem.* **48**, 922 (1956).
18. TOOR, H.L., *Trans. Soc. Rheol.* **1**, 177 (1957).
19. GEE, R.E., and LYON, J.B., *Ind. Eng. Chem.* **49**, 956 (1957).
20. TOOR, H.L., *AIChEJ.* **4**, 319 (1958).
21. FORSYTH, T.H., and MURPHY, N.F., *AIChEJ.* **15**, 758 (1969).
22. SCHECTER, R.S., and WISSLER, E.H., *Nucl. Sci. Engr.* **6**, 371 (1959).
23. WISSLER, E.H., and SCHECTER, R.S., *Chem. Eng. Progr. Symp. Ser.*, 55, (29), 203 (1958).
24. MICHIYOSHI, I., MATSUMATO R., and HOZUMI, M., *Bull. JSME* **6**, 496 (1963).
25. FORABOSCHI, F.P., and DI FEDERICO, I., *Int. J. Heat Mass Transfer* **7**, 315 (1964).
26. GILL, W.N., *AIChEJ.* **8**, 137 (1962).
27. METZNER, A.B., VAUGHN, R.D., and HOUGHTON, G.L., *AIChEJ.* **3**, 92 (1957).
28. METZNER, A.B., and GLUCK, D.F., *Chem. Eng. Sci.* **12**, 185 (1960).
29. OLIVER, D.R., and JENSON, V.G., *Chem. Eng. Sci.* **19**, 115 (1964).
30. KUBAIR, V.G., and PEI, D.C.T., *Int. J. Heat Mass Transfer* **11**, 855 (1968).
31. BEYER, C.E., and DAHL, R.B., *Modern Plastics* **30**, 125 (1952).
32. SCHOTT, H., and KAGHAN, W.S., *Soc. Plastics Eng. J.* **20**, 139 (1964).
33. BIRD, R.B., *Soc. Plastics Eng.* **11**, 35 (1955).
34. VAN LEEUWEN, J., *Kunststoffe* **55**, 441 (1965).
35. VAN LEEUWEN, J., *Polymer Eng. Sci.* **7**, 98 (1967).
36. GRISKEY, R.G. and WIEHE, I.A., *AIChEJ.* **12**, 308 (1966).
37. GRISKEY, R.G., *Proceedings 29th Annual Technical Conference SPE* **XVII**, 98 (1971).

38. COLLINS, E.A., and FILISKO, F.E., *AIChEJ.* **16**, 339 (1970).
39. KIM, H.T., and DARBY, J.P., *Soc. Plastics Engr. J.* **26**, 31 (1970).
40. COLLINS, E.A., and KIM, H.T., *Polymer Eng. Sci.*, *11*, 83 (1971).
41. FORSYTH, T.H., and MURPHY, N.F., *Polymer Eng. Sci.* **9**, 22 (1969).
42. SALTUK, I., SISKOVIC, N., and GRISKEY, R.G., *Polymer Eng. Sci.* **12**, 397 (1972).
43. SALTUK, I., SISKOVIC, N., and GRISKEY, R.G., *Polymer Eng. Sci.* **12**, 402 (1972).
44. GRISKEY, R.G., and CHOI, M.H., and SISKOVIC, N., *Polymer Eng. Sci.* **13**, 287 (1973).
45. KAY, W.M., *Trans. ASME* **II**, 1265 (1955).
46. CHARM, S.E., *Ind. Eng. Chem. Fund.* **1**, 79 (1962).
47. SKELLAND, A.H.P. in *Non-Newtonian Flow and Heat Transfer*, Wiley, New York (1967), p. 421.
48. GRISKEY, R.G., NOTHEIS, P., FEDORIW, W., and VICTOR, S., *Proceedings 33rd Annual Technical Conference SPE* **XXI**, 459 (1975).
49. TIEN, C., *Can. J. Chem. Eng.* **39**, 45 (1961); **40** 130 (1962).
50. SUCKOW, W.H., HRYCAK, P., and GRISKEY, R.G., *Polymer Eng. and Sci.* **11**, 401 (1971).
51. CROZIER, R.D., BOOTH, J.R., and STEWART, J.E., *Chem. Eng. Progr.* **60** (8), 43 (1964).
52. SCHECTER, R.S., Paper presented at 43rd National Meeting American Institute of Chemical Engineers, Tulsa, OK (Sept. 1960).
53. KOZICKI, W., CHAN, C.H., and TIU, C., *Chem. Eng. Sci.* **21**, 665 (1966); *Can. J. Chem. Eng.* **47**, 438 (1969).
54. CARREAU, P., CHAREST, G., and CORNEILLE, J.L., *Can. J. Chem. Eng.* **44**, 3 (1966).
55. GLUZ, M.D., and PAVLECHENKO, I.S., *J. Appl. Chem. (U.S.S.R.)* **21**, 2719 (1966).
56. HAGEDORN, D.W., and SALAMONE, J.J., *Ind. Eng. Chem. Proc. Des. Devel.* **6**, 469 (1967).
57. SANDALL, O.C., and PATEL, K.G., *Ind. Eng. Chem. Proc. Des. Devel.* **9**, 139 (1970).
58. SANDALL, O.C., and MARTONE, J.A., *Ind. Eng. Chem. Proc. Des. Devel.* **10**, 86 (1971).
59. SKELLAND, A.H.P., and DIMMICK, G.R., *Ind. Eng. Chem. Proc. Des. Devel.* **8**, 267 (1969).
60. METZNER, A.B., FEEHS, R.H., RAMOS, H.L., OTTO, R.E., and TUTHILL, J.D., *AIChEJ.* **7**, 3 (1961).
61. WINDING, C.C., DITTMAN, F., and KRANICH, W.L., *Thermal Properties of Synthetic Rubber Lattices*, Report to Rubber Reserve Company, Cornell Univ., Ithaca, NY (1944).
62. ORR, C., and DALLA VALLE, J.M., *Chem. Eng. Prog. Symposium Series*, No. 9, **50**, 29 (1954).
63. THOMAS, D.G., *AIChEJ.* **6**, 631 (1960); *Ind. Eng. Chem.* **55**, 28 (1963).
64. CLAPP, R.M. in *International Developments in Heat Transfer*, American Society of Mechanical Engineers, New York (1963), p. 652.
65. METZNER, A.B., and FRIEND, W.L., *AIChEJ.* **4**, 393 (1958).

66. ACRIVOS, A., *AIChEJ.* **6**, 584 (1960).
67. ACRIVOS, A., SHAH, M.J., and PETERSEN, E.E., *AIChEJ.* **6**, 312 (1960).
68. *Ibid.* **8**, 542 (1962).
69. YAU, J., and TIEN, C., *Can. J. Chem. Eng.* **41**, 139 (1963).
70. NA, T.Y., and HANSEN, A.G., *Can. J. Chem. Eng.* **9**, 261 (1966).
71. BERKOWSKI, B.M., *Inst. Chem. Eng.* **6**, 187 (1966).
72. ACRIVUS, A., *Chem. Eng. Sci.* **21**, 343 (1966).
73. GRANVILLE, P.S., *J. Ship. Res.* **6** Oct. (1962).
74. SKELLAND, A.H.P., *AIChEJ.* **12**, 69 (1966).
75. GRIFFIN, O.M., *Proceedings 29th Annual Technical Conference SPE* **XVII**, 17 (1971).
76. GRIFFIN, O.M., *Chem. Eng. Sci.* **25**, 109 (1970).
77. GRIFFIN, O.M., and SZEWCZYK, A.A., *Proceedings Fourth International Heat Transfer Conference*, Vol. 1, Paris (1970).
78. KASE, S., and MATSUO, *J. Polymer Sci.* **3A**, 2541 (1965).
79. ANDREWS, E.H., *Brit. J. Appl. Phys.* **10**, 39 (1959).
80. BARNETT, T.R., *Appl. Polymer Symposia* **6**, 51 (1967).
81. COPLEY, M., and CHAMBERLAIN, N.H., *Appl. Polymer Symposia* **6**, 27 (1967).
82. WILHELM, G., *Kolloid Z.* **208**, 97 (1966).
83. MORRISON, M.E., *AIChEJ.* **16**, 57 (1970).
84. FOX, V.G., and WANGER, W.H., Paper presented at AIChE Annual Meeting Chicago, IL, Dec. 1970.
85. COLLIER, J.R., DINOS, N., and SIFLEET, W.L., *Proceedings 29th Annual Technical Conference SPE* **XVII**, 12 (1971).
86. WESTOVER, R.F., in *Processing of Thermoplastic Materials*, E.C. BERNHARDT, ed., Reinhold, New York (1965), p. 554.
87. GRISKEY, R.G., and HUBBELL, D.O., *J. Appl. Polymer Sci.* **12** 853 (1968).
88. GRISKEY, R.G., and FOSTER, G.N., *Proceedings 26th Annual Technical Conference SPE* **XIV** (1968).
89. VLACHOPOULOS, J., and KEUNG, J.C.K., *AIChEJ.* **18**, 1272 (1972).
90. METZNER, A.B., and FRIEND, P.S., *Ind. Eng. Chem.* **51**, 879 (1959).
91. MCKELVEY, J.M., in *Processing of Thermoplastic Materials*, E. BERNHARDT, ed., Reinhold, New York (1965), p. 105.
92. PROGELHOF, R.C., THRONE, J.L., and QUINTERE, J., *Proceedings 29th Annual Technical Conference SPE* **XVII**, 112 (1971).
93. KWANT, P.B., ZWANEFELD, A., and DIJKSTRA, F.C., *Chem. Eng. Sci.* **28**, 1303 (1973).
94. MAHALINGAM, R., TILTON, L.O., and COULSON, J.J., *Chem. Eng. Sci.* **30**, 921 (1975).
95. MCKILLOP, A.A., *Int. J. Heat Mass Transfer* **7**, 853 (1964).
96. BASSETT, C.E., and WELTY, J.R., *AIChEJ.* **21**, 699 (1975).
97. FAGHRI, M., and WELTY, J.R., **23**, 288 (1977).
98. POPOVSKA, F., and WILKINSON, W., *Chem. Eng. Sci.* **21**, 1155 (1977).
99. DANG, V.D., *J. Appl. Poly. Sci.* **23**, 3077 (1979).
100. WINTER, H.H., *Poly. Eng. Sci.* **15**, 84 (1975).

101. PEARSON, J.R.A., *Poly Eng. Sci.* **18**, 222 (1978).
102. JOSHI, S.D., and Bergles, A.E., *J. Heat Transfer* **102**, 397 (1980).
103. COX, H.W., and MACOSKO, C.W., *AIChEJ.* **20**, 785 (1974).
104. LIN, S.H., and HSU, W.K., *J. Heat Transfer* **102**, 382 (1980).
105. YBARRA, R.M., and ECKERT, R.E., *AIChEJ.* **26**, 751 (1980).
106. DINH, S.M., and ARMSTRONG, R.D., *AIChEJ.* **28**, 294 (1982).
107. LIN, S.H., and HSIEH, D.M., *J. Heat Transfer* **102**, 786 (1980).

PROBLEMS

4-1 A polymer's refractive index at 25°C is 1.52. Find its thermal conductivity at 175°C.

4-2 If a semicrystalline polymer has a glass temperature of 286 K, what is its thermal conductivity at 573 K?

4-3 According to Fig. 4-2, the generalized curve reaches a thermal conductivity ratio of unity at $T = T_g$. Both above and below T_g, the correlation indicates that the thermal conductivity of the polymer is below its glass temperature value. Why is this so? Provide as logical an explanation as possible.

4-4 In a process, a polymer is injection-molded into the shape of a sphere (radius of 0.025 m). How long should the object be cooled (mold wall temperature of 28°C) if the polymer's heat distortion temperature is 55°C?

Polymer data:

Thermal diffusivity = 1.25×10^{-3} cm^2/s
Heat of fusion = 60 cal/g
Molding temperature = 175°C

4-5 Frequently, polymer chips or pellets are dried in a cylindrical tower equipped with a screw drive to circulate the pellets. Polyester chips are to be dried at a temperature of 190°C. How long will it take a bed (radius of 0.5 m) to reach the drying temperature?

4-6 Nylon 66 is processed in a screw extruder (helical flow channel 0.01 m deep). If the polymer midplane temperature is to reach 140°C in 3.5 min, what should the extruder barrel and screw temperature be?

4-7 What is the thermal diffusivity of a plastic if two sheets of 0.015 m thick are bonded in a half-hour. The press used to bond the sheets has its platens at 225°C. Bonding temperature for the adhesive used in the process is 125°C.

4-8 Several authors [22–24] have assumed uniform viscous heat dissipation in dealing with the solution of the energy equation for non-Newtonian systems. Why is this assumption generally invalid? Are there any situations in which the assumption could be valid or partially valid?

4-9   For an *n* value of 1/3, correlate the (K_b'/K_w) term of Eqs. (4-17) and (4-18) with the ψ (*H*) parameter of Fig. 4-14.

4-10   A polymer solution (*n* of 0.45; *K* of 130 Newtons-s^n/m^2 at 306 K is flowing in a 0.025-m-diam tube. It enters at 311 K and leaves at 327 K. Wall temperature is constant at 367 K. The activation energy for *K* is 13.66 kilojoules/g-mole. Properties are:

ρ = 1050 kg/m^3
C_p = 2.09 kJ/kg-K
k = 1.21 W/m K

What is the flow rate of the solution if the heat exchanger is 1.52 m long?

4.11   The polymer solution in the preceding problem can be processed in one of two different systems. For each case, the solution's mass flow rate (0.5 kg/s) and inlet and outlet temperatures (65.6 and 26.7°C) are the same. Which pipe diameter (0.02 m,wall at 21°C or 0.01 mg, wall at 5°C) will give the lower pressure drop?

4-12   Thermally softened polystyrene is heated in a circular tube (diameter of 0.03 m) that is 3 m long. The mass flow rate of the polymer is 400 g/min. What wall temperature will be needed if the polymer's average inlet and exit temperature are 490 and 513 K? Rheological data for the polymer are: n = 0.22; K = 2.2 $\times$ 10^4 N s^n/m^2.

4-13   Molten polypropylene is to be cooled from 260 to 220°C. If the heat exchanger of the previous problem is available (with the same wall temperature), what mass flow of polypropylene can be accommodated? The *n* and *K* values for the polymer are, respectively, 0.4 and 4.1 $\times$ 10^3 N s^n/m^2.

4-14   A polyethylene at 175°C (ρ_0 of 0.920 g/cm^3) is pumped into a 0.02-m-diam pipe whose wall temperature is 120°C. At what length of pipe will the polyethylene solidify? Flow data for the polyethylene are n = 0.48 and k = 6.5 $\times$ 103 N s^n/m^3. The polymer mass flow rate is 200 g/min.

4-15   Nylon 6 is flowing at an average velocity of 0.02 m/s in a 0.03-m-diam tube 4 m long. The tube wall temperature is 565 K and the polymer exits at 510 K. What is the polymer's inlet temperature? The *n* and *K* values for the nylon are 0.65 and 1.85 $\times$ 10^3 Ns^n/m^2.

4-16   A polymer solution (*n* of 0.42 and *k* of 110 N s^n/m^2) is flowing through a rectangular duct (width, 1 m; height, 0.03 m) for a length of 3 m. The duct's wall temperature is 358 K, and the solution's inlet and outlet temperatures are 310 and 323 K, respectively. What is the solution's average velocity in the tube? Solution properties are:

ρ = 1050 kg/m^3
C_p = 1.19kJ/kgK
k = 1.20 W/m K

4-17   The Kozicki et al. [53] *a* and *b* values [see Eqs. (4-25)–(4-29) are 0.489 and 0.991

for an annulus (two concentric pipes with r inside/r outside of 0.4). A polyethylene melt at 145°C flows into the annular space at a mass flow rate of 400 g/min. Outside diameter is 0.04 m. Wall temperatures are both 175°C. Find the exit temperature of the polymer. K and n values for the polyethylene are 4.5×10^3 N s^n/m^2 and 0.46.

4-18   Repeat Problem 4-17, using an equivalent diameter that is four times the hydraulic radius defined by Eq. (3-38). Compare the results to those of Problem 4-17. Comment on the comparison.

4-19   Find the heat-transfer coefficient for a polymer solution flowing in a 0.015-m-diam tube (wall temperature at 390 K). The solution enters the tube at 330 K with an average velocity of 130 m/s.

Solution properties are:

$\rho = 1040$ kg/m^3
$C_p = 2.1$ kJ/kg K
$k = 1.20$ W/m K
$n = 0.42$
$K = 135$ N s^n/m^2

4-20   Estimate the temperature halfway between the centerline and outside surface of an opaque polymer after an elapsed time of 475 s an elapsed time of the slab is 25°C. Surface temperature is 510°C. Slab thickness is 1.2 cm. Indicate all assumptions.

MINI PROJECT A

Develop a generalized curve for the thermal conductivities of molten polymers similar to the curve in Fig. 4-2.

MINI PROJECT B

Use the heat-transfer data for molten and thermally softened flowing polymer systems (i.e., Figs. 4-26–4-50) to fit the relation

$$Nu = \frac{hD}{k} = C_1(Re)^a(Pr)^b(B^*)^d$$

where C_1 and the superscripts a, b, and d are constants.

MINI PROJECT C

On occasion, polymers are processed as filled systems (i.e., with glass, carbon, or mineral fillers). Develop appropriate methods for the flow and heat transfer of such systems. Some flow data are available in the open literature.

MASS TRANSFER IN POLYMER SYSTEMS

5.1 INTRODUCTION

The goal of all polymer processing operations is to produce a usable object. Basically, these operations involve flow and deformation (rheology) as well as the transfer of energy (thermodynamics and heat transfer). As such, we can state that, for *all* polymer processing operations, necessary areas of knowledge are:

Rheology (Chapter 3)
Thermodynamics (Chapter 2)
Heat transfer (Chapter 4)

In addition, in many instances, it is also necessary to use the concept of mass transfer. Particular cases involving mass transfer include:

1. Devolatilization processes
2. Structural foamed plastic production
3. Reaction injection molding (RIM)
4. Polymer reaction systems
5. Dry or wet spinning of fibers

Mass transfer is analogous in several ways to heat transfer. Both processes are vectorial, that is, they can be described with three components. The fundamental equation that applies to mass transfer is the equation of continuity of species. This equation is essentially the inventory of a compound in a given system. In its most general form, the equation for a species A in cylindrical coordinates is:

$$\frac{\partial c_A}{\partial t} + \frac{1}{r}\frac{\partial}{\partial r}(rN_{A_r}) + \frac{1}{r}\frac{\partial N_{A\theta}}{\partial \theta} + \frac{\partial N_{Az}}{\partial z} = R_A \qquad (5\text{-}1)$$

where

c_A = concentration of species A
N_{Ar}, NA_θ, N_{Az} = components of mass flux N
R_A = the rate of chemical reaction of A per unit volume

If density and diffusivity D_{AB} are constant, then,

$$\frac{\partial c_A}{\partial t} + \left[v_r\frac{\partial c_A}{\partial r} + v_\theta\frac{1}{r}\frac{\partial c_A}{\partial \theta} + v_z\frac{\partial c_A}{\partial z} \right]$$
$$= D_{AB}\left[\frac{1}{r}\frac{\partial}{\partial r}\, r\frac{\partial c_A}{\partial r} + \frac{1}{r^2}\frac{\partial^2 c_A}{\partial \theta^2} + \frac{\partial^2 c_A}{\partial z^2} \right] + R_A \qquad (5\text{-}2)$$

The terms in Eq. (5-2) are qualitatively as follows: $\partial C_A/\partial t$ is an accumulation term; the bracketed term with the velocities represents convection; the term multiplied by the diffusivity D_{AB} deals with molecular diffusion; and the R_A term is for reaction.

5.2   DIFFUSION AND SOLUTION

Fick's first law (written for a one-dimensional case in rectangular coordinates) gives the simplest representation of the molecular diffusion of a species in a system. This expression relates the molar flux J_A of species A to a gradient dC_A/dy by means of a transport coefficient D_{AB}, the diffusivity or diffusion coefficient.

$$J_{A_y} = -D_{AB}\frac{dC_A}{dy} \qquad (5\text{-}3)$$

Note that D_{AB} is a *property* of the system in much the same way that thermal conductivity is for the transfer of heat. Another way of thinking of this is that diffusivity is the ratio of molar flux to concentration gradient whereas thermal conductivity is the ratio of heat flux to temperature gradient.

The mass-transfer process is a kinetic or dynamic one. For diffusion of molecules into a liquid, semisolid, or solid system, there is an important limiting case, which is *solution*. The analogy to this situation is chemical reaction, which is a *kinetic process* that is ultimately limited by *chemical equilibrium*.

Both the diffusivities and solubilities of small molecules are important in polymer processing operations. The rate of transport of small molecules within the polymer being processed is related to the diffusivity. However, solubility can fix the retention of the molecule in the system for a given temperature and pressure. In many cases, therefore, a combination of both diffusivity and solubility is needed for the design or understanding of the given operation.

A third coefficient of some importance is the permeability, which is defined as the amount of a diffusing molecule that passes through a polymer film of unit thickness per second per unit area and a unit difference of pressure. Diffusivity, solubility, and permeability are related by the following:

$$P = D \cdot H \tag{5-4}$$

where D is diffusivity and H the Henry's law constant obtained from the relationship

$$H P_i = X_i \tag{5-5}$$

where P_i is the partial pressure of the diffusing component and X_i the gas concentration in the polymer.

Permeabilities apply mainly to molecular transport in solid polymers, particularly polymer films. Also, diffusivities are usually obtained in conjunction with permeability measurements. Although such behavior is obviously important in many applications, it relates only indirectly to processing that occurs mainly with molten or thermally softened polymers. The principal role of processing with respect to permeabilities and diffusivities in solid polymers is development of the material's structural characteristics which, in turn, affect properties. As such, consideration of permeability will be deferred to Chapter 12, which deals with the effect of processing or structural parameters.

Although discussion will be deferred, it is worthwhile to cite references that consider permeabilities and diffusivities in solid polymers, which can be found in the text *Diffusion in Polymers*, edited by Crank and Park (1). Particular sections of some importance include Chapter 2, by V. Stannett, which summarizes a large amount of diffusivity data at low pressures, and Chapter 8, by J. A. Barrie, which presents water sorption data.

Mass transport of gaseous molecules in molten or thermally softened polymer systems is a particularly pertinent area. This is so because polymer processing operations generally involve such systems. Hence, considerable attention will be directed to diffusivities and solubilities in molten or thermally softened polymers.

TABLE 5-1   Measurement Techniques for Diffusivities and Solubilities of Gas Molecules in Molten or Thermally Softened Polymer Systems

Method	Used for Diffusivity	Used for Solubility	Reference
1. Movement of a color boundary	Yes	Yes	2
2. Strain gage measurements in sintered steel cylinder	Yes	Yes	3–6
3. Special diffusion cell	Yes	Yes	7–9
4. Gas chromatography	Yes	No	10, 11
5. Piezoelectric	No	Yes	12
6. Modified quartz balance	Yes	Yes	13
7. Desorption apparatus	Yes	—	14

Measurement of solubilities and diffusivities have been accomplished by various techniques as shown in Table 5-1.

Diffusion cell data of Table 5-1 are shown in Figs. 5-1 and 5-2. The diffusivity is fit to the data, and the solubility is found when the pressure reading levels off (i.e., equilibrium). Published data can be found in Refs. 2–9, 13, and 15–18.

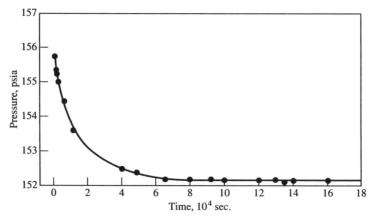

Fig. 5-1   Pressure time for the system argon–polyethylene at 188°C. Diffusivity for the curve is 9.27×10^{-6} cm^2/s [7–9].

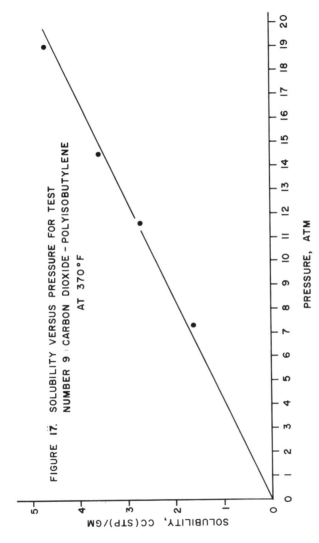

FIGURE 17. SOLUBILITY VERSUS PRESSURE FOR TEST
NUMBER 9 : CARBON DIOXIDE - POLYISOBUTYLENE
AT 370°F

Fig. 5-2   Solubility vs pressure for nitrogen–polyethylene and carbon dioxide–polyethylene [7–9].

227

5.3 GAS DIFFUSIVITIES IN MOLTEN AND THERMALLY SOFTENED POLYMER SYSTEMS

A listing of experimentally determined diffusivities for a variety of gases and polymers is given in Table 5-2 for a temperature of 188°C.

Data for the diffusion of solvents in polystyrene were also determined (13, 15–18). Figure 5-3 presents data for the pentane–polystyrene system.

TABLE 5-2   Diffusivities (cm²/s × 10⁻⁵) in Molten and Thermally Softened Polymers at 188°C

Gas	Diffusivities			
	Polyethylene	*Polypropylene*	*Polyisobutylene*	*Polystyrene*
Helium	17.09	10.51	12.96	—
Argon	9.19	7.40	5.18	—
Krypton	—	—	7.30	—
Monochlorodifluormethane	4.16	4.02	—	—
Methane	5.50	—	2.00	0.42
Nitrogen	6.04	3.51	2.04	0.348
Carbon dioxide	5.69	4.25	3.37	0.39

Source: Methane data from Refs. 3–6. Other polystyrene data from Ref. 2. All other data from Refs. 7–9.

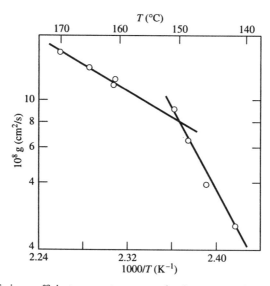

Fig. 5-3   Diffusion coefficients versus temperature for the pentane-polystyrene system (15).

The diffusivities of gases in polymers are related to a variety of parameters: temperature, pressure, the nature of the gas, and the nature of the polymer.

In order, therefore, to deal properly with processing operations, it is necessary to be able to take these parameters into account, which means that correlations are needed. If such correlations are not available, the practitioner will need at least some understanding of the interrelationship of diffusivities with the parameters.

Using diffusivity behavior in general as a guide, we can examine the case for polymers in more detail. For example, it is generally known that diffusivities can be related to *temperature* by an exponential function. The same situation prevails for diffusion of gases into molten or thermally softened polymers where

$$D = D_0 \exp(-E_d/RT) \tag{5-6}$$

The E_d is the activation energy of diffusion (kcal/mole). D_0 is the pre-exponential function (an empirical constant; its units are cm^2/s).

Experimentally determined values of E_d and D_0 are given in Table 5-3.

Values of E_d and D_0 can be estimated for other gases and polymers by using Figs. 5-4 and 5-5. In Fig. 5-4, the logarithm of D_0 is plotted against E_d/R. Figure 5-5 gives a plot of E_d vs a function of the polymer's glass temperature and ϵ/k (a gas molecule parameter; see Table 5-4) for the data of Table 5-3. In order to use the data of Figs. 5-4 and 5-5, first establish the E_d value from Fig. 5-5 for a given polymer (i.e., use the polymer's glass temperature and ϵ/k for the gas). Next, obtain the D_0 value from Fig. 5-4.

The effect of *pressure* on the diffusivity values is negligible *at least* up to 1.01×10^7 Pa (100 atm) and possibly up to 3.03×10^7 Pa (300 atm). As such, the diffusivity values given in Table 5-2 can be used up to the above limits.

TABLE 5-3   Values of E_d and D_0 for Various Gas-Polymer Systems

Polymer	*Gas*	*E_d, kcal/mole*	*$D_0 \times 10^5$, cm^2/s*	*Temperature Range °C*	*References*
Polyethylene	N_2	2.0	53.414	125–188	3–6
	CO_2	4.4	688.135	188–224	7–9
Polypropylene	CO_2	3.0	111.768	188–224	7–9
Polystyrene	H_2	10.1	218.064	120–188	2
	N_2	9.6	21.119	119–188	2
	CH_4	3.6	21.24	125–188	3–6

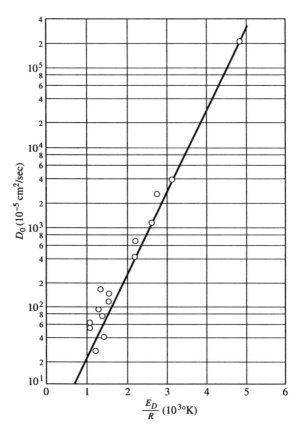

Fig. 5-4 Logarithm of D_0 vs activation energy divided by gas constant.

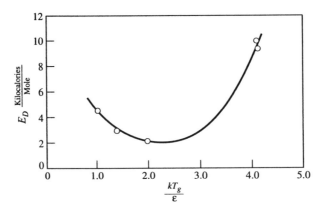

Fig. 5-5 Activation energy versus $\frac{kT_g}{\epsilon}$.

TABLE 5-4 Selected Gas Molecule Parameters

Gas	σ Collision Diameter, Å	ϵ/K, K
H_2	2.915	380
He	2.576	10.2
Ne	2.789	35.7
A_r	3.418	124
K_r	3.498	225
N_2	3.681	91.5
CO	3.590	110
CO_2	3.996	190
Cl_2	4.115	357
CH_4	3.822	137
C_2H_4	4.232	205
C_2H_6	4.418	230
C_3H_8	5.061	254

Source: Data abstracted from Bird et al., *Transport Phenomena* [19].

As mentioned earlier, the diffusivity is related to the molecular size of the diffusing species: The smaller the molecule, the larger the diffusivity. This becomes apparent when a parameter (the logarithm of the diffusivity divided by the square of the diameter of the gas molecule) is plotted against the molecular diameter itself (see Figs. 5-6–5-9). As can be seen, a reasonable correlation can be obtained for a variety of gases with a given polymer. Such plots also make it possible to estimate D for the polymer and a gas not previously measured.

Finally, the question arises as to the effect of the polymer itself on diffusivity for a given gas. Figure 5-10, which plots the diffusivity vs polymer mer weight for the same gas, shows this effect. As evident from Fig. 5-10, the diffusivity decreases with increasing mer weight.

The explanation for the behavior of Fig. 5-10 is not immediately obvious. Chain flexibility is not an appropriate explanation since, as before, the glass temperature of polyisobutylene lies between that for polyethylene and polypropylene whereas the correlation shows the polyisobutylene diffusivity data to be lower than either of the other two polymers. The same holds true for cohesive energy density since the values for polyethylene and polyisobutylene are nearly the same. Once we have considered existing theories of diffusion (20–23) in liquids, we begin to discern a pattern for the correlation. All these theories share one factor in common, namely, that diffusivity is inversely proportional to a

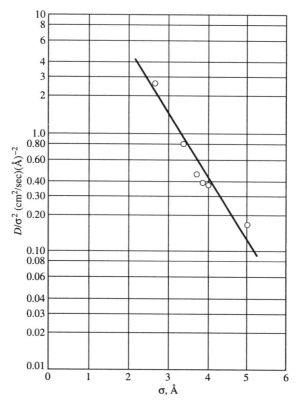

Fig. 5-6   Diffusivity divided by the square of gas collision diameter versus gas collision diameter for polyethylene (8, 9).

viscosity. We cannot properly speak of a viscosity in describing the four polymers of Fig. 5-10 at 188°C. For one thing, polystyrene and polyisobutylene form thermally softened masses rather than melts (as with polyethylene and polypropylene). Furthermore, in polymeric melts, we cannot speak of a viscosity because of the non-Newtonian nature of these materials.

It is possible, however, to speak of relative consistencies of the polymers of Fig. 5-10 or, in another sense, of the relative mobilities of diffusing species in them. If such a distinction is made, the materials could be arranged (in order of increasing consistency or decreasing diffusing species mobility) as polyethylene, polypropylene, polyisobutylene, and polystyrene. This fact checks with Einstein's original relation, which showed diffusivity to be related directly to the mobility of the diffusing species.

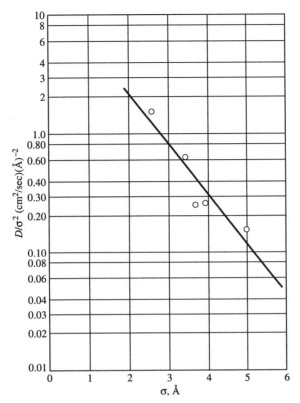

Fig. 5-7 Diffusivity divided by the square of gas collision diameter versus gas collision diameter for polypropylene (8, 9).

Hence, it would seem that the relation indicated in Fig. 5-10 between diffusivity and mer weight is actually between the diffusivity and mobility of the diffusing species since these polymers differ structurally in the pendant groups attached to the basic polyethylene chain. As such, Fig. 5-10 should be used only for polymers whose basic straight-chain structure is that of polyethylene.

Mention should also be made of a diffusion coefficient theory and the model developed by Duda, Vrentas, and their co-workers (18, 24–31). This model, based on free-volume concepts, was suggested by the type of behavior shown in Fig. 5-3 for the pentane–polystyrene system. As can be seen, there are actually two relations between the diffusion coefficient and the reciprocal of the absolute temperature [i.e., at the form of Eq. (5-6)]. One of the relations holds up to 150°C, and the second from 150 to 170°C.

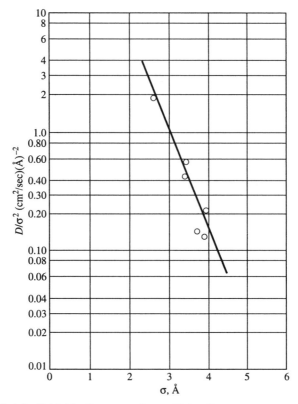

Fig. 5-8 Diffusivity divided by the square of gas collision diameter versus gas collision diameter for polyisobutylene (8, 9).

Alternatively, it is possible to try to fit the entire range of diffusivity–reciprocal of temperature with a curve. This is basically what was done with the free-volume model. The resultant correlation involves two complex equations, together with Eq. (5-6). In order to use the correlation, it is necessary to know:

1. Polymer density data as a function of temperature
2. Density data for the pure solvent as a function of temperature
3. Several values of the polymer-solvent system diffusivity for at least two temperatures
4. Sorption equilibrium data for the polymer solvent system
5. Rheological flow data for the polymer
6. Solvent viscosity data

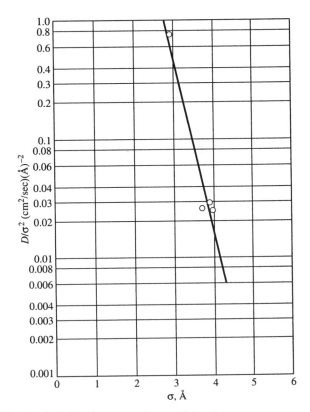

Fig. 5-9   Diffusivity divided by the square of gas collision diameter versus gas collision diameter for polystyrene (8, 9).

Items 1–6 are then used to calculate a number of quantities such as a solvent free-volume parameter, a polymer free-volume parameter, and molar volumes at 0 K for the polymer and solvent. These quantities are then used with the three equation sets to calculate diffusivity.

This method has been applied not to semicrystalline molten polymers (such as polyethylene) but rather to amorphous polymers (such as polystyrene and polymethyl acrylate) that thermally soften. It appears that the simpler generalized techniques described earlier in this chapter are adequate for the higher temperature ranges for both polystyrene and polyvinyl acetate (above 420 K for polystyrene and 358 K for polyvinyl acetate). These temperatures are, respectively, 1.1 and 1.2 times the polymers' glass temperature (in degrees Kelvin). It is, therefore, suggested that the free-volume model be used in the region from the glass temperature up to 1.1 to 1.2 times its value.

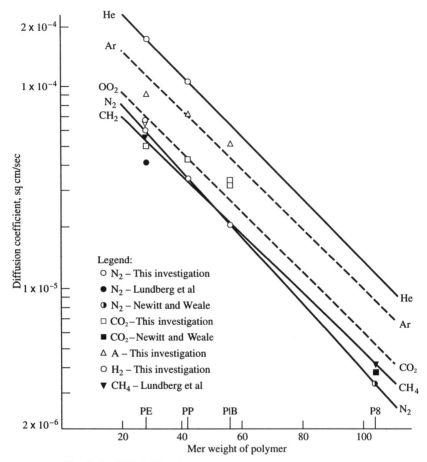

Fig. 5-10   Diffusivities of gases versus polymer mer weights (8, 9).

5.4   DIFFUSION EXAMPLES

Example 5-1

What are the diffusivities of the following systems at 188°C?

Gas	Polymer
Propylene	Polypropylene
Ethylene	Polyethylene
Carbon monoxide	Polyisobutylene
Styrene	Polystyrene

The diffusivity values can be obtained from Figs. 5-6–5-9. In order to determine these values, gas collision diameters (see Table 5-4) are needed. These are:

Gas	Collision diameter Å
Propylene	4.757[a]
Ethylene	4.232
Carbon monoxide	3.590
Argon	3.418

[a]Estimated values; see p. 22 of Ref. 19.

The reduced diffusivity values for each of the systems are 0.165 (Fig. 5-7), 0.340 (Fig. 5-6), 0.370 (Fig. 5-8), and 0.135 (Fig. 5-9). Calculated diffusivity values are:

System	Diffusivity, $cm^2/s \times 10^5$
Propylene–polypropylene	3.73
Ethylene–polyethylene	6.09
Carbon monoxide–polyisobutylene	4.77
Argon–polystyrene	1.58

Example 5-2

What are the diffusivities at 220°C for the following systems:

Methane–polyethylene
Nitrogen–polypropylene
Krypton–polyisobutylene

In order to calculate these values, we will need the E_D for each of the systems. Then, for 493.16 K (220°C), the diffusivity is

$$\frac{D_{493.16}}{D_{461.16}} = \frac{\exp\left(\frac{-E_d}{R(493.16 \text{ K})}\right)}{\exp\left(\frac{-E_d}{R(461.16 \text{ K})}\right)}$$

The E_d values from Fig. 5-5 are obtained by using the data for the systems (T_g and ϵ/K from Table 5-4):

Polymer	$T_g\ K$	Gas	$\epsilon/k\ K$	$E_d\ kcal/mole$
Polyethylene	195	Methane	137	3.000
Polypropylene	255	Nitrogen	91.5	2.787
Polyisobutylene	203	Krypton	225	5.400

The estimated diffusivities are

Polymer	Gas	Diffusivity $cm^2/sec \times 10^5$
Polyethylene	Methane	6.798
Polypropylene	Nitrogen	4.275
Polyisobutylene	Krypton	10.695

Example 5-3

Estimate a value of diffusivity for the system carbon dioxide-polyvinyl-chloride at 200°C.

This can be estimated by two different methods. The first is to get a value for the diffusivity at 188°C (using Figure 5-10). Then take the value to 200°C using the E_d from Figure 5-5.

The second is to obtain E_d from Figure 5-5 and a value of D_0 from Figure 5-4, which are then used to calculate the 200°C diffusivity.

Method one

The repeating mer for polyvinylchloride is CH_2CHCl which gives a mer weight of 62.46. Figure 5-10 for this mer weight indicates a value of 2.2×10^{-5} cm^2/sec for the 188°C diffusivity.

Next we calculate the value of kT_g/ϵ for Figure 5-5. Using the polyvinyl-chloride T_g of 455.16°K and the ϵ/K of 190°K from Table 5-4 yields a value of 2.396. Then, from Figure 5-5, the E_D value is 2.84 kcal/gm mole.

$$D_{473.16\ K} = D_{461.16\ K} \frac{\exp\left(\dfrac{E_D}{R(461.16°K)}\right)}{\exp\left(\dfrac{E_D}{R(473.16°K)}\right)}$$

$$D_{473.16 \text{ K}} = 2.2 \times 10^{-5} \text{ cm}^2/\text{s} \exp \frac{E_D}{R} \frac{(473.16 - 461.16) \text{ K}}{(473.16 \text{ K})(461.16 \text{ K})}$$

$$D_{473.16 \text{ K}} = 2.38 \times 10^{-5} \text{ cm}^2/\text{s}$$

Method 2:

Using Fig. 5-4 for the E_D of 2.84 kcal/g mole, we obtain a D_0 of 60×10^{-5} cm²/s.
Then, D is

$$D = (60 \times 10^{-5} \text{ cm}^2/\text{s}) \exp \frac{-2840 \text{ cal/g mole}}{R(473.16 \text{ K})}$$

$$D = 2.94 \times 10^{-5} \text{ cm}^2/\text{s}$$

The two estimated values are in reasonable agreement. Further, the reader should remember that we are obtaining a diffusivity for a system for which there are no data. As such, the results, at worst, give us at least a lower and an upper limiting value.

Actually, since the polyvinyl chloride point is past the end of the carbon dioxide data (polyisobutylene) in Fig. 5-10, more weight can probably be given to the 2.94 value. In any case, the above methods can be of considerable help to the polymer processing practitioner.

5.5 SOLUBILITIES OF GASES IN MOLTEN AND THERMALLY SOFTENED POLYMERS

As indicated earlier, the limiting effect of the diffusion of a gas into a molten or thermally softened polymer is its solution. Further, such behavior can be expressed in the form of Henry's law [Eq. (5-5)].

Experimentally determined values of H, the Henry's law constant are given in Table 5-5.

As with diffusion, solution is a function of a number of parameters, including temperature, pressure, and the nature of the gases and polymers involved.

The relation of the Henry's law constant to temperature is an exponential one, represented by Eq. (5-7):

$$H = H_0 \exp(-E_s/RT) \tag{5-7}$$

where E_s is the heat of solution in kcal/g mole. Experimental values of E_s are given in Table 5-6.

As evident from Table 5-6, the heat of solution changes with gas structure

TABLE 5-5   **Henry's Law Constants for Various Gas-Polymer Systems**
Henry's Law Constant, cm³ (STP)/g atm

Polymer	Nitrogen	Carbon Dioxide	Monochloro-difluoromethane	Argon	Helium
Polyethylene	0.111	0.275	0.435	0.133	0.038
Polypropylene	0.133	0.228	0.499	0.176	0.086
Polyisobutylene	0.057	0.210	—	0.102	0.043
Polystyrene	0.049	0.220	0.388	0.093	0.029
Polymethyl methacrylate	0.045	0.260	—	0.105	0.066

Additional values: krypton–polyisobutylene 0.114; krypton–polymethyl methacrylate 0.122; neon–polymethyl methacrylate 0.126.
Source: All data in Table 5-5 are from Refs. 7–9.

for a given polymer. This phenomenon was also observed in solid amorphous polyethylene [32] and natural rubber [33], where the heats of solution moved from endothermic to exothermic as the gas collision diameter increased. Figure 5-11 compares data for molten polyethylene to data for solid amorphous polyethylene. The dotted line of Fig. 5-11 can be used to estimate heats of solution for gases other than nitrogen or carbon dioxide in polyethylene.

References in the literature indicate that the Henry's law constant is independent at pressures up to at least 1.01×10^7 Pa [3-6] and possibly up to 3.03×10^7 Pa. The data in Table 5-5 can be used for the preceding pressure range.

The Henry's law constant for a given polymer can be correlated with the ϵ/k factor (see Table 5-4). Such correlations are shown in Figs. 5-12 and 5-13. A more generalized correlation is shown in Fig. 5-14, where the Henry's law constant is correlated with the gas critical temperature for thermally softened polymers (polyisobutylene, polystyrene, and polymethyl methacrylate). Figure 5-14 offers the possibility of estimating a Henry's law constant for a polymer for which there are no data.

TABLE 5-6   **Heat of Solution Values, kcal/g mole**

| Polymer | Gases | | | | Reference |
	H_2	CO_2	N_2	CH_4	
Polyethylene	—	−0.80	0.95	—	7–9, 3–6
Polypropylene	—	1.7	—	—	7–9
Polystyrene	1.9	—	1.7	1.05	2, 3–6

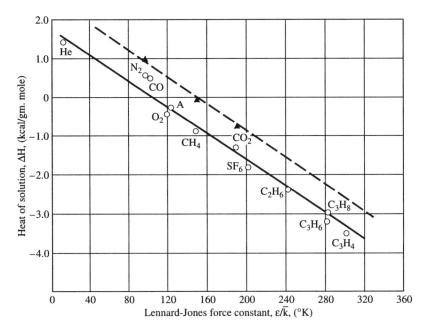

Fig. 5-11   Heats of solution for gases in solid amorphous (solid line, circles) polyethylene from reference (32) and for molten polyethylene (dotted line, triangles) from references (3–9).

5.6   SOLUTION EXAMPLES

Example 5-4

Michaels and Bixler [32] have studied the solution of gases in solid amorphous polyethylene at 298 K. It has been suggested that such data extrapolated to higher temperatures will represent the solution of gases in molten polyethylene. Explore this possibility by using the data of Table 5-5 and Fig. 5-12, along with the following Michaels and Bixler data:

Gas	H, cm^3 (STP)/g (atm)
He	0.0102
N_2	0.0351
A	0.0878
CO_2	0.384

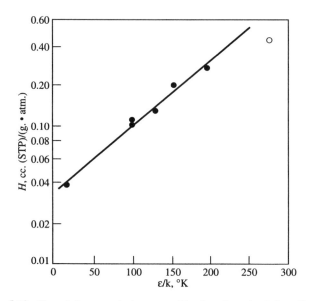

Fig. 5-12   Henry's Law constants versus ϵ/K values for polyethylene (8, 9).

First, find the heat of solution for the appropriate gas from the dotted line of Fig. 5-16. These values for helium, nitrogen, argon, and carbon dioxide, respectively, are 2300, 950, 500, and 800 kcal/g mole. The extrapolated amorphous values are calculated from the relation

$$H_{461.16 \text{ K}} = H_{298.16 \text{ K}} \frac{\exp \dfrac{E_s}{R(298.16)}}{\exp \dfrac{E_s}{R(461.16)}}$$

The extrapolated values are compared to the following experimental values:

Gas	H Experimental	H Extrapolated
He	0.038	0.043
N_2	0.111	0.0619
A	0.133	0.118
CO_2	0.275	0.417

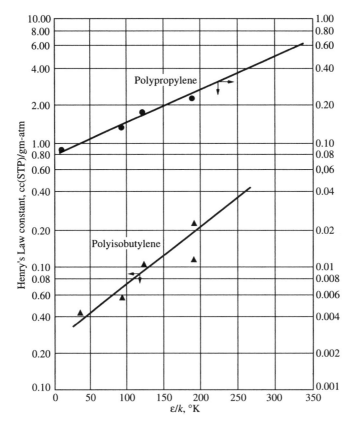

Fig. 5-13   Henry's Law constants versus ε/K for polypropylene and polyisobutylene (8, 9).

The table indicates that the extrapolated values are close for helium and argon and within the same order of magnitude for the other gases. Figure 5-16, which plots the values, shows that, the Henry's law constants, although related, they are not so on a one-to-one basis.

Example 5-5

Estimate the Henry's law constants for carbon monoxide, oxygen, and ethane at 188°C in polyethylene, polypropylene, polyisobutylene, polystyrene, and polymethyl methacrylate.

Use Figs. 5-12–5-14 for the estimation, together with the ϵ/k data of Table 5-4 (the value of ϵ/k for O_2 is 113 K). The estimated Henry's law constants

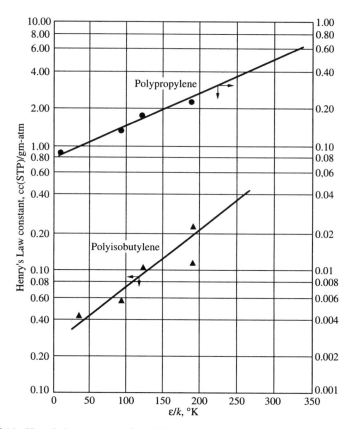

Fig. 5-14   Henry's Law constants for ϵ/K for polystyrene and polymethylmethacrylate (8, 9).

follow:

Polymer	H, cm³(STP)/g (atm)		
	CO	O₂	C₂H₆
Polyethylene	0.11	0.14	0.47
Polypropylene	0.15	0.17	0.37
Polyisobutylene	0.086	0.088	0.35
Polystyrene	0.080	0.082	0.37
Polymethyl methacrylate	0.11	0.13	0.24

Example 5-6

What are the Henry's law constants for nitrogen, krypton, carbon dioxide, and ethane in polyvinyl chloride at 188°C?

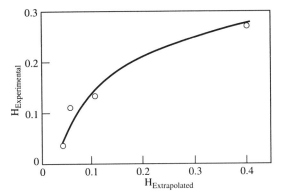

Fig. 5-16 Henry's law constants for molten polyethylene vs extrapolated values of Henry's law constants for amorphous solid polyethylene at 188°C.

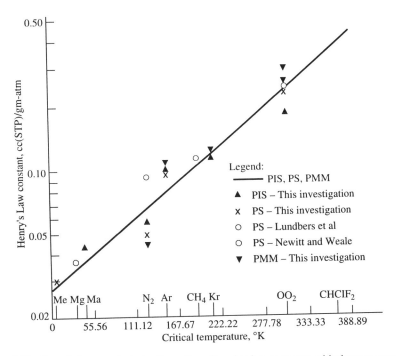

Fig. 5-15 Henry's law constants for thermally softened polymers vs gas critical temperatures [8, 9].

In this case, we estimate the values from Fig. 5-15, which follow:

Gas	H, $cm^3(STP)/g$ (atm)
N_2	0.067
Kr	0.140
CO_2	0.270
C_2H_6	0.275

REFERENCES

1. CRANK, J., and PARK, G.S. (eds.), *Diffusion in Polymers*, Academic Press, New York (1968).
2. NEWITT, D.M., and WEALE, K.E., *J. Chem. Soc. (London)*, 1541 (1948).
3. LUNDBERG, J.L., WILK, M.B., and HUYETT, M.J., *J. Appl. Phy.* **31**(6), 1131 (1960).
4. LUNDBERG, J.L., WILK, M.B., and HUYETT, M.J., *J. Polymer Sci.* **57**, 275 (1962).
5. LUNDBERG, J.L., WILK, M.B., and HUYETT, M.J., *Ind. Eng. Chem. Fundamentals* **2**, 37 (1963).
6. LUNDBERG, J.L., MOONEY, E.J., and RODGERS, C.E., *J. Polymer Sci.* A-2, **7**, 947 (1969).
7. GRISKEY, R.G., and DURILL, P.L., *AIChE J.* **12**, 1147 (1966).
8. Ibid., **15**, 106 (1967).
9. GRISKEY, R.G., *Modern Plastics* **54**(6), 158 (1977).
10. GIDDINGS, J.C., SEAGER, S.L., STUCKI, S.R., and STEWART, G.N., *Analyt. Chem.* **32**, 867 (1960).
11. PURNELL, J.H., and BOHEMAN, J., *J. Chem. Soc.* **360** (1961).
12. BONNER, D.C., and CHENG, Y.T., 67th Annual Meeting AIChE, Washington, DC (1974).
13. DUDA, J.L., KIMMERLEY, G.K., SIGELKO, W.L., and VRENTAS, J.S., *I & EC Fundamentals* **12**, 133 (1973).
14. MATSUKO, T., HOMMA, Y., and KARASAWA, M., *Sen-i Gakkaishi* **34**, T-137 (1978).
15. DUDA, J.L., and VRENTAS, J.S., *J. Polymer Sci.*, A-2 **6**, 675 (1968).
16. DUDA, J.L., NI, Y.C., and VRENTAS, J.S., *J. Appl. Polymer Sci.* **22**, 689 (1978).
17. DUDA, J.L., VRENTAS, J.S., JU, S.T., and LIU, H.T. *AIChE J.* **28**, 279 (1982).
18. DUDA, J.L., JU, S.T., and VRENTAS, J.S., *I & EC Prod. Res. Dev.* **20**, 330 (1981).
19. BIRD, R.B., STEWART, W.E., and LIGHTFOOT, E.N., *Transport Phenomena*, Wiley, New York (1960).
20. POWELL, R.E., ROSEVEARE, W.E., and EYRNG, H., *Ind. Eng. Chem.* **33**, 430 (1941).
21. OTHMER, D.F., and THANKAR, M.S., *Ind. Eng. Chem.* **45**, 589 (1953).
22. WILKE, C.R., and CHANG, P., *AIChE J.* **1**, 264 (1955).
23. SCHEIBEL, E.G., *Ind. Eng. Chem.* **46**, 2007 (1954).
24. VRENTAS, J.S., and DUDA, J.L., *J. Polymer Sci.* **15**, 403 (1977).
25. *Ibid.*, 417, 441.
26. VRENTAS, J.S., and DUDA, J.L., *J. Appl. Polymer Sci.* **21**, 1715 (1977).
27. Ibid., **22**, 2325 (1978).

28. VRENTAS, J.S., and DUDA, J.L., *AIChE J.,* **25**, 1 (1979).
29. VRENTAS, J.S., and DUDA, J.L., *J. Polymer Sci., Phys.* **17**, 1085 (1979).
30. VRENTAS, J.S. LIU, H.T., and DUDA, J.L., *J. Appl. Polymer Sci.* **25**, 1297 (1980).
31. VRENTAS, J.S., LIU, H.T., and DUDA, J.L., *J. Appl. Polymer Sci.* **25**, 1793 (1980).
32. MICHAELS, A.S., and BIXLER, H.J., *J. Polymer Sci.* **50**, 393 (1961).
33. VAN AMERONGEN, G.J., *J. Polymer Sci.* **5**, 307 (1950).

PROBLEMS

5-1 Compare values of the diffusivity of *n*-pentane in polystyrene at 188°C found from Figs. 5-3 and 5-9. Discuss the results.

5-2 The polystyrene polymerization process uses styrene as a monomer. As such, the diffusivity of styrene in polystyrene is an important property. Find a value of such a diffusivity at 200°C.

5-3 What is the diffusivity of nitrogen in polycarbonate at 195°C?

5-4 Which gas (2,2 dimethylbutane or 2,3 dimethylbutane) has the higher diffusivity in molten polypropylene (at 188°C)? Discuss the results.

5-5 Find the diffusivity of chlorine in polychlorotrifluoroethylene at 188°C.

5-6 A polymer has values of D_0 and E_D of 1.8 cm^2/s and 9400 cal/g mole, respectively, for nitrogen diffusion. What is the polymer?

5-7 Authors have suggested that families of polymers such as polyolefins could give a common line on plots of the type of Fig. 5-6. Test this hypothesis by replotting the data from Figs. 5-6–5-8 in a single graph. Could a reasonable correlation line be derived by ignoring data for a given gas? If so, explain why this can be done.

5-8 What would be the diffusivity of carbon monoxide in a copolymer of ethylene and propylene (5.0 methyl groups per 100 C atoms)? Justify all assumptions.

5-9 Compare the data of Figs. 5-12 and 5-13. Is any generalized correlation for the polyolefins possible? Are any correlations of any type possible for any two of the polymers? In each case, discuss positive or negative results.

5-10 Estimate heat of solution values for ethane in molten polypropylene and oxygen in thermally softened polystyrene. Justify the methods used.

5-11 What is the Henry's law constant of nitrogen in polyvinyl chloride at 210°C?

5-12 Estimate a Henry's law constant for nitrogen in an ethylene–propylene copolymer (4.1 methyl groups per 100 C atm). Justify all assumptions.

5-13 What will the concentration of argon be in molten polyethylene at a pressure twice atmospheric and a 205°C temperature?

5-14 A molten polypropylene system (saturated with nitrogen at 220°C and 1 atm pressure) is reduced to 0.3 atm. What will the nitrogen concentration be after the pressure reduction?

5-15   Two samples of the same molten polyethylene (at 200°C) are placed in 0.03-m-diam cylindrical vessels. In one case, an oxygen gas is placed above a melt 0.3 m high. In the second case, ethane gas contacts a melt 0.52 m high. In which system would the first gas molecule transverse the melt's axial length (i.e., from top to bottom)? Which sample would have the highest gas concentration?

MINI PROJECT A

A 50–50 molten blend of polyethlene and polypropylene is to be processed in a system requiring contact with various gases. Develop a method for determining diffusivities and solubilities in molten or thermally softened gases. Use it to compute the diffusivity and Henry's law constant for carbon monoxide at 220°C and atmospheric pressure.

MINI PROJECT B

Chapter 5 has presented data and correlations for various situations involving take-up of a gas in a molten or a thermally softened polymer. Using approaches developed elsewhere, such as thermodynamics and transport processes, modify one of the correlations for the case in which more than one gas is present. Justify any and all assumptions.

MINI PROJECT C

Water is commonly found in many processing operations. Provide a detailed and clear explanation of how you would handle a steam-molten polymer system in terms of diffusion and solution. Explain the differences, if any, in the behavior of steam-molten 6-10 nylon. Estimate diffusivities and the Henry's law constant for systems.

CHAPTER 6

CHEMICAL REACTION KINETICS IN POLYMER SYSTEMS

6.1 INTRODUCTION

In the earlier chapter on polymer fundamentals, the concepts of addition and condensation polymerizations were introduced. The recognition that differences in polymerizations could be categorized was due initially to Wallace Carothers. His perception related to the formation of a small by-product molecule in the case of condensation polymerization and the lack of such a molecule in addition polymerization.

Later, Flory focused on the actual mechanisms occurring in addition and condensation polymerizations. In the condensation case, the mechanism was a stepwise reaction whereas, in the addition case, a chain reaction occurred.

The other distinctions between addition and condensation are given in Table 1-2.

In this chapter, attention will be directed to both types of polymerization, as well as polymerization processes and reactors.

6.2 CONDENSATION POLYMERIZATION

One of Carothers' concepts, that of *functionality*, relates directly to condensation polymerizations. This concept dealt with the average number of reactive groups that occurred in each monomer molecule. The effect of functionality is shown in Table 6-1.

TABLE 6-1   Functionality and Products

Functionality	Product
Mono- (one group)	Low molecular weight material
Bi- (two groups)	Linear polymer
Poly- (more than two groups)	Branched or cross-linked polymers

There are a number of ways that a stepwise reaction can take place. These include direct reaction, interchange, and acid chloride/anhydride. Direct reactions include formation of polyesters and polyamides. Typical interchange reactions involve acetol–alcohol, amine–amide, and amine–ester. The last-named are cases in which an anhydride or acid chloride are reacted with a glycol or amine.

Table 6-2 lists some typical polycondensation reactions.

TABLE 6-2   Polycondensation Reactions

Reactants	Polymer Product	Chain Linkage	By-Product
Dibasic acid–dialcohol	Polyester	O $\parallel$ $-C-O-$	H_2O
Dihydroxy acids	Polyester	O $\parallel$ $-C-O-$	H_2O
Dibasic acid–diamine	Polyamide	O $\parallel$ $-C-NH-$	H_2O
Alpha amino acids	Polyamide	O $\parallel$ $-C-NH-$	H_2O
Phosgene–diols	Polycarbonate	O $\parallel$ $-O-C-NH-$	HCl
Diisocyanate–diol	Polyurethane	O $\parallel$ $-N-C-O-$	—
Dichloride–sulfide (Na_2S)	Polysulfide	$-S-$	NaCl

6.3 CONDENSATION POLYMERIZATION STATISTICS AND KINETICS

Consider a condensation reaction that forms a linear polymer. For this system, we define the following:

p = probability that a unit has reacted or extent of reaction at some time

$(1 - p)$ = probability of finding an unreacted unit at some time

N_0 = original number of units

N = number of unreacted units at any time

The number-averge degree of polymerization or the average number of monomer units per chain, X_n, is then

$$X_n = \frac{N_0}{N} = \frac{1}{(1 - p)} \qquad (6\text{-}1)$$

Likewise, the weight-average degree of polymerization is

$$X_w = \frac{(1 + p)}{(1 - p)} \qquad (6\text{-}2)$$

The ratio of the degrees of polymerization (the breadth of the molecular weight) distribution curve is

$$\frac{X_w}{X_n} = 1 + p \qquad (6\text{-}3)$$

For a large extent of reaction ($p \to 1$),

$$\frac{X_w}{X_n} \to 2.0 \qquad (6\text{-}4)$$

Polycondensations usually involve both forward and reverse reactions. At equilibrium, these forward and reverse rates of reaction are equal. Such a condition results after a long time has elapsed. In the early stages of a polymerization, the reverse rate of reaction is slow. As a result, only the forward reaction need be considered for reasonable times.

Consider a second-order polycondensation reaction with constant volume and initial reactant group concentration being equal. Then,

$$\frac{-d(E)}{dt} = k(E)(F) \tag{6-5}$$

where k is specific reaction rate and E and F are concentrations. Since $E_0 = F_0$,

$$\frac{-d(E)}{dt} = k(E)^2 \tag{6-6}$$

Then, since at $t = 0$, $E = E_0$ and at $t = t$, $E = E$,

$$\frac{1}{(E)} - \frac{1}{(E_0)} = kt \tag{6-7}$$

$$p = \frac{(E_0) - (E)}{(E_0)} = \frac{(E_0)\, kt}{1 + (E_0)} \tag{6-8}$$

and

$$X_n = 1 + (E_0)\, kt \tag{6-9}$$

Hence, X_n is linear with time as Fig. 6-1 shows.

For a polycondensation that forms a polyester from a glycol and a dibasic acid, the kinetic result, if a strong acid is added, is

$$\frac{-d\,(COOH)}{dt} = k\,(COOH)^2\,(OH) \tag{6-10}$$

where the terms in parentheses are concentrations and k is the specific reaction rate.

If $(COOH)$ and (OH) are equal (as c) and if p, the extent of reaction, is used, then

$$2\,C_0^2\, kt = \frac{1}{(1 - p)} + const \tag{6-11}$$

which means that a plot of $1/(1 - p)^2$ should be linear with time. This is shown in Fig. 6.2.

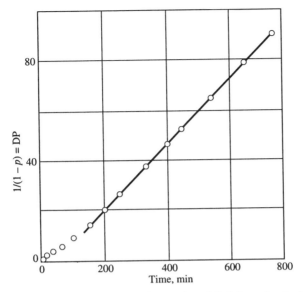

Fig. 6-1   Degree of polymerization vs time for reaction of diethylene glycol with adipic acid (catalyst of 0.4 mole per unit *p*-toluene sulfonic acid) [1].

6.4   POLYFUNCTIONAL CONDENSATION POLYMERIZATIONS—THE GEL POINT

As noted earlier, functionalities higher than 2 introduce the possibility of branching or cross-linking. In the latter case, it is ultimately possible to attain molecular weights that approach infinity. Such materials are termed *gels*, with the remainder of the reaction mixes being *sols*. The sol, which is soluble, can be separated from the insoluble gel. If, however, the amount of gel increases past a given point, the mix becomes unprocessible. This occurs because the viscous thermoplastic mass is transformed into a thermoset (an elastic solid with a very high viscosity).

The critical point at which the amount of gel begins to increase rapidly is called the gel point.

In order to determine this point, we begin by specifying α, a branching coefficient, which is the probability that a functional group is connected to another branch unit. For a reaction system in which the bifunctional groups A–A and B–B are present along with polyfunctional units A_f (of functionality f), the polymer will be

$$A_{f-1} - A[B - BA - A]_i - BA - A_{f-1} \qquad (6\text{-}12)$$

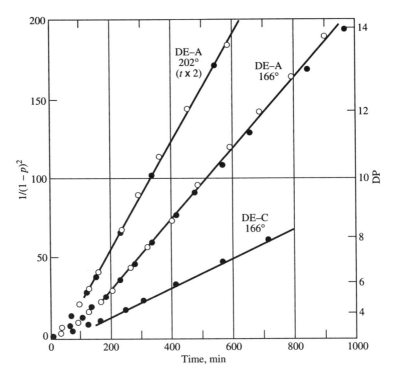

Fig. 6-2   Reaction of diethylene glycol with adipic acid (two top lines) and caproic acid (bottom line); times at 202°C are multiplied by 2 [1].

For Eq. (6-12), the gel formation criterion is that at least one of the $f - 1$ units is connected to another branch unit. Such a case has a probability of $1/(f - 1)$. Hence, the critical α value is

$$\alpha_c = \frac{1}{(f - 1)} \tag{6-13}$$

If we relate α to the extent of reaction, we find that

$$\alpha = \frac{r p_A^2 \, \rho}{1 - r p_A^2 \, (1 - \rho)} = \frac{p_B^2 \, \rho}{r - p_B^2 \, (1 - \rho)} \tag{6-14}$$

where $r = N_A/N_B$ and ρ is the ratio of A units on branches compared to all A units.

Some special cases are given in Table 6-3.

In experiments with glycerol and dibasic acids, gelation occurred at $p = 0.765$. If $\alpha_c = p^2$ from Table 6-3, then $p = 0.58$. This compares to a theoretical value of 0.5 from Eq. (6-13).

Such behavior, namely, α_c values higher than theoretical, was found in other cases as well. This is accounted for by the possibility that some functional groups form only intramolecular links.

6.5 FREE-RADICAL ADDITION POLYMERIZATION

Addition polymerization can take place by a number of routes. A particular type that holds for many cases is *free-radical polymerization*. In this case, the reaction is initiated by a free radical, which then continues as a chain reaction composed of initiation, propagation, and termination steps.

The initiation step involves the creation of free radicals, which can then add monomer in the propagation step. Production of initiator can be done through physical energy (thermal, photochemical means) or chemical energy (the initiator reacts with the monomer).

Some typical cases of initiation are:

Photochemical:

$$M + h\nu \rightarrow M\cdot \tag{6-15}$$

TABLE 6-3 Special Gelation Cases

Condition	α
$r = 1$ ($p_A = p_B$)	$\dfrac{p^2 \rho}{1 - p^2 (1 - \rho)}$
$\rho = 1$	$r\, p_A^2$ or $\dfrac{p_B^2}{r}$
$r = \rho = 1$ only branch units	$\dfrac{p^2}{p}$

Thermally induced:

$$2M \rightarrow M - M\cdot \tag{6-16}$$

Initiator-monomer reaction:

$$I + M \rightarrow RM + \text{remainder} \tag{6-17}$$

Initiator decompositions:

$$I \rightarrow 2R\cdot \tag{6-18}$$

$$I + h\nu \rightarrow 2R\cdot \tag{6-19}$$

Where I is the initiator, M is the monomer, R a free radical, h Planck's constant, and ν frequency.

Terminations can occur by a number of ways. These include: monomer addition, mutual termination, spontaneous reaction, and disproportionations (two inactive polymers are produced).

A typical free-radical chain polymerization is depicted below:

Initiation:

$$I \xrightarrow{k_1} 2R\cdot \tag{6-20}$$

$$R + M \xrightarrow{k_2} M_1\cdot \tag{6-21}$$

Propagation:

$$M_1\cdot + M \xrightarrow{k_p} M_2\cdot \tag{6-22}$$

and

$$M_2\cdot \ll M \xrightarrow{k_p} M_3\cdot \tag{6-23}$$

or, in general,

$$M_x\cdot + M \xrightarrow{k_p} M_{x+1}\cdot \tag{6-24}$$

The k_p (reaction rate constants for propagation) are all the same.

Termination:

By combination,

$$M_m\cdot + M_n\cdot \xrightarrow{k_t} M_{m+n} \tag{6-25}$$

or disproportionation,

$$M_m\cdot + M_n\cdot \xrightarrow{k_t^1} M_m + M_n \tag{6-26}$$

Then, the rate of initiation, r_i, is given by

$$r_i = \frac{d(M\cdot)}{dt}\bigg|_i = 2\alpha k_1(I) \tag{6-27}$$

where α is the fraction of radicals formed that initiates the chain reaction of Eq. (6-21).

Likewise, the rate of termination, r_t, is given by

$$r_t = -\frac{d(M\cdot)}{dt}\bigg|_t = 2\,k_t(M\cdot)^2 \tag{6-28}$$

Frequently, free-radical concentration $(M\cdot)$ becomes a constant early in the reaction since radicals are created and removed at the same rates.

Hence,

$$r_i = r_t \tag{6-29}$$

and

$$(M\cdot) = \left[\frac{\alpha k_1(I)}{k_t}\right]^{1/2} \tag{6-30}$$

and the rate of propagation, r_p, is the same as the overall rate of monomer disappearance. This occurs since the monomer of Eq. (6-21) is small relative to the monomer involved in Eqs. (6-22–6-24). Hence,

$$r_p = -\frac{d(M)}{d_t} = k_p(M)(M\cdot) \tag{6-31}$$

or

$$r_p = k_p(M) \left(\frac{\alpha k_1(I)}{k_t} \right)^{1/2}$$

(6-32)

Equations 6-31 and 6-32 show that the overall polymerization rate should be proportional to the initiator concentration raised to the 1/2 power. Figure 6-3 confirms this since the logarithm of r_p plotted against the logarithm of initiator concentration yields straight lines with a slope of 1/2.

If we divide the equation for the rate of initiation (6-27) by the rate of propagation (6-32), we obtain

$$\frac{k_p^2}{k_t} = \frac{2r_p^2}{r_i(M)^2}$$

(6-33)

Also, the kinetic chain length v^1 can be related to the rate of propagation:

$$v^1 = \frac{k_p^2(M)^2}{2 \, k_t r_p}$$

(6-34)

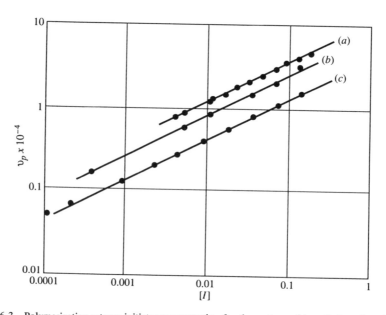

Fig. 6-3   Polymerization rate vs initiator concentration for the systems: (a) methyl methacrylate with azobisisobutylnitrile; (b) methyl methacrylate with benzoyl peroxide; (c) styrene with benzoyl peroxide [5–7].

The degree of polymerization x_n is equal to twice v^1 for combination termination and equal to v^1 for disproportionation terminator. Deviations that occur are due to chain transfer reactions in which the following occurs:

$$M_x + XP \rightarrow M_xX + P \qquad (6\text{-}35)$$

$$P + M \rightarrow PM \qquad (6\text{-}36)$$

Chain transfer agents can be used to control degree of polymerization (see Fig. 6-4).

In a heterogeneous reaction system (as in emulsion polymerization), the kinetics change considerably. Basically, such a system uses a soap (i.e., sodium or potassium salts of sulfates or organic acids) that allows the creation of micelles (agglomerations of anions). These micelles serve as accumulations of monomer and ultimately as polymerization sites. It can be shown that

$$r_p = k_p \frac{N}{2A} (M) \qquad (6\text{-}37)$$

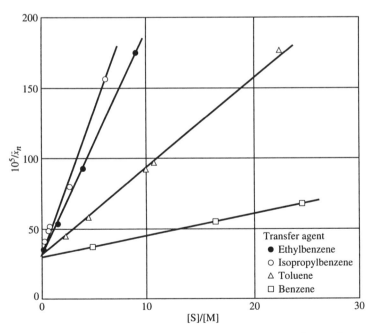

Fig. 6-4 Effect of chain transfer agent concentration (*S*) on polystyrene degree of polymerization [8].

where N is the number of particles per volume of reaction mass and A is Avogadro's number (or radicals per mole radical). In Eq. (6-37), note that increasing soap concentration will increase N and, hence, increase the rate of reaction.

6.6 NONRADICAL ADDITION POLYMERIZATION

One form of nonradical addition polymerization is *cationic polymerization.* In this case, strong electron acceptors in the form of strong Lewis acids initiate the chain reaction. These catalysts include $SnCl_4$, H_2SO_4, $AlCl_3$, $AlBr_3$, and BF_3. A characteristic of this form of polymerization is a high rate at low temperatures. As an example, consider the polymerization of isobutylene industrially at a temperature of $-100°C$ (using BF_3 or $AlCl_3$). A polymer of molecular weight up to several million is obtained. In contrast, both rate and molecular weight are much lower at higher temperatures.

A mechanism for cationic polymerization A is the catalyst RH; a cocatalyst, usually water, is shown below:

$$A \rightarrow A - RH \tag{6-38}$$

$$A - RH + M \xrightarrow{k_i} HM^+ {}^-R \tag{6-39}$$

$$HM_x^+ {}^-AR + M \xrightarrow{k_p} HM_{x+1} {}^-AR \tag{6-40}$$

$$HM_x^+ {}^-AR \xrightarrow{k_t} M_x + A - RH \tag{6-41}$$

For the cationic case above, the rate of polymerization, r_p, is

$$r_p = k_p(M^+)(M) = \frac{k_p k_i}{k_t}(C)(M)^2 \tag{6-42}$$

where (C) is the catalyst concentration. Also,

$$x_n = \frac{k_p}{k_t}(M) \tag{6-43}$$

If the addition polymerization is initiated by anions, it is categorized as *anionic polymerization.* The catalysts used for this type of polymerization include the alkali metals and organic alkali metal salts.

An example of anionic polymerization kinetics [9] for the synthesis of polystyrene (using styrene, potassium amide, and liquid ammonia) is shown below:

$$KNH_2 \rightarrow K^+ + N_2^-$$ (6-44)

$$NH_2^- + M \xrightarrow{k_i} NH_2M^-$$ (6-45)

$$NH_2M_x^- + M \xrightarrow{k_p} NH_2M_{x+1}^-$$ (6-46)

$$NH_2M_x^- + NH_3 \xrightarrow{k_{tr}} NH_2M_xH + NH_2^-$$ (6-47)

where k_{tr} is the reaction rate constant for transfer.
The equations for the rate of propagation and α_n are

$$r_p = \frac{k_p k_i}{k_{tr}} \frac{(NH_2^-)}{(NH_3)} (M)^2$$ (6-48)

$$\alpha_n = \frac{k_p}{k_{tr}} \frac{(M)}{(NH_3)}$$ (6-49)

In anionic polymerization, $r_i \cong r_p$, unlike free-radical polymerization, where $r_i \ll r_p$. Hence, chains start growing rapidly. Also, for very pure systems, termination does not occur. As a result, the chains can grow until the monomer is totally used. Adding more monomer causes the reaction to continue.

These systems are known as "living" polymers [10] and yield essentially monodisperse products. Adding a proton-donating compound (water) will "kill" the polymer by terminating the reaction.

Another type of nonradical addition polymerization is *coordinate* or *complex catalyst* polymerization. This approach stems from the pioneering work of Ziegler and co-workers [11] and Natta [12]. The polymerization is carried out with a suspension of a reaction between a metal alkyl (usually Al, Zn, and Mg) and a transition metal halide (V, Zr, and Ti).

Although a number of kinetic schemes have been proposed for coordination polymerization, none have really been substantiated. Even though this is the case, the process itself has attained considerable industrial usage because of its ability to produce isotactic and syndiotactic polymers, as well as those of very high molecular weights.

As might be suspected, polymers can be synthesized by using different polymerizations. Table 6-4 illustrates this for various monomers.

TABLE 6-4   Alternative Polymerization Methods

Monomer	Radical	Cationic	Anionic	Coordination
Acrylic esters	Yes	No	Yes	No
Vinylidene esters	Yes	No	Yes	No
Acrylonitrile derivatives	Yes	No	Yes	No
Ethylene	Yes	Yes	Yes	Yes
Butadiene	Yes	Yes	Yes	No
Styrene	Yes	Yes	Yes	No
α-Methylstyrene	Yes	Yes	Yes	No
Methyl vinyl ketone	Yes	Yes	Yes	No
N-Vinylcarbozole	Yes	Yes	No	No
N-Vinylpyrrolidone	Yes	Yes	No	No
Methacrylic esters	Yes	No	Yes	No

6.7   COPOLYMERIZATIONS

As mentioned earlier, copolymerizations involve two or more monomers in the synthesis process (remember that polymers that have regular repeating units involving two monomers are considered homopolymers).

For a system involving two monomers M_1 and M_2, we can write four equations describing the potential reactions:

$$P_1 + M_1 \xrightarrow{k_{11}} P_1 \tag{6-50}$$

$$P_1 + M_2 \xrightarrow{k_{12}} P_2 \tag{6-51}$$

$$P_2 + M_2 \xrightarrow{k_{22}} P_2 \tag{6-52}$$

$$P_2 + M_1 \xrightarrow{k_{21}} P_1 \tag{6-53}$$

In Eqs. (50-53), P_1 is a growing chain radical with an end unit of M_1, and P_2 is a similar material with an end unit of M_2. Note that P_1 is created in Eq. (6-53) and eliminated in Eq. (6-51). This means

$$k_{12}(P_1)(M_2) = k_{21}(P_2)(M_1) \tag{6-54}$$

with the monomer rates of consumption given by

$$-\frac{d(M_1)}{dt} = k_{11}(P_1)(M_1) + k_{21}(P_2)(M_1) \tag{6-55}$$

$$-\frac{d(M_2)}{dt} = k_{12}(P_1)(M_2) + k_{22}(P_2)(M_2) \tag{6-56}$$

Also, the copolymer's composition is given by

$$\frac{d(M_1)}{d(M_2)} = \frac{(M_1)}{(M_2)}\left[\frac{r_1(M_1) + (M_2)}{(M_1) + r_2(M_2)}\right] \tag{6-57}$$

where $r_1 = k_{11}/k_{12}$ and $r_2 = k_{22}/k_{21}$.

If F_2 and F_1 are the mole fraction of monomers M_1 and M_2 in the copolymer and f_1, f_2 the mole fractions in the feed, then

$$F_1 = \frac{r_1 f_1^2 + f_1 f_2}{r_1 f_1^2 + 2f_1 f_2 + r_2 f_2^2}$$

$$= \frac{(r_1 - 1)f_1^2 + f_1}{(r_1 + r_2)f_1^2 + 2(1 - r_2)f_1 + r_2} \tag{6-58}$$

Figures 6-5 and 6-6 show the ideal copolymerization case ($r_1 r_2 = 1.0$) and various r_1/r_2 ratios.

TABLE 6-5   Effect of Reactivity Ratios and Copolymerizations

r_1, r_2 Values	Result
$r_1 = r_2 = 0$	$F_1 = 0.5$ (perfectly alternating)
$r_1 = r_2 = \infty$	Mixture of homopolymer 1 and homopolymer 2
$r_2 r_1 = 1.0$	Ideal copolymerizations, $F_1 = \dfrac{r_1 f_1}{f_1 (r - 1) + 1}$ (see Fig. 6-5)
$r_1 = r_2 = 1.0$	$F_1 = f_1$
$r_1 > 1.0, r_2 > 1.0$	$F_1 = f_1 = \dfrac{(1 - r_2)}{(2 - r_1 - r_2)}$

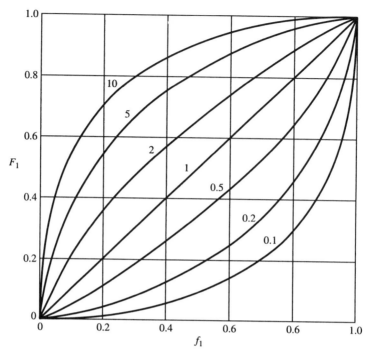

Fig. 6-5 F_1 as a function of f_1 when $r_1 r_2 = 1.0$ [14].

6.8   POLYMERIZATION PROCESSES

Actual polymerizations are carried out in a variety of ways. These can be categorized as homogeneous and heterogeneous systems. Homogeneous systems include bulk and solution polymerization, whereas suspension polymerization, emulsion polymerization, bulk polymerization with precipitate, interfacial polycondensation, and solid-state polycondensation constitute the heterogeneous systems.

Bulk polymerization is perhaps the most straightforward polymerization technique. Essentially, a mass of monomer is reacted to form the polymer. These polymerizations are used most often for condensation reactions since such systems have low exotherms. Free-radical polymerizations are particularly difficult to handle because such reactions are highly exothermic.

Another aspect of the problem with free-radical cases is the phenomenon known as autoacceleration, or the *Tromsdorff effect* (see Fig. 6-7). This occurs because the reaction is controlled by the rate of diffusion of the reactants, which lowers the termination rate.

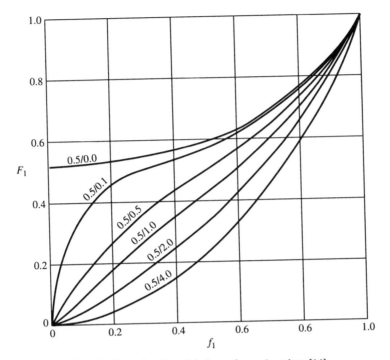

Fig. 6-6 F_1, as function of f_1, for various r_1/r_2 values [14].

A modification of bulk polymerization for free-radical systems that over-comes these problems is the use of a continuous rather than a batch reactor. Such a reactor gives much better heat-transfer control and narrower molecular weight distributions.

Solution polymerization is a homogeneous process that overcomes the prob-lems of bulk polymerization by the use of an inert solvent. Essentially, the solvent acts as a heat sink and also reduces the viscosity of the reaction mass.

Ionic reactions are usually carried out in solutions. The same is true for most coordination polymerizations.

Suspension polymerization, a heterogeneous process, uses the monomer as a dispersed phase with a continuous phase that is usually water. This dispersion in the form of droplets (0.01–5 cm in diameter) is brought about by a combi-nation of a suspending agent and agitation. Typically, the agent is a water-soluble polymer. Sometimes, finely divided insoluble solids are also used.

The initiator for suspension polymerization is in the monomer phase. Kinetics are those of bulk polymerization.

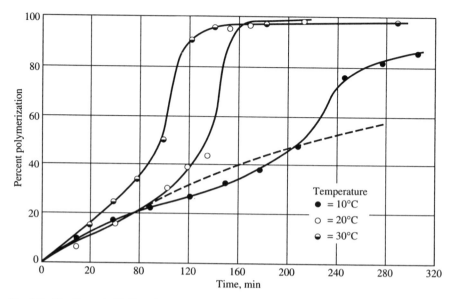

Fig. 6-7 The Tromsdorff effect (autoacceleration) for polymethyl methacrylate polymerization [15].

The suspension polymerization process is particularly dependent on agitation between conversions of 20%–70%. Product polymer is in the form of beads that can be molded directly. In addition the beads can be directly used for ion-exchange resins.

Emulsion polymerization, mentioned earlier during the discussion of free-radical chain reactions, is a process that uses a soap to generate an emulsion. The resultant micelles in the emulsion serve as centers for reaction.

The initiator is in the aqueous phase. Particles are much smaller (about 10 times) than those in suspension polymerization (where the catalyst is monomer-soluble and the kinetics are similar to bulk-type polymerizations). The emulsion polymerization kinetics are governed by Eq. (6-37), which differs considerably from bulk polymerization kinetics. Figure 6-8 shows the relation of polymerization to soap concentration.

A modification of bulk polymerization that is a heterogeneous system is one in which the polymer product precipitates. Typical polymers that can be produced by this technique are polyacrylonitrile and polyvinyl chloride. The polymerization rate for polyvinyl chloride is given by

$$r_p = k_p \left(\frac{\alpha k_i(I)}{k_t} \right)^{1/2} [(M) + f(P)] \qquad (6\text{-}59)$$

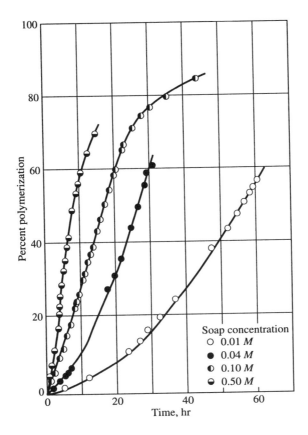

Fig. 6-8   Effect of soap concentration on emulsion polymerization of isoprene [16].

In Eq. (6-59), $f(P)$ is a term related to the polymer (precipitate) concentration that varies from polymer concentration (P) at short times to $(P)^{2/3}$ at higher conversions.

Interfacial polycondensation is a low-temperature, low-pressure process involving two liquid phases. One of the monomers involved in the reaction is dissolved in water, whereas the other is usually in some organic chloride. Polymer forms at the interface of the two phases.

This form of polymerization is highly dependent on mass transfer to the extent that this rate takes precedence over the chemical reaction rate. Figure 6-9 illustrates the importance of mass transfer on interfacial polycondensation.

Solid-state polycondensation is a process that uses an inert gas sweep through a bed of solid particles. Essentially, the process removes the small by-product

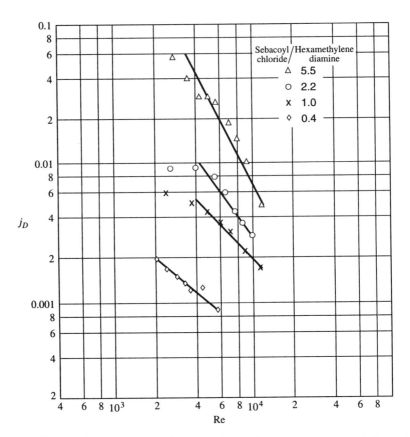

Fig. 6-9   Relation of j_D (mass-transfer function) vs Reynolds number. Trends are similar to those for flow past cylinders in packed beds and other systems [17].

molecules generated in the polycondensation reaction. The result is that the polycondensation equilibrium

$$P_1 + P_2 \rightleftarrows P_3 + X \qquad (6\text{-}60)$$

(where P_1, P_2 are shorter chains, P_3 is a longer chain, and X is the by-product molecule) is driven to the right (i.e., X is removed). This causes the polymer's degree of polymerication to increase.

Figure 6-10 shows some typical curves for 66 nylon (polyhexamethylene adipamide) for temperatures well below the polymer's melting point.

Table 6-6 compares the advantages and disadvantages of the homogeneous

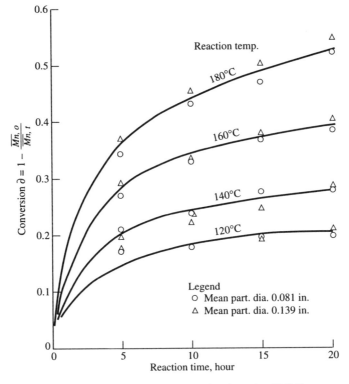

Fig. 6-10   Conversion vs reaction time for nylon 66 [18].

TABLE 6-6   Comparison Homogeneous Polymerization Processes

Process	Advantages	Disadvantages
Bulk (batch reactor)	Simple; minimum contamination; greatest yield per unit volume	Highly exothermic; broad molecular weight distribution; difficult to control; hard to remove last traces of monomer
Bulk (continuous reactor)	Better heat transfer; gives narrower molecular weight distribution	Needs separation of product
Solution	Exotherms better controlled; can use solution directly; theoretical kinetics apply, which simplifies design of reaction	Lower rate and average chain length; solvent handling and removal; lower yield per unit volume

TABLE 6-7   Comparison Heterogeneous Polymerization Processes

Process	Advantages	Disadvantages
Suspension	Excellent heat transfer and control; product can be used immediately	Requires continuous agitation; low yield per unit volume; possible stabilizer contamination; washing, drying may be needed
Emulsion	Excellent control of both viscosity and heat; product can be directly usable; residue monomer low; attains high molecular weight and narrow distribution	Emulsifier contamination; washing and drying may be needed
Interfacial polycondensation	Low temperature and pressure; can produce polymers not available by other means	Product separation problems; cost of monomers
Solid-state polycondensation	Can attain high molecular weights; can be used for formed objects	Requires inert gas sweeps

polymerization processes, and Table 6-7 compares the various heterogeneous systems.

REFERENCES

1. FLORY, P.J., in *High Molecular Weight Organic Compounds*, Frontiers in Chemistry VI, R.E. BURK and O. GRUMMETT, eds., Interscience, New York (1949), pp. 221–283.
2. KIENLE, R.H., VAN DER MEULEN, P.A., and PETKE, F.E., *J. Am. Chem. Soc.* **61**, 2258, 2268 (1939).
3. KIENLE, R.H., and PETKE, F.E., *J. Am. Chem. Soc.* **62**, 1053 (1940).
4. Ibid **63**, 481 (1941).
5. ARNETT, *J. Am. Chem. Soc.* **74**, 2027 (1952).
6. MAYO, F.R., GREGG, R.A., and MATHESON, M.S., *J. Am. Chem. Soc.* **73**, 1691 (1951).

7. SCHULZ, G.V., and BLASCHKE, F.Z., *Physik. Chem.* **B51**, 75 (1942).
8. GREGG, R.A., and MAYO, F.R., *Disc. Faraday Soc.* **2**, 328 (1947).
9. HIGGINSON, W.C.E., and WOODING, N.S., *J. Chem. Soc.* 760 (1952).
10. SZWARC, M., *Makromal. Chem.* **35**, 132 (1960).
11. ZIEGLER, K., HOLZKAMP, E., BREIL, H., and MARTIN, H., *Agnew Chem.* **67**, 426, 541 (1955).
12. NATTA, G., *J. Am. Chem. Soc.* **77**, 1708 (1955).
13. NATTA, G., *J. Polymer Sci.* **16**, 143 (1955).
14. BILLMEYER, F.W. JR., *Textbook of Polymer Science*, Wiley, New York (1966), p. 314.
15. NAYLOR, M.A., and BILLMEYER, F.W. JR., *J. Am. Chem. Soc.* **75**, 2181 (1953).
16. HARKINS, W.D., *J. Am. Chem. Soc.* **69**, 1428 (1947).
17. GRISKEY, R.G., HUNDIA, S.R., and SISKOVIC, N., *J. Appl. Polymer Sci.* **18**, 30 (1974).
18. GRISKEY, R.G., CHEN, F.C., and BEYER, G.H., *AIChE J.* **15**, 680 (1969).

PROBLEMS

6-1 The reaction rate constant (at 160.5°C) for the polycondensation reaction of 12-hydroxy stearic acid

$$HO{\rm-}CH{\rm-}(CH_2)_{10}COOH$$
$$|$$
$$C_6H_{13}$$

is

$$k = 2.47 \times 10^{-4} \text{ L/g mole s}$$

(with a catalyst level of 3.31×10^{-2} g mole/L of *p*-toluene sulphonic acid).

From this information develop a plot of the extent of reaction against time for various initial concentrations of the stearic acid.

6-2 Two monomers, *E* and *F*, participate in a polycondensation reaction. Find the relations that tie together conversion and number-average chain length, assuming equal moles of the monomers.

6-3 We want to reach a degree of polymerization of 96 for a polycondensation reaction where equal moles of the monomers are used. What conversion would be needed? If only half that conversion is attained what is the corresponding degree of polymerization? Comment on the product in this latter case.

6-4 Two moles of ethylene glycol (A) are reacted with one mole of phthalic anhydride (B) and one mole of BTDA (C).

$$HO—CH_2—CH_2—OH \qquad \text{(A)}$$

(B)

(C)

Will a gel form?

6-5 In polycondensation systems at higher temperatures in the melt, interchange reactions can take place. These reactions can alter the molecular weights of the polymer system since two molecules having the same degree of polymerization (D.P.) can react to yield one of greater D.P. and one of lesser D.P. Free interchange of this type will yield the most probable distribution.

Consider an experiment that mixes 100 g of one polymer ($Mn = 10{,}000$; $Mw = 20{,}000$) with 100 g of another ($Mn = 30{,}000$; $Mw = 60{,}000$).

What conclusions can you reach for the following three results? Prove your answers.

Result 1: Products Mn and Mw are, respectively, 15,000 and 40,000.

Result 2: Products Mn and Mw are, respectively, 15,000 and 30,000.

Result 3: Products Mn and Mw are, respectively, 15,000 and 34,000.

6-6 For the system pentaerythritol (P), phthalic anhydride (PA), and tricarballylic acid (TA), how many moles of P would cause gelling?

Compound	*Code*	*Moles*
$C(CH_2-OH)_4$	P	?
	PA	1.0
$CH_2 - CH - CH_2$ $\quad\;$ | $\qquad$ | $\qquad$ | COOH $\;\;$ COOH $\;\;$ COOH	TA	1.0

6-7   Acetic acid is frequently used to "cap" (react with the end groups on the polymer chains) for the reaction between hexamethylene diamine (HMD) and adipic acid (AA) to form 66 nylon.

      What relative ratios of moles of these reactants and acetic acid have to be used to produce polymer chains capped at both ends for an average D.P. of 96?

$$NH_2-(CH_2)_6-NH_2 \qquad HMD$$

$$COOH-(CH_2)_4-COOH \qquad AA$$

6-8   In the polycondensation reaction of 66 nylon, water is formed as a by-product:

$$H_2N-(CH_2)_6-NH_2 + HO\overset{O}{\overset{\|}{C}}-(CH_2)_4-\overset{O}{\overset{\|}{C}}-OH$$

$$\longrightarrow [NH-CH_2)_6-NH-\overset{O}{\overset{\|}{C}}-(CH_2)_4-\overset{O}{\overset{\|}{C}}]_n + H_2O$$

The energy in kcal/g mole needed to break the following bonds N—H, C—O, C—N, and H—O are, respectively, 103, 86, 73, and 119.

      Is it also possible to form 66 nylon by forming NH_3 gas as a by-product instead of water?

6-9   It is proposed to make a condensation polymer from ethylene glycol $HO(CH_2)_2OH$ and acetone

$$CH_3-\overset{}{\underset{\overset{\|}{O}}{C}}-CH_3$$

with water as a by-product.

      Is this possible? Also, how many kg of water are formed per kg of mer?

Some additional data: Energies to break the C—H and C—C bonds are 104 and 88 kcal/g mole (also see data from Problem 6-8).

6-10 The rate constants k_i, k_p, and k_t can be represented by the Arrhenius equation

$$k_i = A_i\, e^{-Ei/RT}$$

$$k_p = A_p\, e^{-Ep/RT}$$

$$k_t = A_t\, e^{-Et/Rt}$$

Values of the A and E for the polymerization of styrene in benzene (using azo-bis-isobutylonitrile (AZBN) as a source of radicals) are: 5.6×10^{14} s^{-1} and 30 kcal/g mole for initiation; 2.2×10^7 L/g mole s 7.8 kcal/g mole for propagation; and 2.6×10^9 L/g mole s and 2.4 kcal/g mole for termination.
Use these data to obtain a single Arrhenius expression for the polymerization of styrene.

6-11 For the reaction of the previous problem, consider the reaction temperature to 375 K and the initiator (AZBN) concentration to be 0.002 g mole/L.
What is the maximum concentration of the growing chains? How much time will be required for this to occur? What is the significance of this parameter?

6-12 It is possible to polymerize styrene (normal boiling point of 146°C; MW of 104) in the bulk or in the solution. The heat of polymerization is 68.65 kJ/g mole.
Compare bulk and solution polymerization of styrene. Use a 20% styrene (by mass) in the solution. Assume that the C_p values for both styrene and the solvent are 2.09 J/g K.

6-13 A polyvinyl acetate latex contains 10^{20} particles/m^3. It is desired to obtain both a higher concentration of polymer and a larger particle size. In order to do this, an additional 3.6 parts (by weight) of monomer per part of polymer will be added (no soap will be added). If the reaction is taken to 82% conversion, what will be the time needed for the reaction the rate of heat removal (J/m^3 of the original system) for an isothermal (330 K) reaction. Also, at what conversion would the monomer droplets disappear and the rate become no longer constant? Pertinent data are

$$k_p = 3.7 \text{ m}^3/\text{g mole s}$$

$$\Delta H_p = 87.9 \text{ kJ/g mole monomer}$$

Polymer density = 1200 kg/m^3

Monomer density = 800 kg/m^3

6-14 At higher temperature in free-radical polymerization systems, a reverse reaction (depropagation) takes place. For such a situation, the disappearance of monomer is given by

$$-\frac{d(M)}{dt} = kp(M)(M) - k_{dp}[M]$$

where k_{dp} is the specific reaction rate for depolymerization or depropagation.

The E_{dp} (activation energy) is greater than the E_p with the difference between them being the heat of polymerization. For a given monomer concentration, there will be a "ceiling temperature" (temperature at which propagation and depropation rates are equal).

Estimate this value for polystyrene with a monomer concentration of 1.2 mole/L. The heat of polymerization is 68.65 kJ/g mole. Use a value of 10^{13} s^{-1} for A_{dp} (Arrhenius equation frequency factor).

6-15 You, as a new hire, have inherited the process responsibilities for several projects. In each of these cases, some laboratory notebooks are available for the projects. Unfortunately, they are in poor shape. However, some data relating monomer concentration to time are available for polymers A and B. Both these polymers are known to be addition polymers. Determine whether these polymers have been synthesized by radical, cationic, or anionic methods.

Polymer A

Monomer, mole/L	Time, min
1.0	0
0.875	15
0.764	30
0.669	45
0.585	60

Polymer B

Monomer, mole/L	Time, min
1.0	0
0.769	15
0.625	30
0.526	45
0.455	60

6-16 Obtain plots of F_1 (copolymer composition) vs f_1 (monomer composition) for the system (all at 60°C) given below.

Monomer 1	r_1	Monomer 2	r_2
Butadiene	1.39	Styrene	0.78
Vinyl acetate	0.01	Styrene	55
Maleic anhydride	0.002	Isopropenyl acetate	0.032

Do any of the above systems match ideal copolymerization? Are there any azeotropes? What will the directions of composition drift be with conversion?

6-17   A free-radical polymerization is carried out with a monomer concentration of M_0, initiator of I_0, and at an absolute temperature of T_0. What will be the effects on rate of polymerization and the number-average degree of polymerization for the following system changes:

a. Using $5M_0$, $0.5\ I_0$, and $1.2\ T_0$
b. Using $3M_0$, $2.0\ I_0$, and $1.1\ T_0$
c. Using $6M_0$, $1.5\ I_0$, and $1.17\ T_0$

6-18   Styrene is polymerized at 60°C using an initiator concentration of 1×10^{-3} g mole/L (azodi-isobutyronitrile). The fraction of radical formed that initiates the chain reaction is 0.6, while k_1 is $0.85 \times 10^{-5}\ s^{-1}$. Termination k_t value is 1.8×10^7 L/mole s.

What is the average radical lifetime and the steady-state radical concentration? Liquid styrene density is 909 kg/m³.

6-19   The following laboratory data are available.

Temperature, °C	Conversion, %	Time	Initial Monomer M_0 mole/L	Initial Initiator I_0 mole/L
60	40	10	0.80	0.0010
80	75	11.67	0.50	0.0010
60	50	8.33	1.00	0.0025

Based on these data, find the time for 50% conversion at 60°C with an M_0 of 0.25 mole/L and an I_0 of 0.0100 Mole/L. Also, find the Arrhenius expression for the polymerization.

6-20   Two hundred liters of methyl methacrylate

$$(CH_2 = C-COOCH_3)$$
$$|$$
$$CH_3$$

is reacted at 330 K with 10.5 moles of an initiator. Values of k_i, k_p, and k_t are $1.1 \times 10^{-5}\ s^{-1}$, 5.5 L/mole s, and 25.5×10^6 L/mole s. The monomer's density is 940 kg/m³.

If the fraction of radicals formed that initiates the chain reaction is 0.3, find both the kinetic chain length and the amount of polymer produced in 6.2 h.

6-21   The activation energy for the melt polymerization of 66 nylon is 12.96 kcal/g mole. How does this value compare to the activation energy for solid-state poly-condensation (Fig. 6-10).

6-22 Two monomers, A and B, are copolymerized. If A is taken as monomer 1 and if $r_1 = 0.96$, $r_2 = 0.55$, $f_2 = 2.0f_1$, find (1) which of the two is found to a greater extent in the initially produced polymers, and (2) what the situation is after 10% conversion and 30% conversion.

CHAPTER 7

POLYMER PROCESSES: EXTRUSION

7.1 INTRODUCTION

The goal of all polymer processing operations is to produce a usable object. Basically, these operations involve flow and deformation as well as the transfer of energy. As such, we can state that, for *all* polymer processing operations, two necessary parameters are:

1. Flow or deformation or both (i.e., rheology)
2. Transfer of heat and thermal behavior

In some situations, the aspects of

3. Transfer of mass
4. Chemical reaction

are also necessary to the process. For example, mass transfer is an important parameter in producing structural foams. Chemical reaction is intimately involved in reaction injection molding.

In a sense, many chemical and petroleum processes also involve the parameters listed above. One very different aspect of polymer processing, however, is that changes in polymer structure (crystallinity, orientation, etc.) can and do occur during the operations themselves. Such changes exert a direct influence on a polymer's end properties (mechanical, electrical, etc.).

Table 7-1 shows the effect of changes in a polymer's structure on properties. The structural characteristics were those discussed in Chapter 1. A plus sign indicates that an increase in a given characteristic will increase the property whereas a minus sign indicates the opposite. A question mark means that the effect of the structural property change is unclear.

The skilled polymer-processing practitioner will always be aware of the important processing–structural characteristics–properties relationships since these very often can control the success of the processing operation.

In a broad sense, all polymer-processing operations can be characterized as batch, semicontinuous, or continuous. Table 7-1A shows this for a number of important industrial operations.

Another method of describing the semicontinuous (and occasionally the batch processes such as transfer molding) is as a *cyclical process.*

The sections that follow will treat each of the operations of Table 7-1A in detail.

The principles and fundamentals presented in these chapters will be used to develop the basic understanding of the various polymer processes so that operations, analysis, design, and development can be properly treated in an engineering sense.

7.2 EXTRUSION: BACKGROUND

Extrusion can be defined as the act of shaping a material by forcing it through a die. In the case of polymers, the devices used to carry out extrusions are called extruders. These devices are expected to be pulse-free pumps capable of delivering a thermally homogeneous polymer melt at a uniformly high rate. As such, the extruder is the essential part not only of operations using dies (extrusion, blow molding) but also of cases in which the melt is injected into a mold (injection molding).

The starting point for the process of extrusion will therefore be to develop an understanding at the extruder itself. This understanding will obviously be useful for the discussion at other processing operations (blow molding, injection molding) in which the extruder also plays a central role.

Extruders are basically helical screw pumps (see Fig. 7-1) that convert solid polymer particles) to a melt, which is delivered to a die or a mold. The screw extruder can be classified as either a single-screw or a twin-screw device. For the present, we will confine our discussion to a single-screw device (see Fig. 7-1).

Polymeric material (in the form of pellets) is fed into the extruder's hopper and thence to the screw channel. The screw, driven by a motor through a gear reducer, rotates in a hardened barrel. A thrust bearing absorbs the rearward thrust

TABLE 7-1 Relation of Polymer Properties to Structure

Characteristic Property	D.P.	Branching	Cross-Linking	Polar Structures	Chain Flexibility	Crystallinity	Size of Spherulites	Orientation	Molecular Weight Dist.
Ultimate tensile strength	+	?	?	+	?	+	?	+	?
Modulus of elasticity	+	?	+	?	−	+	?	+	?
Impact strength	+	?	−	?	−	+	Smaller +	+	Narrow +
Elongation at break	+	?	−	?	+	−	Smaller −	?	Narrow +
Range of reversible extensibility	+	?	−	?	+	−	?	?	?
Surface hardness	+	?	+	+	+	?	?	+	?
Temperature resistance	+	?	+	+	−	+	Large +	+	High D.P. +
Electrical resistance	?	?	?	−	−	+	?	+	?
Dielectric constant	?	?	?	+	+	?	?	?	?
Swelling resistance	+	?	+	+	+	−	Smaller −	+	High D.P. +
Creep or cold flow resistance	+	?	+	+	−	+	?	+	?
Moisture resistance	+	+	+	?	−	+	?	+	?
Alkali resistance	+	+	+	?	−	+	?	+	?
Acid resistance	?	?	+	?	?	+	?	+	?
Adhesive power	?	+	−	+	+	−	?	?	Broad +

Adapted from Mark (13).

280

TABLE 7-1A   Categorization of Polymer-Processing Operations

Batch	*Semicontinuous*	*Continuous*
Casting	Blow molding	Calendering
Compression molding	Injection molding	Extrusion
Sheet forming	Rotational molding	Pultrusion
Thermoforming		Fiber spinning
Transfer molding		

of the screw. Thermal energy is supplied by the internal heat generation of the flowing polymer or external heaters or both. In some cases, the internal heat becomes so great that external cooling is actually needed. The plastic granules are melted as they are conveyed and forced through a breaker plate–screw combination ultimately to a die. Sometimes a back-flow ball valve is positioned between the breaker plate and adapter to control the process.

There are three zones or regimes in the extruder (see Fig. 7-2). These are:

1. Solids-conveying, or feed, zone

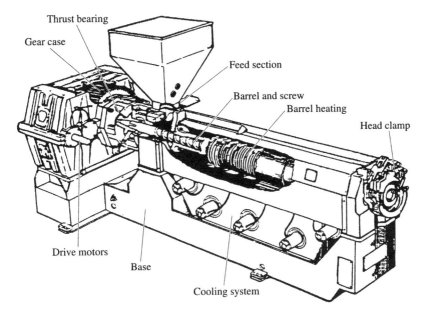

Fig. 7-1   Single-screw extruder [1].

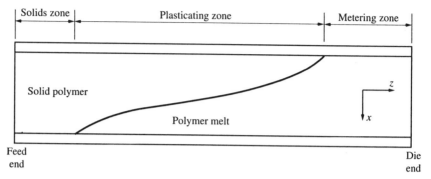

Fig. 7-2   Schematic diagram of zones in the screw extruder.

2. Transition, or plasticating, zone
3. Melt, or metering, zone

In the first zone, pellets are conveyed to the main segment of the extruder. It is essential that the conveying capacity is equal to the extruder's melting and pumping capacity.

The theoretical approaches for this zone are not well defined. Instead, a semi-empirical approach is used that considers the pellets to behave as a solid plug. This plug advances with little deformation, and its rate of movement depends on both the back pressure in the extruder and the frictional forces on the screw flight and barrel.

The frictional forces are functions of screw geometry and the nature of the extruder surfaces. A particularly critical factor is the helix angle.

In essence, the *transition*, or *plasticating, zone* connects the *feed* and *metering zones*. The length of this zone varies with the materials being extruded. Generally, the flow channel cross section is reduced in this zone. An exception is for rubbery materials for which the pitch angle is changed so that overworking can be avoided.

The treatment of the *plasticating zone* requires an analysis that combines flow, heat transfer, and mixing. Effective techniques require the assumption of models that enable usable techniques to be developed.

The material in the *metering zone* is a melt. As such, the treatment combines flow and heat transfer. The metering section, while still complex, lends itself more readily to technical analysis than the other two zones. Before undertaking the technical analysis at each of the zones, it is appropriate to consider the various important aspects of screw extruder geometry. Figure 7-3 is a schematic of a screw and a barrel (double-flighted screw with two flow channels in parallel).

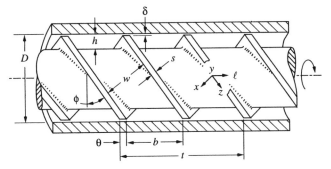

Fig. 7-3   Geometric aspects of an extruder [3].

The various geometric parameters shown in Fig. 7-3 are defined as:

B = Axial distance between flights
D_B = Barrel diameter
D_S = Screw diameter
W = Width of the flow channel
b = Screw flight width in the axial direction
e = Width of the screw flight perpendicular to the flight
$h = (D_B - D_S)$
δ = Clearance between the screw flight edge and the barrel
ϕ = Helix angle

Channel width can be further specified:

$$\text{Width at barrel surface} = \frac{L}{P} \cos \phi_B - e \qquad (7\text{-}1)$$

where P is the number of channels in parallel ($P = 2$) for double-flighted screws.

$$\text{Width at screw} = \frac{L}{P} \cos \phi_S - e \qquad (7\text{-}2)$$

The diverse nature of polymers that can be extruded as well as a wide variety of operating conditions combine to create the possibility of many different screws. A sampling of such screws used commercially is shown in Fig. 7-4.

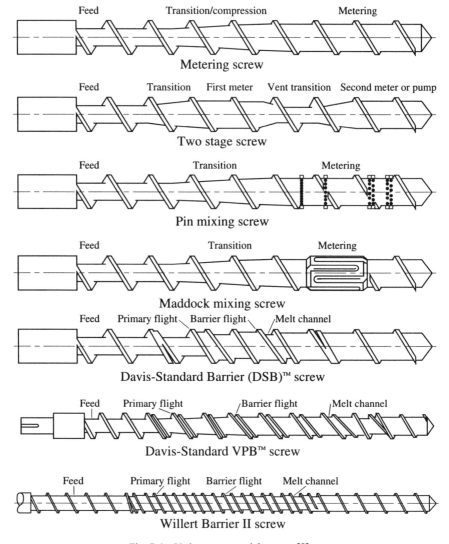

Fig. 7-4   Various commercial screws [8].

Polymer conveyed through the extruder ultimately passes through a die (or a mold). Typical ranges of pressures at a die are given in Table 7-2.

In contrast, pressures in injection-molding operations can vary from 14 to 210 MPa. The reader should, however, recall (Table 7-1A) that extrusion is a continuous process whereas injection molding is semicontinuous.

TABLE 7-2 Extruder Die Pressures

Product	Pressure in MPa at the die
Blown film	6.9–34.5
Cast film	1.4–10.4
Sheet	1.5–10.4
Pipe	2.8–10.4
Wire coating	6.9–34.5
Filament	6.9–20.7

Source: Adapted from Ref. 9.

7.3 EXTRUDER SOLIDS-CONVEYING SECTION

The movement of the solid particles in the feed section is best treated as that of a solid plug that advances with little deformation. The progress of the plug depends on back pressure and the frictional forces of the barrel and screw, which are functions of screw geometry and the nature of the surface.

The basic flow equation for such movement is

$$Q_S = \pi^2 \, NhD_B(D_B - h) \frac{\tan \theta \tan \phi_B}{\tan \theta + \tan \phi_B} \frac{W}{W + e} \tag{7-3}$$

where W the average channel width, ϕ a complicated function of geometry, and N the screw speed.

For the case in which the frictional effects of barrel and screw are the same and there is no appreciable pressure differential,

$$\cos \theta = K \sin \theta + \frac{W_S}{W_B} \sin \phi_B \, K + \frac{D_S}{D_B} \cot \phi_B \tag{7-4}$$

where W_S and W_B are channel widths at the screw and barrel, respectively.

The K factor is

$$K = \frac{\overline{D}}{D_B} \frac{\sin \overline{\phi} + \mu_S \cos \overline{\phi}}{\cos \overline{\phi} \, \mu_S \sin \overline{\phi}} \tag{7-5}$$

and $\overline{D}$ is the mean diameter, with $\overline{\phi}$ the corresponding helix angle.

A somewhat simpler case is one in which there is no frictional effect and negligible pressure. This gives

$$Q_S = \pi^2 \, NhD_B \, (D_B - h) \sin \phi_B \cos \phi_B \frac{\overline{W}}{\overline{W} + e} \tag{7-6}$$

The case for consideration of pressure and of all possible friction forces gives a much more complicated expression for θ. This can be found elsewhere.

7.4 EXTRUDER MELTING, OR PLASTICATING, SECTION

The polymer pellets are converted from a solid to a melt in the plasticating section. Here we have a two-phase system in which the proportions of solid and molten material change.

Two melting mechanisms (Fig. 7-4A) can take place. If the screw is above the melting temperature, the solid is surrounded on three sides by a melt film and is adjacent to a melt pool. When the screw is below the polymer melting temperature, the film does not exist on two of the solid bed's sides.

Development of appropriate equations to describe the plasticating section involves combining the hydrodynamics, heat transfer, and melting phenomena. If the channel depth h is taken to be constant, the following equations are

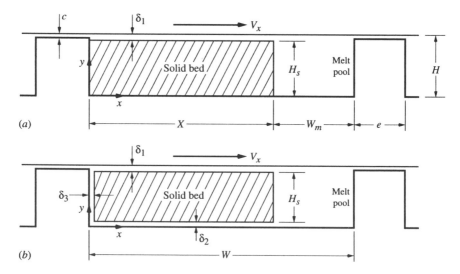

Fig. 7-4A   Polymer melting mechanisms [4].

derived.

$$\frac{X}{W} = \left(1 - \frac{C_2 Z}{2\rho_s V_{sz} hW^{1/2}}\right)^2 \tag{7-7}$$

$$Z_m = \frac{2\rho_s V_{sz} hW^{1/2}}{C_2} = 2\pi_s V_{sz} h\left(\frac{2\lambda W}{C_1 \rho m V_x}\right)^{1/2} \tag{7-8}$$

$$C_1 = \left[\frac{\mu_1 V_r^2}{2} + km (T_B - T_m)\right] \tag{7-9}$$

$$C_2 = \left(\frac{C_1 \rho m V_x}{2\lambda}\right)^{1/2} \tag{7-10}$$

where X is the width of the solid bed at any helical length Z; ρ_s and ρm are the solid and melt densities, T_B and T_m are the barrel and melt temperatures, V_{sz} is the solid-bed downstream velocity, V_x is the cross-channel velocity, and V_r is a resultant *relative* velocity; k_m is the melt thermal conductivity, and λ is the heat of fusion.

Equations (7-7) and (7-8) make it possible to estimate both the helical length required for melting and the profiles of the solid/melt interface for the plasticating section.

More detailed and complicated treatments are given in the literature.

7.5 MELT, OR METERING, SECTION

When the polymer is finally a melt, its flow can be treated by a hydrodynamic analysis. This analysis shows that there are three principal flow types that can interact in this section.

These are:

1. Drag flow: The melt is dragged forward in the screw channels by the action of the screw.
2. Pressure flow: This is flow that results in backward flow because of the pressure differential in the extruder (high pressure at the die and low pressure in the feed section).
3. Leakage flow: This flow can occur backward in the space between the edge of the flights and the barrel.

Drag and pressure flows usually predominate. Leakage flow can occur if the screw flights have eroded or if the extruder is badly designed, with an excessive clearance between the flight edges and the barrel.

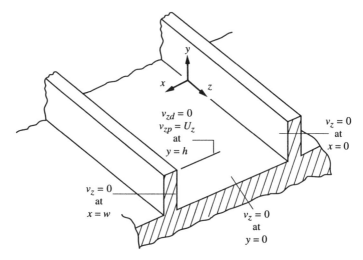

Fig. 7-5   Extruder flow channel [5].

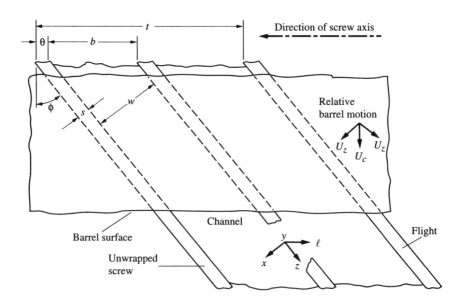

Fig. 7-6   Extruder flow simulation [3].

It is usual, therefore, to consider only the drag and pressure flows so that

$$q_{\text{Total}} = q_{\text{Drag}} + q_{\text{Pressure}} \tag{7-11}$$

The flows are complicated functions at the extruder geometry. A semiempirical approach ultimately yields

$$q_{\text{Total}} = F_D \propto N - F_P \frac{\beta}{\mu}\left(\frac{\partial P}{\partial L}\right) \tag{7-12}$$

where

$$\propto = \frac{\pi^2 D^2 h[(1 - ne)/t]\ \sin^2\ \phi}{2} \tag{7-13}$$

$$\beta = \frac{\pi D h^3\ [(1 - ne)/t]\ \sin^2\ \phi}{12} \tag{7-14}$$

F_D and F_P are shape factors given by Fig. 7-7.
Physically, F_D and F_P represent the effect of flow distortion caused by the flight edges.

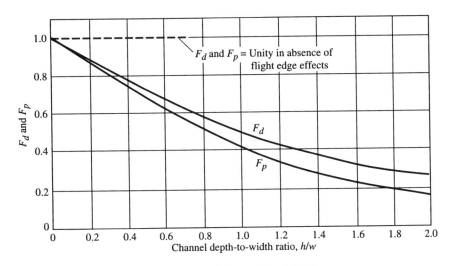

Fig. 7-7   Geometric factors F_d and F_p [5].

7.6 SOME DESIGN AND OPERATING PRINCIPLES FOR EXTRUDERS

The proper design of an extruder requires that all three sections, solids-conveying, melting, and metering, be properly matched. If they are not, faulty operation will result. Figure 7-8 illustrates this point for three cases. In case A, the melting capacity exceeds the capacity of the metering section while, in case C, the melting capacity is too low. In case B, both the melting and metering capacities are matched. The optimum operation is represented by case B. Excess of melting capacity causes surges in operation. Too low a melting capacity, on the other hand, "starves" the extruder.

In designing an extruder, two limiting cases can apply. One of these, isothermal operation, assumes that the temperature is balanced (i.e., constant) in the metering section. The other case is adiabatic operation.

Equations (7-11–7-14) are used for the isothermal situation. These equations, together with an energy balance and a temperature relationship for viscosity, are used for the adiabatic case.

Adiabatic operation will give a lower output than isothermal operation at a given extruder speed and pressure differential. The difference between the outputs increases with increasing pressure differential.

Less elegant design procedures are available for either the design or the analysis of extruder operations. For example,

$$\text{Power} = 0.00053 \; C \; q(T_E - T_F) \tag{7-15}$$

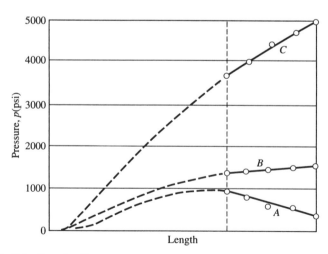

Fig. 7-8 Extruder die pressure behavior as a function of process variables [5].

where power is in hp (horsepower), q is in lbm/h, c is an average specific heat in Btu/lbm °F, and T_E, T_F are, respectively, the extrudate and feed temperatures.

If the surface speed (in feet/minutes) and horsepower are known, the minimum screw diameter D min. is

$$D \text{ min.} \cong 4.2 \, \frac{\text{hp}}{V_s} \qquad (7\text{-}16)$$

where V_s is surface speed.

Finally, the length of the extruder is in a ratio (L/D) of 16–24. Longer barrels are favored because they provide

1. Better mixing action
2. Uniformity of extrudate
3. More uniformity at higher rates

7.7 TWIN AND MULTIPLE SCREWS

It is also possible (as mentioned earlier) to have extruders that use twin or multiple screws. Such extruders are utilized in certain operations because they offer specific advantages. These include: (1) increased output at low screw speed, (2) improved pumping control over a wide range of operating conditions, (3) decreased viscous dissipation and internal heat generation, (4) ability to handle materials that are difficult to feed, and (5) lower power requirements (see Fig. 7-9).

Twin screw extruders can be operated with either counter- or corotating screws. Additionally, the screws can be fully intermeshing, partially intermeshing, or nonintermeshing. Figure 7-10 illustrates the various cases.

One additional method of categorizing twin screws relates to the size and shape of the screw channels and flights. This method uses the categories of nonconjugated and conjugated screws. Nonconjugated screws have flights that fit loosely into the other screws' channels and have ample flow passages. Conjugated screws each have flights with similar sizes and shapes that snugly fit the other screws' channels with negligible clearance. Table 7-3 lists possible configurations for fully intermeshing counter- and corotating screws.

Flow in twin-screw units is made up of drag flow and leakage flow. Overall flow is, therefore, the difference between drag flow and leakage flow. The effect of leakage flow is expressed in terms of a percentage of drag flow. These percentages range from 50% to 65% for counterrotating units and from 10% to 15% for corotating units. Drag flows for counterrotating and corotating units are

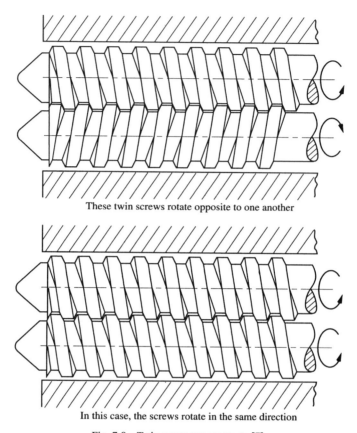

These twin screws rotate opposite to one another

In this case, the screws rotate in the same direction

Fig. 7-9   Twin-screw arrangements [7].

given in Eqs. (7-17) and (7-18):

$$q_d = \pi\, Dh\, \sin\, \phi(\pi D - 2Dh)N \qquad (7\text{-}17)$$

$$q_d = -\, \pi^2 D^2 Nh\, \tan\, \phi \qquad (7\text{-}18)$$

Based on the foregoing, mass flow outputs w for counterrating [Eq. (7-17)] and corotating units [Eq. (7-18)] would be

$$0.35\, q_d\rho < w < 0.50\, q_d\rho \qquad (7\text{-}19)$$

$$0.85\, q_d\rho < w < 0.90\, q_d\rho \qquad (7\text{-}20)$$

where ρ is the melt density of polymer being processed.

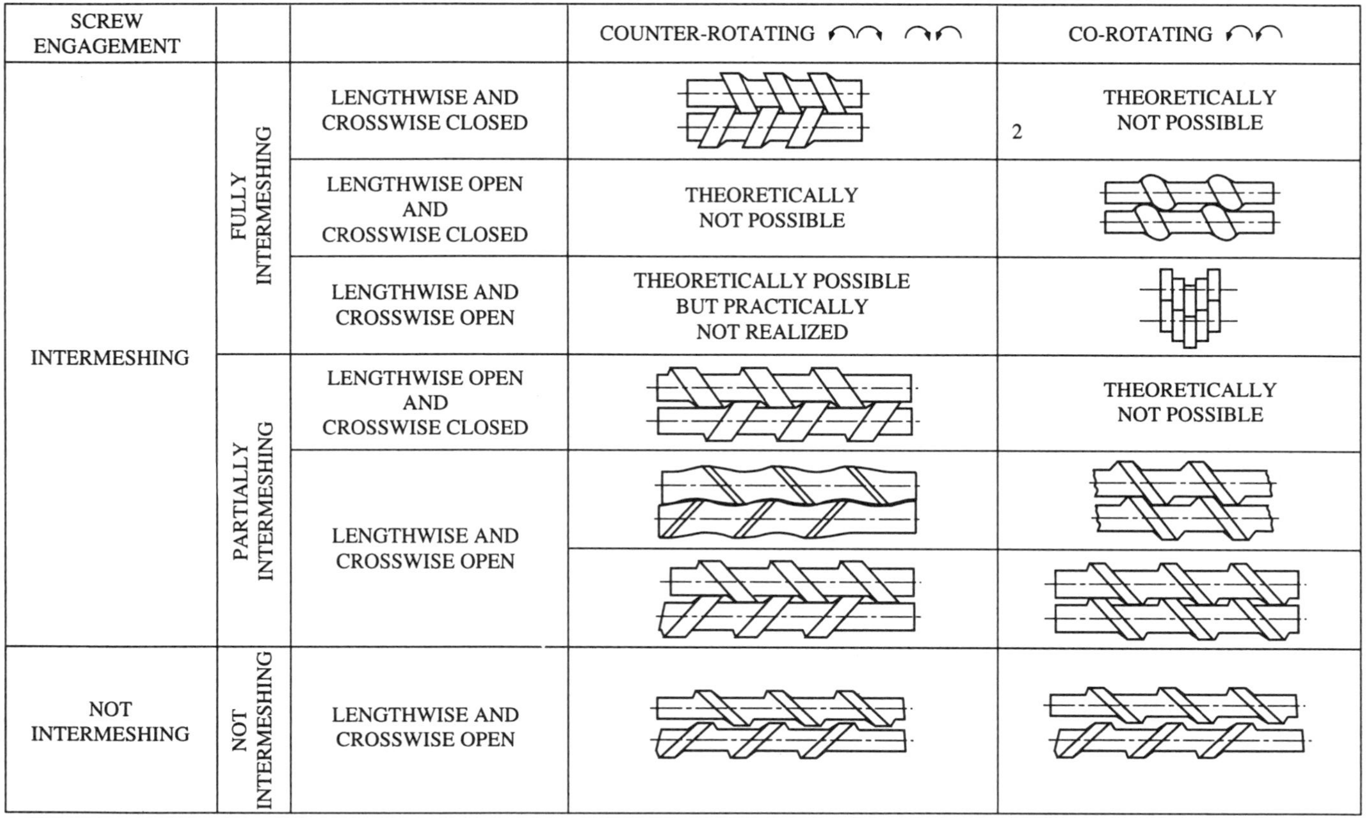

SCREW ENGAGEMENT		COUNTER-ROTATING	CO-ROTATING
INTERMESHING	FULLY INTERMESHING	LENGTHWISE AND CROSSWISE CLOSED		THEORETICALLY NOT POSSIBLE (2)
		LENGTHWISE OPEN AND CROSSWISE CLOSED	THEORETICALLY NOT POSSIBLE	
		LENGTHWISE AND CROSSWISE OPEN	THEORETICALLY POSSIBLE BUT PRACTICALLY NOT REALIZED	
	PARTIALLY INTERMESHING	LENGTHWISE OPEN AND CROSSWISE CLOSED		THEORETICALLY NOT POSSIBLE
		LENGTHWISE AND CROSSWISE OPEN		
NOT INTERMESHING		LENGTHWISE AND CROSSWISE OPEN		

Fig. 7-10 Comparison of commercial twin-screw arrangements [8].

293

TABLE 7-3   Conjugated and Nonconjugated Systems

Conjugated?	Counterrotating	Corotating
Yes	Lengthwise closed; crosswise closed	Lengthwise open; crosswise closed
No	Lengthwise open; crosswise closed	
No	Lengthwise open; crosswise open	Lengthwise open; crosswise open

Source: Adapted from Ref. 10.

7.8   VENTED EXTRUDERS

At times, polymers being processed must be devolatilized or degassed. Such an operation can be carried out in a vented extruder (see Fig. 7-11). Such extruders can have two vents (as shown) but usually have a single vent.

The pressure profile in the vented extruder is such that pressure first reaches a low value (because of the change in screw geometry) and then actually falls below atmospheric pressure (caused by a vacuum applied at the vent) in the devolatilization section. Figure 7-12 shows the altered screw geometry at the vent openings, which starve flow in the given section.

Venting can be carried out in either single- or twin-screw extruders. In either case, the removal of the material to be vented involves a complex interaction of heat, mass, and momentum transfer. Analyses based on various assumptions exist for both single- and twin-screw systems.

Single-screw systems use mass-transfer penetration theory [10], together with certain geometric assumptions, to develop an expression for a film efficiency E_F:

$$E_F = 1 - \exp(-E_x) \tag{7-21}$$

and

$$E_x = \frac{k_f S_f + k_p S_p}{Q} \tag{7-22}$$

where k_f and k_p are mass-transfer coefficients for the film on the barrel surface and the polymer pool, respectively; S_f and S_p are surfaces of the film and pool; and Q is the volumetric flow rate.

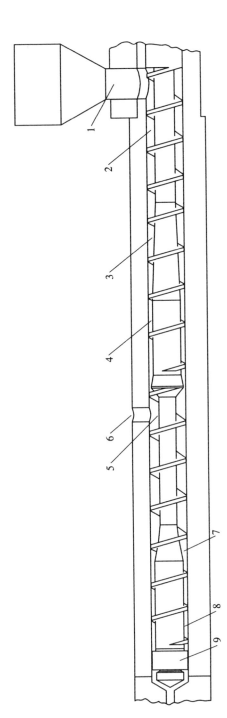

Fig. 7-11 Vented extruder [8]. (1) Wet material enters from a conventional hopper. (2) The pellets are conveyed forward by the screw feed section and are heated by the barrel and by some frictional heating. Some surface moisture is removed here. (3) The compression or transition section does most of the melting. (4) The first metering section accomplishes final melting and evens the flow to the vent section. (5) Resin is pumped from the first metering section to a deep vent or devolitizing section. This vent section is capable of moving quantities well in excess of the material delivered to it by the first metering section. For this reason, the flights in the vent section run partially filled and at zero pressure. It is here that volatile materials such as water vapor, and other nondesirable materials, escape from the melted plastic. The vapor pressure of water at 500°F is 666 psi. These steam pockets escape the melt, and travel spirally around the partially filled channel until they escape out the vent hole in the barrel. (6) Water vapor and other volatiles escape from the vent. (7) The resin is again compressed, and pressure is built in the second transition section. (8) The second metering section evens the flow and maintains pressure so that the screw will be retracted by the pressure in front of the nonreturn valve. (9) A low-resistance, sliding-ring, nonreturn valve works in the same manner as it does with a nonvented screw.

295

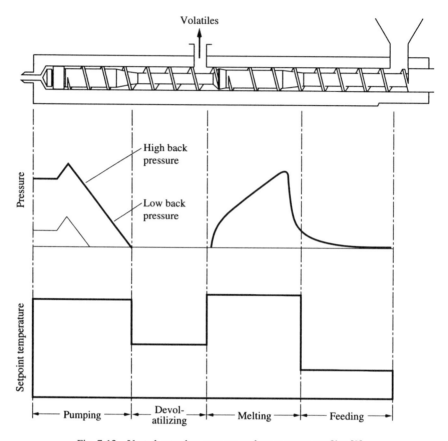

Fig. 7-12   Vented extruder: pressure and temperature profiles [9].

E_f increases with the screw speed N when

$$- \ln(1 - E_F) - C_1\left(\frac{D}{N}\right)^{1/2} \qquad (7\text{-}23)$$

and decreases if

$$- \ln(1 - E_F) = C_1(N + C_3)^{1/2} + C_2(N)^{1/2} \qquad (7\text{-}24)$$

the C values are constants.

The twin-screw case [11] uses a model based on a series of stagewise units and a semi-infinite mass-transfer diffusion model.

The results are

$$\frac{C_n}{C_0} = (1 - F)^n \qquad (7\text{-}25)$$

and

$$F = \frac{b}{(v \, Q)^{-1/2}} \frac{4Dt_e}{\pi t_s} \qquad (7\text{-}26)$$

where v is the hold-up volume in each stage, b is a geometric parameter, and t_e and t_s are, respectively, the residence time in a stage and the overall residence time.

In both the single- and twin-screw cases, the equations are not design equations but are rather models that can be used to compare the given performances of particular extruders.

7.9 EXTRUDER EXAMPLES

Example 7-1

Consider a screw extruder with the following dimensions in the solids-conveying section.

$D_B = 5.06$ cm
$D_S = 3.49$ cm
$L = 5.08$ cm
$e = 0.508$ cm
$\phi = 17.7°$

For this extruder, with nylon as the polymer, find the mass flow per revolution.

First, calculate the various items needed (i.e., H, W):

$$h = \frac{D_B - D_S}{2} = 0.787 \text{ cm}$$

$$D = D_B - h = 4.275 \text{ cm}$$

$$\phi_B = \tan^{-1} \frac{L}{\pi D_B} = 17°44^1$$

$$\phi_S = \tan^{-1} \frac{L}{\pi D_S} = 24°50^1$$

$$\phi = \tan^{-1} \frac{L}{\pi D} = 20°43^1$$

$$W = L \cos \phi - e = 4.244$$

$$W_B = L \cos \phi_B - e = 4.331$$

$$W_S = L \cos \phi_S - e = 4.102$$

Determine Q_S/N for the case in which there is *no frictional effect* and negligible pressure [i.e., use Eq. (7-6)].

$$\frac{Q_S}{N} = \pi^2 h \, D_B(D_B - h) \sin \phi_B \cos \phi_B \frac{W}{w + e}$$

Substituting the various values gives

$$\frac{Q_S}{N} = 43.246 \text{ cm}^3/\text{rev}$$

The nylon density is 0.475 g/cm^2. Hence, the mass output per revolution is

$$\frac{w_s}{N} = 43.246 \text{ cm}^3/\text{rev} \; 0.475 \text{ g/cm}^3$$

$$\frac{w_s}{N} = 20.54 \text{ g/rev}$$

Next, use the case in which the *frictional forces are equal* (μ_s, the coefficient of friction is 0.25) and pressure is negligible.

For this case,

$$\cos \phi = K \sin \theta + \frac{W_S}{W_B} \sin \phi_B + \frac{D_S}{D_B} \cot \phi_B$$

$$K = \frac{D}{D_B} \frac{\sin \phi + \mu_s \cos \phi}{\cos \phi - \mu_s \sin \phi}$$

$$K = 0.586$$

and

$$\cos \theta = 0.586 \sin \phi + 0.5996$$

$$\theta = 28^\circ 28^1$$

$$\tan \theta = 0.5422$$

Then, using

$$\frac{Q_S}{N} = \pi^2 h \, D_B (D_B - h) \frac{\tan \theta \tan \phi_B}{\tan \theta + \tan \phi_B} \frac{W}{w + e}$$

gives

$$\frac{Q_S}{N} = 4.648 \text{ cm}^3/\text{rev}$$

and

$$\frac{w_s}{N} = 14.24 \text{ g/rev}$$

A comparison of the calculated and experimental values are:

Case	w_s/N *g/rev*
1 (Nonfriction; negligible pressure)	20.54
2 (Some frictional effects; negligible pressure)	14.54
3 (Actual experiment)	14.90

Note that the agreement for case 2 with experiment is quite good.

Example 7-2

Analyze the melting profile, and find the length required for total melting of polyethylene in an extruder that has 12 turns for the plasticating segment. The 5.635-cm-diam extruder is single-flighted, with the flight width 0.635 cm and the helix angle 18°. Channel depths in the feed and metering sections are 1.270 and 0.3175 cm, respectively. Extruder output is 0.0152 kg/s (N = 82 rpm; T_B = 150°C).

In the plasticating section, there are 18.34 cm/turn and the following parameters:

Helical length of plasticating section = 220.1 cm
Channel width = 5.43 cm
Barrel velocity (V_B) = 27.25 cm/s

The solid-bed velocity, using a density of 920 kg/m³, is

$$V_S = \frac{(0.152 \text{ kg/m}^3)}{(0.0543 \text{ m}) \ (0.0127 \text{ m}) \ (920 \text{ kg/m}^3)}$$

$$V_S = 0.024 \text{ m/s} = 2.4 \text{ cm/s}$$

Relative velocity difference between barrel and solid is

$$V_r = (27.25 - 2.4) \text{ cm/s} = 24.85 \text{ cm/s}$$

Assuming the average temperature to be the average of T_B and T_{melting} (109°C), and allowing some viscous heating, gives a T_{AV} of 134°C. For this temperature, the melt properties are:

$$k_m = 1.82 \ \frac{w}{m°C}$$

$$\rho_m = 790 \text{ kg/m}^3$$

$$(C_p)m = 2510 \text{ J/kg °C}$$

$$\lambda = 1.297 \times 10^5 \text{ J/kg}$$

Average shear rate (assuming a melt film thickness of 0.01 cm) is

$$\gamma_{\text{AV}} = \frac{24.85 \text{ cm/s}}{0.01 \text{ cm}} = 2485 \text{ s}^{-1}$$

at 134°C, the μ_1 is 1700 poise.

An estimate of the temperature rise due to viscous heating can be obtained by using

$$\Delta T_{\text{VD}} = \frac{\mu_1 V_r^2}{12 \text{ km}}$$

Substitution gives a ΔT_{VD} that confirms the assumed average temperature of 134°C.

Now, we can use Eqs. (7-7), (7-9), and (7-10):

$$\frac{X}{W} = \left(1 - \frac{C_2 Z}{2\rho_s V_{sz} H W^{1/2}}\right)^2$$

$$C_1 = \frac{\mu_1 v_r^2}{2} + km\,(T_B - T_m)$$

$$C_2 = \frac{C_1 \rho m V_x}{2\lambda}^{1/2}$$

Substituting the various quantities gives

$$\frac{X}{W} = (1 - 0.002893Z)^2$$

where Z is in centimeters.

For the solids-conveying section,

$$\text{cm/turn} = \pi\,\frac{(5.635 - 1.27)\text{cm}}{\cos 18°} = 16.77 \text{ cm/turn}$$

After the last turn of the solids-conveying section, Z will be 16.77 cm and X/W will equal 0.905.

Next, calculate the Z_m from Eq. (7-8):

$$Z_m = \frac{2\rho_s V_{sz} h W^{1/2}}{C_2} = 2\rho_s V_{sz} h\,\frac{2\lambda W^{1/2}}{C_1 \rho m V_x}$$

$$Z_m = 236.6 \text{ cm}$$

Repeating the X/W calculation for Z/Z_m values yields

Z/Z_m	Turns	X/W
0	0	0.905
0.2	2.58	0.779
0.4	5.16	0.625
0.6	7.74	0.435
0.8	10.32	0.195
0.9	11.61	0.074
1.0	12.90	0

The total number of turns for melting are then $(1.0 + 12.9)$ or 13.9. Actually, an experiment showed that 12 turns were needed.

Example 7-3

Compare the outputs for the extruder of Example 3-1 and one in which h, the channel depth, is increased from 0.100 to 0.200 in.

In Example 3-1, the equation for q, which was shown to be

$$q = 0.775N - 3.13 \times 10^{-4} \frac{\Delta P}{\mu}$$

came from Eqs. (7-11–7-14) and Fig. 7-7.

$$q_{Total} = q_{Drag} + q_{Pressure}$$

$$q_{Total} = F_D \propto N - F_P \frac{\beta}{\mu} \left(\frac{\partial P}{\partial L} \right)$$

where

$$\propto = \frac{\pi^2 D^2 h[(1 - ne)/t] \sin^2 \phi}{2}$$

$$\beta = \frac{\pi D h^3 [(1 - ne)/t] \sin^2 \phi}{12}$$

From Fig. 7-7,

$$F_D = 0.93$$

$$F_P = 0.93$$

The resultant q equation is

$$q = 1.55N - 2.50 \times 10^{-3} \frac{\Delta P}{\mu}$$

A comparison of the output vs pressure for both screws is shown in Fig. 7-13. Initially, the deeper (0.200-in.) screw has the higher output. However, above a 5000-psi pressure differential, the shallower screw has the higher output.

Example 7-4

A twin-screw devolatilization extruder processes 1.136 kg/s of polymer. The original percent volatiles are 4.5%, which are reduced to 0.45%. What will the final volatile percent be if the amount processed increases to 1.705 kg/s?
We first use Eq. (7-24) to get the F value (assuming 25 stages):

$$\frac{C_n}{C_0} = \frac{0.45}{4.50} = (1 - F)^{25}$$

$$F_1 = 0.088$$

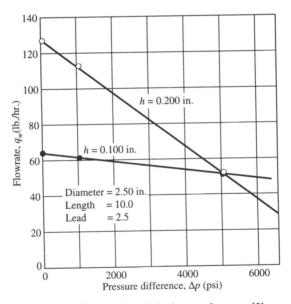

Fig. 7-13   Effect of channel depth on performance [5].

Next, the change in amount processed will change the volumetric throughput in the extruder (assuming that density is relatively constant). This means that, from Eq. (7-25), we obtain

$$F_2 = F_1 \frac{Q_1}{Q}$$

$$F_2 = 0.088 \ [0.667] = 0.072$$

and

$$C_{n_2} = C_0(1 - F_2)^{25}$$

$$C_{n_2} = (4.5)(1-0.072)^{25} = 0.70\%$$

Example 7-5

Find the power required to operate the 0.10-in. and 0.200-in. channel depth extruders of Example 7-3 adiabatically at a flow rate of 60 lbm/h using polyethylene. Feed and extrudate temperatures are 80°F and 400°F, respectively.

Consider two cases using Eq. (7-15) and the first law of thermodynamics.
Case 1 [Eq. (7-15)]: Let C = 0.8 Btu/lbm °F

Power = 0.00053 $Cq(T_E - T_F)$
Power = (0.00053)(0.8)(60)(400 − 80)
Power = 8.14 hp

Note that the power required is predicted to be the same for both channel depths.

Case 2 (First law of thermodynamics): No kinetic or potential energy changes.

Adiabatic:

$$Q - W_S = m\left[\Delta h + \frac{\Delta\mu^2}{2\ gc} + \frac{g}{gc}\ (Z) \right]$$

$$- W_S = m\Delta h$$

In the foregoing, Q is heat input/unit time, W_S power, m flow rate, Δh enthalpy change, $\Delta\mu^2/2\ gc$ kinetic energy change, and $(g/gc)\Delta Z$, potential energy change.

Now, in order to get the Δh value, we must also know the final pressure. In the 0.100-in. channel extruder, pressure is seen (Fig. 7-13) to be 140° psi, while the final pressure in the 0.200-in. channel extruder is 4500 psi.

The enthalpy values can be read from Fig. 2-17. These yield

$$\Delta h = (350 - 21) \frac{\text{Btu}}{\text{lbm}}$$

$$\Delta h = 329 \frac{\text{Btu}}{\text{lbm}}$$

for the 0.100-in. channel, and

$$\Delta h = (358 - 21) \frac{\text{Btu}}{\text{lbm}}$$

$$\Delta h = 337 \frac{\text{Btu}}{\text{lbm}}$$

for the 0.200-in. channel.

The lower value (21 Btu/lbm) in the above is for polyethylene at 80°F and atmospheric pressure.

Therefore, the power required for the 0.100-in. channel is

$$- W_S = 60 \frac{\text{lbm}}{\text{h}} \frac{\text{h}}{3600 \text{ s}} \, 329 \frac{\text{Btu}}{\text{lbm}}$$

$$- W_S = 5.483 \frac{\text{Btu}}{\text{s}} \frac{(1.34102 \text{ hp})}{(0.94783 \text{ Btu/s})}$$

$$- W_S = 7.758 \text{ HP}$$

and for the 0.200-in. channel,

$$- W_S = 60 \frac{\text{lbm}}{\text{h}} \frac{\text{h}}{3600 \text{ s}} \, 337 \frac{\text{Btu}}{\text{lbm}}$$

$$- W_S = 5.616 \frac{\text{Btu}}{\text{s}}$$

$$- W_S = 7.947 \text{ HP}$$

The preceding calculation shows that Eq. (7-15) gives a reasonable estimate of the power required since it deviates by only 4.92% and 2.42%, respectively, for the 0.100-in. and 0.200-in. channels.

Note that the difference in the first law calculation is due to the pressure difference of the two cases (1400 vs 4500 psi).

REFERENCES

1. *Modern Plastics Magazine.*
2. DONOVAN, R.C., *SPE ANTEC* **16**, 561 (1970).
3. SQUIRES, P.H., *Soc. Plast. Engr. Trans.* **4**, 1 (1964).
4. FENNER, R.T., *Principles of Polymer Processing*, Chemical Publishing, New York (1980).
5. BERNHARDT, E. (ed.), *Processing of Thermoplastic Materials*, Reinhold, New York (1959).
6. TADMOR, Z., and KLEIN, I., *Engineering Principles of Plasticating Extrusion*, Van Nostrand Reinhold, New York (1970).
7. MARTELLI, F.G., *Twin-Screw Extruders: A Basic Understanding*, Van Nostrand Reinhold, New York (1983).
8. ROSATO, D.V., and ROSATO, D.V., *Plastics Processing Data Handbook*, Van Nostrand Reinhold, New York (1990).
9. ROSATO, D.V., and ROSATO, D.V., *Injection Molding Handbook*, Van Nostrand Reinhold, New York (1986).
10. BIESENBERGER, J.A., *Devolatilization of Polymers*, MacMillan, New York (1983).
11. MIDDLEMAN, S., *Fundamentals of Polymer Processing*, McGraw-Hill, New York (1977).
12. FENNER, R.T., and WILLIAMS, *J. Mech. Eng. Sci.* **13**, 65 (1971).
13. MARK, H., Ind. Eng. Chem. *34*, 1343 (1942).

PROBLEMS

7-1 What would be the maximum output of the extruder of Example 3-1?

7-2 An extruder (0.05-m-diam screw; channel depth of 0.01 m) has a helix angle of 26°. If all surfaces have the same coefficient of friction, what is the Q/N value for the extruder's solids-conveying section?

7-3 In the extruder of Example 7-2, what would be the optimum helix angle value if no ΔP occurred in the solids-conveying section and if the frictional terms had the same value?

7-4 If the space between the flight edge and the extruder barrel becomes large, then flow opposed to the drag flow takes place. This flow, called *leakage flow*, can be described by the equation

$$q_{\text{leakage}} = \frac{(ne\ \tan\ \phi)\delta^3}{12\ \mu}\left(\frac{\partial P}{\partial L}\right)$$

Plot leakage flow divided by total flow (ratios from 0.02 to 0.15) vs e values. Assume that the extruder is that of Example 3-1.

7-5 Rework Example 3-1 for a polypropylene melt (n of 0.27; K of 3×10^4 N s^n/ m^2).

7-6 Which melt extruder system—a deep channel with a large die opening or a shallow channel with a small die opening—has the higher flow rate? Assume that you have the choice of either a deep or shallow channel combined with either a large or small die opening, for example, deep channel–small die. What combination of channel and die will require the lowest pressure drop? What combination will give the highest pressure drop? Prove your answers.

7-7 Rework Example 3-3 for 6 nylon at 500 K. The rheological data for the polymer are: $n = 0.65$ and $K = 2 \times 10^3$ N s^n/m^2.

7-8 Polyethylene is extruded at the rate of 0.54 kg/min with a mechanical power input of 5.62 kW. Find the rate of heat input in kW if the exit temperature and pressure of the polymer are, respectively, 428 K and 1.21×10^7 N/m^2. Comment on your results.

7-9 The output of an extruder can be changed by changing ϕ, the helix angle. Devise a method to determine the value of the optimum helix angle (value that gives the highest output). Show equations, plots, etc., used for this method.

7-10 Pumping efficiency (E) of a screw extruder is defined as

$$E = \frac{Q \Delta P}{\text{Power}}$$

where the denominator is the total power input to the extruder.
Develop a curve of E vs helix angle for an isothermal extruder.

7-11 Small-scale (0.05-m-diam) and large-scale (0.5-m-diam) extruders that are mechanical power–controlled give the following results:

	Small	Large
Out (kg/s)	3.16×10^{-4}	0.303
Power (kW)	0.216	182.7
Extrudate temperature (K)	442	429
Extrudate pressure (N/m^2)	1.28×10^7	1.10×107
Screw speed (rpm)	15	15

Deduce appropriate scaling factors (i.e., diameter ratio raised to a power) for output, power, temperature, and pressure. Hint: Powers should be whole numbers or zero.

7-12 Would you expect the scaling factors to be any different if thermal input was controlling the process? If so, how would these values compare to the ones of Problem 7-12? In other words, are they larger, smaller, or the same?

7-13 Fenner and Williams [12] found power data as a function of a dimensionless group A:

$$A = \frac{Q}{(\pi D_s N) \cos \phi \, hw}$$

The system was non-Newtonian fluid (n of 0.81 and K of 117.2 N s$_n$/m^2) processed in an extruder, where

$\phi = 36.6°$
$D_s = 0.0381$ m
$L = 0.305$ m
$h/w = 0.036$
$\delta/h = 0.053$

The experimental data are:

A (dimensionless)	Power (kW)
0.004	0.082
0.055	0.079
0.080	0.075
0.120	0.067
0.155	0.064

Is the system isothermal or adiabatic?

7-14 A molten polymer (n of 0.48 and K of 2.76 × 10^4 N s^n/m^2) has an output of 0.051 kg/s when processed at 464 K in an extruder whose dimensions are:

$D_s = 0.0089$ m
$\theta = 17.7°$
$h = 0.01$ m
$L = 0.51$ m
$N = 120$ rpm

Find the pressure developed in the extruder.

7-15 The melting section of an extruder has the following dimensions and operating parameters:

$D_B = 0.064$ m
$w = 0.053$ m
$h = 0.0094$ m
$N = 60$ rpm

Polymer data are:

C_{pm} = 2.6 kJ/kg K
k_m = 0.19 W/m°C
ρ_m = 860 kg/m³
λ_m = 130 kJ/kg
n = 0.34 (at 110°C)
K = 5.6 × 104 N sⁿ/m² (at 110°C)

Apparent viscosities can be corrected to higher temperatures by using the factor

$$\exp[-0.01(T - 110°C)]$$

Six turns are allowed to melt the polymer. Each turn has 4.4×10^{-2} m.
   If 0.019 kg/s of polymer is processed, are the six turns sufficient for melting?

7-16   Find x/w as a function of turn number in the preceding example.

7-17   Solid polyethylene pellets are fed to an extruder with the following parameters:

D_B = 0.076 m
h = 0.005 m
ϕ = 20°
w = 0.05 m
N = 75 rpm

The extruder is equipped with a 0.012-m-diam die 0.04 m long.
   Calculate pressure drop across the die. Clearly indicate all assumptions.

7-18   What would the maximum discharge pressure be for the extruder of Example 3-1.

7-19   If a polystyrene polymer (n = 0.26; K = 3 × 10⁴ N sⁿ/m²) is processed as in Example 3-2, what would the results be?

7-20   Which one of the following polymers would have the highest maximum output if processed in the extruder of Example 3-1, assuming that each is a melt?

Polymer	n	$K \, N \, s_n/m^2$
Polycarbonate	0.70	9 × 10²
Polymethyl methacrylate	0.22	7 × 10³
Polypropylene	0.40	3 × 10³

MINI PROJECT A

Using the approaches of Chapter 4, analyze the viscous dissipation effects found in an extruder. Select one of the extruders described in the Examples in the text or in the Problems of Chapter 7. What do you conclude as a result of your analysis?

MINI PROJECT B

A method of synthesizing a polymer is to carry out the polymerization in a screw extruder. Supposing that we carry out a polycondensation reaction (i.e., creating a small molecule by-product) in such a system, how would you go about designing such an extruder? Remember that chemical reactions depend on time and temperature and that the presence of the by-product molecule will retard reaction (i.e., reaction reaches an equilibrium that requires by-product removal). Remember, too, that the small molecule by-product will have a solubility in the polymer mass.

It is suggested that your first cut be an attempt to set up all aspects. Next, simplify your approach but only enough to facilitate a design. Assume second-order kinetics for the polymerization.

MINI PROJECT C

In some systems that are processed, such as vented extruders, volatiles are purged at various points along the extruder length. Carry out a mass-transfer analysis for small-molecule volatiles in a polymer that is being processed. Can such an analysis be useful in siting the venting systems? If so, illustrate how this can be done.

MINI PROJECT D

Polymer blends are frequently processed in screw extruders. Develop appropriate designs for such extruders if the blends are compatible (i.e., such as polyethylene–polypropylene). For example, consider a 50–50 blend.

MINI PROJECT E

Filled polymer systems (glass, carbon, inorganic) are frequently processed in screw extruders. How would you alter the design equations of Chapter 7 to handle such systems. Give a detailed and coherent response.

CHAPTER 8

INJECTION-MOLDING SYSTEMS

8.1 INJECTION MOLDING—A CYCLIC PROCESS

Although considered by some to be a batch process, injection molding is really a semicontinuous process. Further, it is also a cyclic process in which plastic granules are heated until melted or thermally softened, forced into a mold and then cooled to a desired shape.

Figure 8-1 shows a typical injection molding process unit. A portion of the unit is the driving system which forms a molten or thermally softened polymer mass and then forces it through a path shown schematically in Fig. 8-2.

There are four principal types of injection molding that are used to heat the plastic and then force the molten or thermally softened polymer into the mold. These include:

1. Single-stage plunger
2. Two-stage plunger-plunger
3. Two-stage screw-plunger
4. Reciprocating screw

The single-stage plunger (Fig. 8-3) is the oldest of all of the systems in use. It is activated by hydraulic pressure (up to 2.1×10^5 kPa). A heated torpedo is used to spread the plastic and to assist in softening or melting it. The plunger forces the resultant softened or molten plastic into the mold.

The two-stage plunger-plunger (Fig. 8-4) system uses one device to create

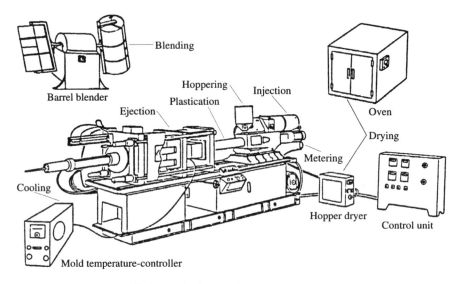

Fig. 8-1   Injection-molding process unit [1].

the molten or softened polymer mass and then force it into an injection cylinder. Stage two, the second plunger, then forces the mass into the mold.

Figure 8-5 shows a screw-plunger system, which uses a screw extruder to produce moldable plastic and move it to an injection chamber. The plunger position then forces the softened or molten mass into the mold.

The reciprocating screw unit (Fig. 8-6) is fed solid plastic through a hopper. When the mold is closed off, the plastic is melted, and a considerable back pressure is generated. This back pressure moves the screw to the rear and leaves a reservoir of molten plastic ahead of the screw.

The rearward motion of the screw sets off a switch, which causes the screw to move forward and force the melt into the mold.

As noted, the single-stage plunger was the earliest system used for injection molding. Initially, there were a large number of problems, such as leakage,

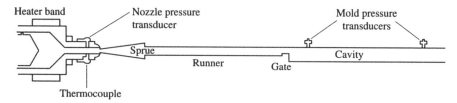

Fig. 8-2   Path of injection molded plastic [2].

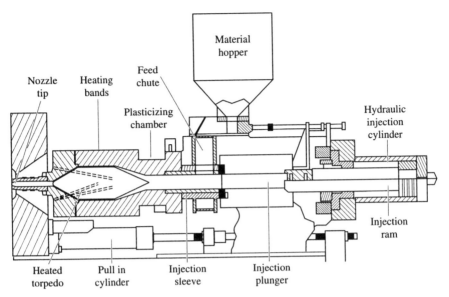

Fig. 8-3   Single stage plunger [3].

excessive pressure drops in the chamber, and difficulties with temperature problems. Most of these were cleared up by improved designs and controls. However, the single-stage plunger still had defects, such as long cycle times, low injection rates (volume/time), and high mold-clamping requirements.

The introduction of the plunger-plunger system alleviated all these difficulties, resulting in shorter cycle times, higher injection rates, and lower clamping requirements. Even so, color changes and problems with heat-sensitive materials remained until they were overcome with the introduction of the screw-plunger system.

Ultimately, the reciprocating screw system was developed. Adaptation of such units led to improved rates of softening or melting of the plastic, closer tolerances on short size (amount of plastic to mold), better control of temperatures and more reliable overall performance. They are also highly effective for vented operation.

8.2   THE PLASTIC FLOW PATH

As shown in Fig. 8-2, the plastic to be molded successively flows through a nozzle, sprue, runner, and gate on its way to the cavity (mold).

The first part of the flow system is the *nozzle*. This ultimately connects the

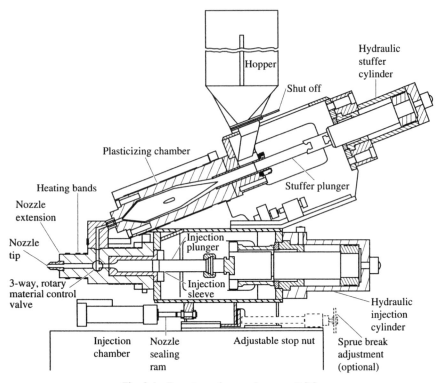

Fig. 8-4   Two-stage plunger-plunger unit [3].

injection chamber with the mold. Three basic types of nozzles can be used. These are:

1. A valve that can be closed from an outside source (i.e., the cylinder)
2. An open flow channel
3. A check valve that is kept closed by a spring (external or internal) and opened by pressure

Some nozzle types are shown in Figs. 8-7–8-9. Figure 8-7 is a version of a standard nozzle that is essentially a channel for flow. The device is equipped with a heating band to ensure that the plastic mass remains molten or thermally softened. These units can be made in more than one piece (i.e., with a removable tip).

Figure 8-8 depicts a needle shutoff valve that utilizes an external spring. When the injection pressure builds up, the needle is forced back, allowing the plastic to flow through the remaining channels (sprue, runner, gate).

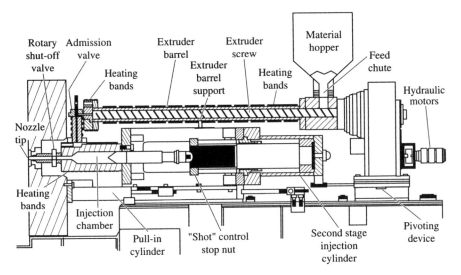

Fig. 8-5   Two-stage screw-plunger system [3].

A special type of nozzle is the one used for nylon (Fig. 8-9), which has a problem with drooling. This unit has a reverse taper that is combined with a carefully monitored energy input (heating band).

The *sprue*, which is a flow channel connecting the nozzle and runner (see

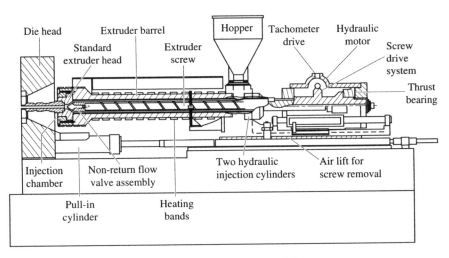

Fig. 8-6   Reciprocating screw [3]

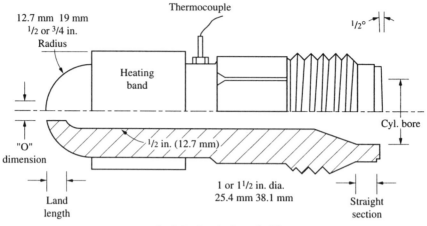

Fig. 8-7 Standard nozzle [3].

Fig. 8-10), is kept to as small a size as possible. This is done to minimize cycle times, which are extended by the longer cooling times required by large-diameter sprues. Generally, the sprue should involve only a small portion of the overall pressure drop in the nozzle-sprue-runner-gate-cavity system.

The *runner* or *runner system* is designed to allow for both rapid mold filling and minimum pressure loss. Various types of cross sections are used. Preferred cross sections are circular (full round) or trapezoidal since they both have minimum surface-to-volume ratios, which minimize heat losses and pressure drops. Figure 8-11 shows the relation between the dimensions of a trapezoidal runner.

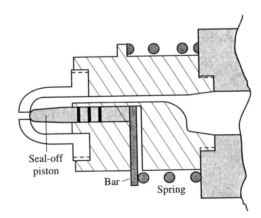

Fig. 8-8 Needle type shut off nozzle [3].

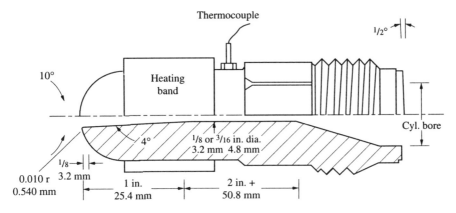

Fig. 8-9   Nylon nozzle [3].

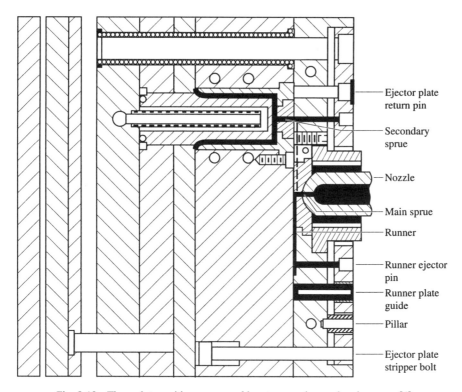

Fig. 8-10   Three plate positive runner mold, note secondary and main sprues [4].

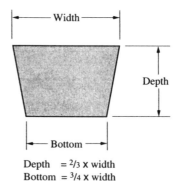

Depth $= 2/3$ x width
Bottom $= 3/4$ x width

Fig. 8-11   Trapezoidal runner cross section [4].

Multicavity units use runner systems (i.e., many runners). Such systems must be carefully designed to ensure that all mold cavities are filled at the same rate. If this is not done, short shots (incomplete mold filling) will occur. Some typical runners and cavity systems are shown in Fig. 8-12. Figure 8-13 contrasts effective and ineffective runner systems.

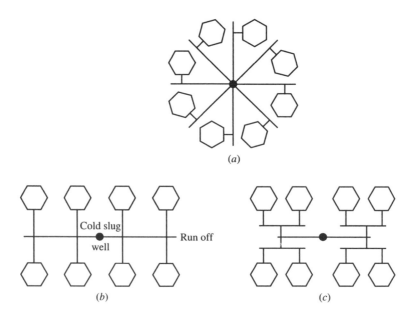

Fig. 8-12   Runner systems; (a) radial; (b) standard; (c) H design [4].

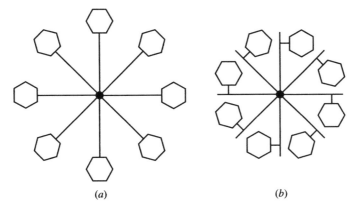

(*a*)   (*b*)

Fig. 8-13   Comparison radial systems; (a) ineffective; (b) effective [4].

The runner systems can be operated in either an insulated or heated mode. Insulated runner systems are preferred for larger objects, whereas heated systems are used to attain more rapid cycles.

Gating and *gates* are a very important part of the flow systems since they present a high resistance to flow. Proper gates allow the melt or thermally softened mass to reach the mold cavity quickly while minimizing energy and pressure losses. Further, the gate must allow the mold cavity pressure to be low at first and to increase rapidly. Another requirement is that the plastic in the gate rapidly solidifies once the mold is filled and up to pressure.

Figure 8-14 shows a variety of gate configurations. Table 8-1 summarizes the uses and advantages of each gate.

The mold cavity itself represents a blend of engineering and art. One important aspect is that there be sufficient venting capacity to allow trapped gases to escape during the filling of the cavity. Failure to do so will result in poor shaping or outright burning of the object.

Another important technical task is to design an appropriate cooling capacity for the mold. The object itself and its unsteady-state cooling are part of what must be considered. In addition, the convective and conductive heat transfer from the mold cavity must be dealt with correctly.

8.3   INJECTION-MOLDING PROCESS ANALYSIS

As mentioned previously, the molding process is a cyclic one that involves variation of both temperature and pressure with respect to time. Figure 8-15 shows a schematic of pressure with time for the injection-molding cycle.

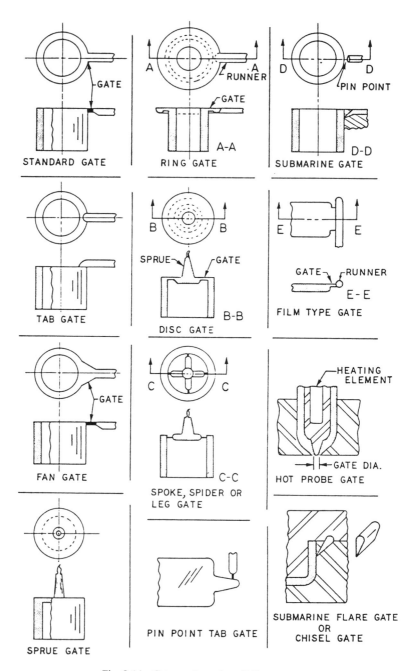

Fig. 8-14   Gate configurations [21]

TABLE 8-1 Gate Types

Gate	Uses/Advantages
Edge	For top, side, or bottom of part
Submarine	Allows automatic degating of part from runner system during ejection
Pinpoint	Permits automatic ejection
Disk	For objects with large cut-out areas; eliminates weld lines
Center	Similar to pinpoint but is larger; also gate extension is left in molded part
Fan	Useful for fragile sections and large-area objects
Ring	For cylindrical shapes
Tab	A small gate area that enhances frictional heating; useful for acrylics, ABS, and polycarbonates

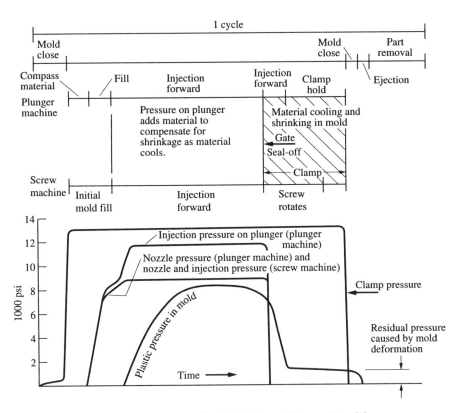

Fig. 8-15   Pressure profiles in injection molding machines [3].

The schematic shows four separate pressure–time traces for each of the plunger and screw machines. These traces include nozzle, injection clamp, and mold pressures as functions of time.

As can be seen, in both types of machines, the pressure in the mold does not begin to increase until the mold is filled. The plastic pressure then rises rapidly, levels off, and decreases only when the gate is sealed. During this latter part of the cycle, the pressure decreases rapidly as the material cools and shrinks. A small amount of residual pressure exists because of the deformation of the mold.

If the process is viewed from the aspect of the polymer, we first see the filling and packing of the mold. After this takes place, the plastic object cools and solidifies.

A proper and detailed analysis would require correctly following the flow of the molten or thermally softened polymer through the nozzle-sprue-runner-gate system into the cavity. This treatment is a complex and formidable task for the following reasons:

1. The polymer is non-Newtonian and possibly viscoelastic.
2. Viscous dissipation could be a sizable factor.
3. The system is nonisothermal and represents convective heat transfer.
4. Pressure and temperature change with time.
5. Polymer compressibility behavior will affect the process.
6. Rheological and thermal property (specific heat, thermal conductivity) behavior must be related to temperature and pressure.

Further, flow into the cavity itself represents an unsteady flow case with all of the foregoing still applicable.

In addition, analysis of flow and heat transfer in the nozzle-sprue-runner-gate system requires a knowledge of the dimensions of these elements. Such knowledge is, of course, not available if a system is being designed. The result is an intricate relation with the cavity filling aspect.

Even so, it is possible to use empirical or simplified approaches to gain understanding of the process. One such expression is shown in Eq. (8-1) [4], which allows filling times to be determined by using an approach based on the flow in a given geometry.

$$t_f = \beta \eta_0 e^{k(T-T^*)}(P_B)^{-\alpha} \tag{8-1}$$

where t_f is fill time, α and β are constants determined for the machine, T and T^* are the temperature and reference temperature, P_B the nozzle pressure, and k the constant for temperature effect on viscosity.

Another approach to filling is to relate the equation to the gate and mold cavity. An expression for case a in Fig. 8-16 is

$$L = \frac{4QT}{\pi D^2} \tag{8-2}$$

and, for case b,

$$R = \left(\frac{Qt}{\pi H}\right)^{1/2} \tag{8-3}$$

where L, R, and H are the indicated dimensions, Q is the volumetric flow, and t is the time.

Another simplified approach [7–9] was based on penetration speed into a straight cylindrical cavity. The results of these studies are shown in Fig. 8-17, which plots penetration speed against time. These data show that V penetration speed is, empirically,

$$V = V_0\, e^{-t/\beta} \tag{8-4}$$

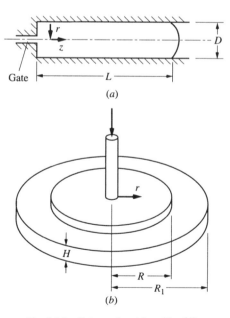

Fig. 8-16   Gates and mold cavities [6].

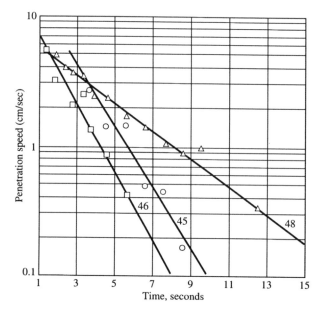

Fig. 8-17   Penetration speed versus time [9].

where V_0 and β depend on the mold design, operating conditions, and polymer properties.

For infinite time, the final penetration X would be

$$X = V_0\beta \tag{8-5}$$

Other attempts to describe mold filling quantitatively were the studies [10–13] that treated the flow as if it were isothermal. One set of results [10–12] yielded calculated results that gave curves of pressure vs injection rate that were displaced from experimental results. The other study [13] gave curves of pressure cavity thickness that matched experimental results. In addition, calculated penetration velocity–time data matched the results shown in Fig. 8-17.

The flow of the molten or thermally softened polymer mass into the mold cavity is, as previously indicated, a complicated matter. A schematic diagram of the flow pattern is shown in Fig. 8-18. Several aspects of the process merit discussion. First of all, the flow is complicated by the formation of a solidified polymer layer [15–17] at and near the mold wall. Next, the molten or thermally softened polymer flows from the cavity centerline to the front of the mass and then outward toward the wall. This behavior is termed a "fountain effect" [14,15].

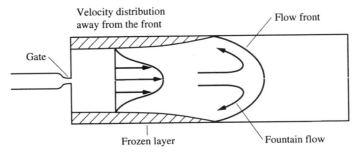

Fig. 8-18   Fountain flow [14].

In addition, the point of maximum shear stress occurs not at the interface of the solid polymer and flowing mass but rather a short distance away [14]. This results because the flowing system's highest temperature at the center gives the lowest apparent viscosity value. Cavity maximum shear rates often are in the range of 8000–15,000 s^{-1} [14].

Various studies have been directed to the filling and packing of molds. One such study that combined both theory and experiment was the work of Kamal and Kenig [18]. Their work involved the solution of the dimensionless equation [19] of continuity, motion, and energy by different methods for a power law fluid.

Figures 8-18 and 8-20 compare predicted (theoretical) and experimental data for the filling of a 180° disk cavity with edge flow. In Fig. 8-19, theoretically determined values of positions of the melt front show the same relative behavior with time as the experimental values (transducer and photographic). However, as can be seen, the theoretical values are considerably larger. Pressure changes with time and flow rate are given as functions of time in Fig. 8-20. The general theoretical prediction of pressure behavior qualitatively matches the trend of the experimental results but once more doesn't agree completely in a quantitative sense.

The import of the foregoing is that while such an analysis at the mold filling will yield a reasonable picture of the events, it does not completely satisfy the need for a quanitative evaluation.

Cooling of the plastic part in the mold is essentially a problem in unsteady state heat transfer. The equation describing this (if density and thermal conductivity are constant) is

$$\rho \, C_p \, \frac{\partial T}{\partial t} = k\left(\frac{\partial^2 T}{\partial x^2} + \frac{\partial^2 T}{\partial y^2} + \frac{\partial^2 T}{\partial z^2}\right) + A_0 \tag{8-6}$$

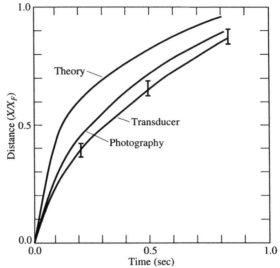

Fig. 8-19   Relative melt front positions as a function of time [18].

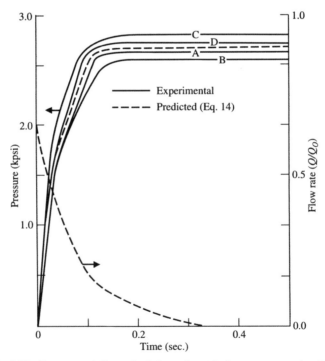

Fig. 8-20   Pressure variation and relative volumetric flow rate versus time [18].

for rectangular coordinates. Equation (8-6) is for the case of semicrystalline polymers, such as polyolefins and polyamides. If the polymer is amorphous, such as polystyrene or polymethyl methacrylate, A_0 (the term dealing with heat of fusion) can be neglected. This means that the unsteady-state analysis already used in Chapter 4 is applicable.

In discussing the cooling process, mention should also be made of another study by Kenig and Kamal [20], which combined a theoretical analysis and experimental data. The theoretical analysis was essentially an energy balance between the mold coolant, the mold, and the polymer. This approach included a term for polymer solidification (crystallization). The equations used (8-7–8-10), which follow, propose that there are four series resistances to heat transfer. First, the recirculating fluid (at constant temperature T_f) cools the molten polymer cylinder. Thus, there is heat transfer at the coolant/steel interface:

$$\left(\rho\, C_p\, \frac{\partial T}{\partial t}\right)_s = \left(k\, \frac{\partial^2 T}{\partial r^2}\right)_s + h_0(T_f - T_s) \tag{8-7}$$

Next, conduction occurs through the steel:

$$\left(\rho\, C_p\, \frac{\partial T}{\partial t}\right)_s = \left(k\, \frac{\partial^2 T}{\partial r^2}\right)_s \tag{8-8}$$

Convective heat transfer involves energy transfer through the steel/polymer interface

$$\left(\rho\, C_p\, \frac{\partial T}{\partial T}\right)_p = \left(k\, \frac{\partial^2 T}{\partial r^2}\right)_p + h_i(T_p - T_s) \tag{8-9}$$

and conduction in the plastic mass

$$\left(\rho\, C_p\, \frac{\partial T}{\partial t}\right)_p = \left(k\, \frac{\partial^2 T}{\partial r^2}\right)_p + \left(k'\, \frac{\partial^2 T}{\partial r^2}\right)_c \tag{8-10}$$

where the subscripts S and P stand for steel and polymer, respectively. The subscripted term C is a crystallization term to correct for any energy liberated by the crystallization.

Figure 8-21 shows the temperature profiles as functions of time and distance. These results show excellent agreement between the experimental and analytical results.

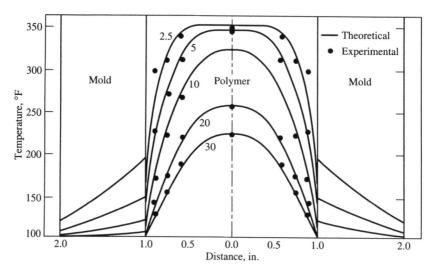

Fig. 8-21   Temperature profiles in molded object [20].

8.4   INJECTION-MOLDING PRODUCT PROBLEMS

Injection molding can give rise to a number of product problems. These include: short shots (incomplete filling of a mold), voids and sinks (holes in molded parts or insufficient material), weld lines and flow marks (e.g., seams and welding of cooler material around projections), sticking problems (mold or material adhesion), warping (bending or distortion of parts), burning, and shrinkage (reduction in size of molded parts).

Table 8-2 presents in chart form a troubleshooting guide for molding problems.

The problem of shrinkage is one that can be dealt with in a more quantitative manner. Basically, percent shrinkage is given by the expression

$$\begin{matrix} \text{Percent} \\ \text{shrinkage} \end{matrix} = S = \frac{V_0 - V_f}{V_0}(100) \qquad (8\text{-}11)$$

where V_0 is the volume of the total mass of the molded part at filling conditions (temperature and pressure) and V_f is the volume of the cooled part as it is removed from the mold (i.e., at final temperature and pressure).

For the above, the mass m of polymer injected is

$$m = \frac{V_0}{v_0} \qquad (8\text{-}12)$$

TABLE 8-2 Troubleshooting Guide for Molding Problems: +, Increase Variable; −, Decrease Variable

Solution Problem	Cylinder Temp.	Ram Pressure	Injection Time	Cycle Time	Mold Temp.	Gate Size	Cure Time	Other
Short shot	+	+	+	+	+			Open gates
Shrinkage	−	+	+	+	−	+		
Voids	−	+	+	+	+	+		Quench 100–130°F
Sinks	−	+	+	+	−	+	+	Increase injection speed
Weld lines, flow marks, poor surface	+	+			+			Increase venting; clean cavity
Cavity sticking		−			−		+	
Part sticking		−			+		−	
Warping	−	+	+	+	−	+	+	Increase ram-in time; increase injection speed
Burning	−	−						Increase venting; decrease feed
Poor additive dispersion	(Increase injection pressure)	(Decrease ram-in time)						

329

where v_0 is the specific volume of polymer at the filling temperature and pressure. Likewise the V_f is given by

$$V_f = m \, v_f \qquad (8\text{-}13)$$

where v_f is the specific volume of polymer when the mold is opened.

Hence, percent shrinkage is

$$S = \frac{v_0 - v_f}{v_0} (100) \qquad (8\text{-}14)$$

This means that percent shrinkage can be predicted from molding conditions, pressure-volume-temperature data, or use of an equation of state.

8.5   SPECIALIZED INJECTION-MOLDING PROCESSES

Two specialized injection-molding processes will be considered in some detail. These processes are structural foam and reaction injection molding. In the case of structural foam, a gas is distributed throughout the molded plastic to form a foam. When cooled, the product is a structural thermoplastic foam. The reaction injection-molding process involves mixing two reactive components and injecting them into a mold, where the product is formed.

Both of these processes share several advantages, including (1) the capability of producing large parts, (2) low process energy requirements, and (3) greater product flexibility.

Structural foam and reaction injection molding are alike in another respect: Both are especially complicated processes. For example, the structural thermoplastic foams can involve the combination of a chemical reaction (the decomposition of a chemical blowing agent), mass transfer (the diffusion of the gases generated from the decomposition of the blowing agent), and heat transfer from a viscoelastic non-Newtonian fluid that undergoes a complicated shear history. Systems that use physical blowing agents eliminate the chemical reaction step but still involve a complex amalgam of mass transfer, heat transfer, and rheology.

Structural foam molding can be accomplished with either a low- or high-pressure process. In the low-pressure process, a combination of a chemical blowing agent and the plastic are injected as a short shot into a mold. The gases derived from the blowing agent expand the system until the polymer fills the mold. Pressure is generated by the gases (200–600 psi). These molding operations can be carried out in conventional, two-stage, and special devices as shown in Figs. 8-22–8-24, respectively.

In the high-pressure process (Fig. 8-25), a full shot of the resin and blowing

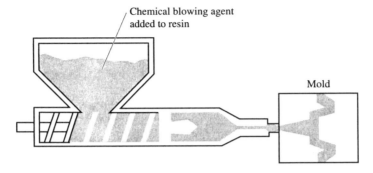

Chemical blowing agent
added to resin

Mold

Fig. 8-22 Conventional foam injection unit [21].

agent are charged to the mold. After a skin of solid plastic is formed, the mold is expanded to allow the gas to force the unsolidified plastic to the larger shape. The advantage of the high-pressure process is that it forms a better surface, with less splay and improved detail.

The pressure-temperature-time relation in the mold controls both the size and dispersion of the gases generated from the blowing agent. In the mass-transfer sense, both diffusion and solution play a role; diffusion in the movement of the gases in the resin, and solution in fixing the amount of gas retained in the object. A typical density profile for a structural foam is shown in Fig. 8-26.

Structural foams have significant density reductions, as great as 40%. They are more rigid and have a higher strength-to-mass ratio. The acoustical and insulating characteristics of the objects are also enhanced. However, tensile and impact strengths are reduced.

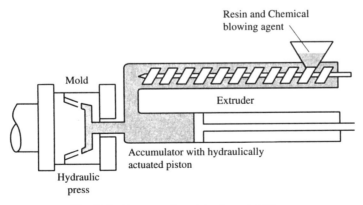

Resin and Chemical
blowing agent

Mold

Extruder

Accumulator with hydraulically
actuated piston

Hydraulic
press

Fig. 8-23 Two stage foam injection unit [21].

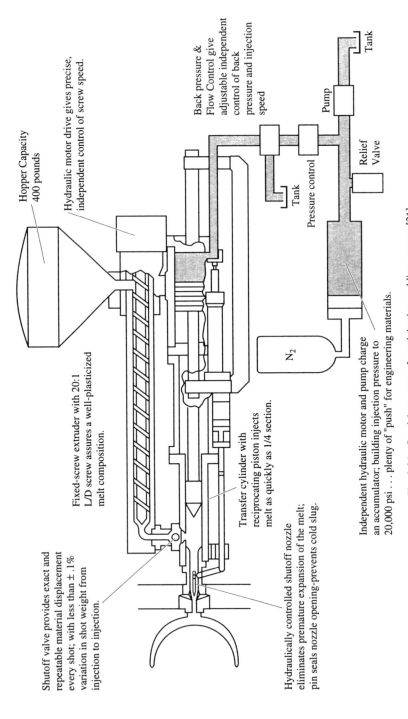

Hopper Capacity
400 pounds

Hydraulic motor drive gives precise,
independent control of screw speed.

Fixed-screw extruder with 20:1
L/D screw assures a well-plasticized
melt composition.

Back pressure &
Flow Control give
adjustable independent
control of back
pressure and injection
speed

Pump

Tank

Relief
Valve

Tank
Pressure control

N₂

Transfer cylinder with
reciprocating piston injects
melt as quickly as 1/4 section.

Shutoff valve provides exact and
repeatable material displacement
every shot; with less than ± .1%
variation in shot weight from
injection to injection.

Hydraulically controlled shutoff nozzle
eliminates premature expansion of the melt;
pin seals nozzle opening–prevents cold slug.

Independent hydraulic motor and pump charge
an accumulator; building injection pressure to
20,000 psi . . . plenty of "push" for engineering materials.

Fig. 8-24 Special two stage foam injection molding process [21].

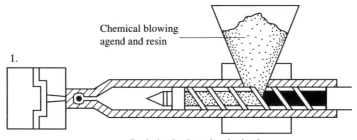

1.

Chemical blowing
agend and resin

Resin is plasticated and mixed
with chemical blowing agent.

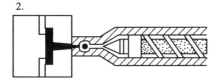

2.

Mold filled–full pressure–full shot temperature
rise to decompose chemical blowing agent as
melt passes through nozzle.

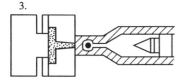

3.

Platen retracted–chemical blowing agent
expands resin to fill enlarged cavity.

Fig. 8-25   High pressure foam operation [21].

As mentioned previously, the reaction injection-molding process combines two precisely metered and well-mixed reactive streams. An example is the process that reacts catalyzed highly reactive streams of urethane components. One stream contains a polyether backbone, a catalyst, and a cross-linking agent. The other stream has an isocyanate. In addition, a blowing agent is included in one of the streams.

A typical process (see Fig. 8-27) uses a high-pressure metering system that combines the streams in a mixing head by impingement. Static mixers in the runner system to the mold produce additional mixing. After the materials are injected into the mold, the blowing agent expands the shot to fill the mold.

The polymer principally used in reaction injection molding is thermoset (cross-linked) polyurethane. Nylon (thermoplastic) and polyesters and epoxies (both thermosets) are also processed in this manner.

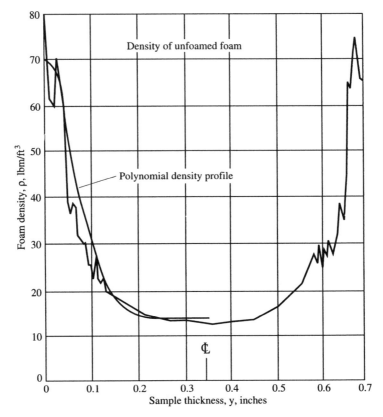

Fig. 8-26   Structural foam density profile [22].

Inherent complexities in both structural foam and reaction injection molding (e.g., chemical reaction, heat and mass transfer, complicated rheological behavior, pressure and temperature changes) preclude entirely satisfactory technical analyses of the processes. Qualitative or semiquantitative approaches, however, can be used to produce suitable products.

8.6   INJECTION MOLDING EXAMPLES

Exhibit 8-1

A polymer used in an injection-molding process has the following operating data for a given mold (P_B is nozzle pressure, η_0 is the polymer Newtonian

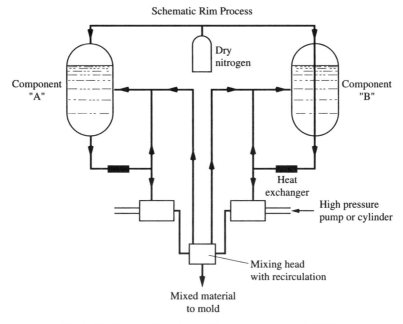

Fig. 8-27   Reaction injection molding process schematic [21].

viscosity at the temperature immediately prior to the mold, and t_f is full time).

P_B, psi	η_0, poise	t_f, s
14,000	62,500	6.7
12,000	12,960	2.7

What would the fill times be for the following P_B and η_0?

P_B, psi	η_0, poise
19,000	38,600
19,000	210,000
16,000	80,500
14,000	14,500
10,000	12,960

In order to estimate the fill times, we can use a modified form of Eq. (8-1):

$$t_f = \beta \eta_0 \, P_B^{-\alpha}$$

This equation does not use the temperature-dependent term of Eq. (8-1). Generally, in process work, the α (exponent for P_B) is found to be between 4 and 5.

Writing the expressions for the fill times and other data gives:

$$6.7 = \beta(62,500)(14,000)^{-\alpha}$$

$$2.7 = \beta(12,960)(12,000)^{-\alpha}$$

Dividing one by the other eliminates β and gives

$$\frac{6.7}{2.7} = \left(\frac{62,500}{12,960}\right)\left(\frac{14,000}{12,000}\right)^{-\alpha}$$

The α term can then be found. In this case, $\alpha = 4.3$. Next, we can determine β. This value is found to be 7.22×10^{13}.

Fill times can now be estimated for the nozzle pressure–viscosity data. The results follow, together with observed fill times (all t_f are in seconds):

P_B psi	η_0 poise	t_f (Calc.)	t_f (Obs.)
19,000	38,600	1.05	1.20
19,000	210,000	5.75	6.70
16,000	80,500	4.60	4.70
14,000	14,500	1.49	1.20
10,000	12,960	5.88	6.70

The results show that the calculated values, while generally below the observed values, are still reasonably close (average deviation of 13%). As such, the modified equation offers a reasonable method for estimating fill time.

Example 8-2

Process data [21] are available for injection molding of 66 nylon (see Fig. 8-28). For the case of a 2000-psi cavity pressure, check the moldability relationship, that is, ln fill velocity vs fill time, discussed earlier [Eqs. (8-4) and (8-5) and Fig. 8-17].

To obtain the needed data, we proceed in the following manner: First, we select a part thickness that gives a fill time directly. Next, we read across to the 2000-psi curve and obtain the t/L ratio (thickness of part) to cavity length. The fill velocity can be estimated by dividing cavity length by fill time.

As an example, let the part thickness be 0.04 in. The t/L ratio is obtained by

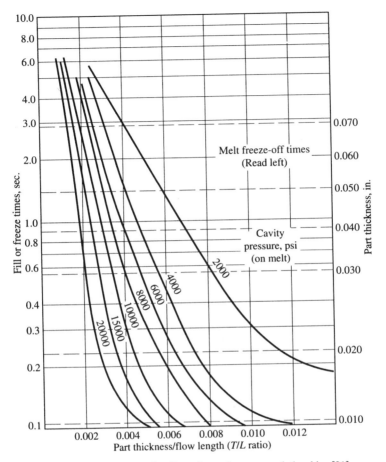

Fig. 8-28   Fill time, cavity dimension and pressure relationships [21].

reading to the 2000-psi curve. This gives a value of 0.007. Then,

$$L = \frac{t}{0.007} = \frac{0.04 \text{ in.}}{0.007} = 5.714 \text{ in.}$$

The fill time corresponding to a 0.04-in.-thick part is 0.9 s. This allows a calculation of fill velocity:

$$\text{Fill velocity} = \frac{5.714 \text{ in.}}{0.9 \text{ s}} = 6.349 \text{ in./s}$$

Repeating the process gives a table of data (fill velocities and fill times).

Fill Velocity v, in./s	ln$v$	Fill Time t, s
6.39	1.855	0.3
6.67	1.898	0.6
6.35	1.848	0.9
5.95	1.783	1.5
5.88	1.772	2.0
5.77	1.753	3.0

Figure 8-29 plots the ln$v$ vs t. As shown, the plot verifies the moldability relations described earlier.

Example 8-3

Estimate the shrinkages for objects molded from polystyrene and polymethyl methacrylate for the following molding conditions:

Mold Pressure, psi	Mold Temperature, °F
5,000	500
10,000	500
20,000	500

Assume final conditions to be 5°C below the polymer heat-distortion temperatures and atmospheric pressure. The heat-distortion temperatures are 91°C for polymethyl methacrylate and 89°C for polystyrene.

To determine the volumetric behavior, we use the equation of state for amor-

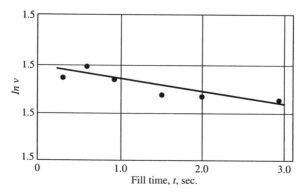

Fig. 8-29   Logarithm fill velocity versus time.

phous or solid semicrystalline polymers:

$$V = \frac{(0.01205)}{(\eta_0)\ 0.9421}\ P^{n-1}\left(\frac{T}{T_g}\right)^{m+1} R$$

given in Chapter 2.

Note that the ratio of final volume to molded volume is given by

$$\frac{V_{\text{final}}}{V_{\text{molded}}} = \frac{(P_{\text{final}})^{n-1}}{(P_{\text{molded}})^{n-1}}\ \frac{(T_{\text{final}}/T_g)^{m+1}}{(T_{\text{molded}}/T_g)^{m+1}}$$

The terms involving m and n cannot be combined because these exponents are functions of pressure.

Doing a sample calculation for polystyrene (T_g = 373 K) at the molding condition of 20,000 psi and 500°F, we first obtain n and m from the plots given in Chapter 2.

Condition	n	m
Molding	0.9954	−0.942
Final	1.00	−0.914

Therefore,

$$\frac{V_{\text{final}}}{V_{\text{molded}}} = \frac{(1)^{\circ}}{(1360)^{0.0046}}\ \frac{(367/373)^{0.086}}{(533/373)^{0.058}} = 0.9462$$

$$\% \text{ Volume shrinkage} = S = 100(1-0.9462) = 5.38\%$$

This calculation can be repeated for the other molding conditions. For polymethyl methacrylate, the only difference will be T_{final} (86°C) and T_g (378°C).

Volume ratios for each polymer are given below for each molding pressure. (Remember that molding temperatures do not vary.)

Mold Pressure, psi	Polystyrene, $V_{\text{final}}/V_{\text{molded}}$	Polymethyl Methacrylate, $V_{\text{final}}/V_{\text{molded}}$
5,000	0.9648	0.9654
10,000	0.9600	0.9603
20,000	0.9462	0.9463

These values are similar because the polymer glass and heat-distortion temperatures are very close in value.

Example 8-4

A high-density ($\rho_0 = 0.958$ g/cm^3) polyethylene is used for injection molding. The polymer is molded at 1800 psi and 481°F. If the mold temperature is 90°F, what is the volume shrinkage?

In this case, the polymer differs from those in the preceding example since polyethylene is semicrystalline and not amorphous. The volumes can be determined by two different techniques. The first is to use available pressure-volume-temperature data (see Figs. 2-1–2-8). Another method is to calculate the volumes from an equation of state and the volume charge correlation of Chapter 2.

Method 1:

The specific volume on filling (at 1800 psi, 481°F) can be obtained from Fig. 2-1 for a pressure of 122.5 atm and a temperature of 250°C.

$$v_0 = 1.371 \text{ cm}^3/\text{g}$$

Likewise, for the 90°F case (and atmospheric pressure),

$$v_f = 1.087 \text{ cm}^3/\text{g}$$

The percent volume shrinkage is

$$S = \frac{1.371 - 1.087}{1.371} 100 = 20.7\%$$

Method 2:

The v_f can be obtained using the solid-region equation of state [Eq. (2-22)]. If we divide the expression for v_f by that for the v_1 (at 298 K and 1 atm), we obtain

$$v_f = v_1 \left(\frac{T_f}{T_1}\right)^{m+1}$$

$$v_f = (1.0438) \frac{305.6}{298}^{0.086}$$

$$v_f = 1.046 \text{ cm}^3/\text{g}$$

Next, to get v_0, we first need the v_m (the solid volume at the polymer melt temperature of 139°C):

$$v_m = v_1 \left(\frac{T_m}{T_1} \right)^{m+1}$$

$$v_m = (1.0438) \frac{412}{298}^{0.086}$$

$$v_m = 1.073 \text{ cm}^3/\text{g}$$

The change in volume on melting is given by Eq. (2-24).

$$\Delta v_m = 0.19 \frac{T_m}{298} V_w$$

The V_w value for polyethylene is 20.46 cm³/mole. Using this with a value of 28 g/mole (polyethylene mer weight), we get a Δv_m of 0.1919 cm³/g.

This gives a v_2 in the melt of $(1.073 + 0.1919)$ cm³/g, or 1.2649 cm³/g.

Then, the fill volume is found from using the equation of state for molten polymer systems (2-23).

$$V = K \left(\frac{T}{T_g} \right)^x P^y$$

The v_0 is

$$v_0 = v_2 \frac{(T_f/T_g)^{x1}}{T_2/T_g} P^x \frac{(P_f)^{y1}}{(P_2)^y}$$

The primed x and y values indicate that they are different because of the difference in pressures for v_0 (122.5 atm) and v_2 (1 atm).

Using values of x and y from the charts in Chapter 2 ($x^1 = 0.262$, $y^1 = 0$, $x = 0.28$, $y = 0.0002$). The y values essentially made the p^y terms unity.

$$v_0 = 1.2649 \text{ cm}^3/\text{g} \frac{(523/330)^{0.212}}{(412/330)^{0.280}}$$

$$v_0 = 1.341 \text{ cm}^3/\text{g}$$

The percent volume shrinkage is then

$$S = \frac{1.341 - 1.046}{1.341} 100 = 22\%$$

Shrinkage is sometimes expressed as a linear dimension (i.e., inches/inch). In order to obtain linear shrinkage (L.S.) we do the following:

$$\text{L.S.} = \left(1 - \frac{v_f}{v_0}\right)^{1/3} = 0.074$$

For the two cases considered,

$$\text{L.S.} = \left(1 - \frac{1.087}{1.371}\right)^{1/3} = 0.074$$

$$\text{L.S.} = \left(1 - \frac{1.046}{1.341}\right)^{1/3} = 0.079$$

That is, the linear shrinkage would be 0.074 to 0.079 in./in.

The variation in the two methods is slight, which makes possible reliable estimates of shrinkage for polymers for which there is no experimental *P-V-T* data.

Example 8-5

A high-density ($\eta_0 = 0.958$ g/cm^3) polyethylene is molded at 1800 psi and 525°F. What is the cooling time required in the mold for a 0.2-cm-thick slab (the other two dimensions are 5.08 cm)? Assume that the polymer centerline temperature is 90°F. Mold temperature is 80°F.

This is a situation involving a semicrystalline polymer. If we assume the molded object to be a large slab, then the governing equation is

$$\rho \, C_p \frac{\partial T}{\partial t} = k \frac{\partial^2 T}{\partial x^2} + A_0$$

where x is the dimension measured from the slab centerline (0.1 cm from the outside); x at the centerline is zero.

The solution of the preceding equation requires that the relationship of A_0 to pressure, temperature, and time be known. Such information is not available. As a result, the preceding equation cannot be solved directly.

It is possible, however, to carry out an analysis that would convey the impact of the solidification term. The analysis will involve the following steps:

1. Find the time required to bring the average temperature of the molten mass to 281.6°F (the polymers T_m).
2. Determine the time for the solidification of the mass at 281.6°F.
3. Calculate the time needed to bring the solidified polymer's centerline temperature to 90°F.

For the first case, using Fig. 4-9,

$$y = \frac{80 - 281.6}{80 - 525} = 0.453$$

Assuming that $m = 0$,

$$x = 0.28 \frac{\alpha\theta}{r^2 m}$$

$$\theta = \text{time} = \frac{(0.28)(0.1 \text{ cm})^2}{1.169 \times 10^{-3} \text{ cm/s}^2}$$

$$\theta = 2.40 \text{ s}$$

In case 2, we use a limiting solution for solidification, presented in the following form [21]:

$$Z = 2\lambda(\alpha t)^{1/2}$$

$$\lambda e^{\lambda^2} \text{ erf } \lambda = \frac{C_p T_m}{L\pi^{1/2}}$$

where Z is total slab thickness, L the heat of fusion, α thermal diffusivity, and T_m the melting temperature; λ is a mathematical function. Figure 8-30 gives a relation between λ and λe^{λ^2} erf λ. All properties are for the solid.

Using the value of heat of fusion from Fig. 2-17 and the C_p for polyethylene, we obtain a value of 0.71 for λ from Fig. 8-22:

$$t = \frac{Z^2}{2\lambda} \frac{1}{\alpha}$$

$$t = 9.61 \text{ s}$$

Finally, for the solid polymer (case 3), using Fig. 4-7,

$$y = \frac{80 - 90}{80 - 281.6} = 0.0496$$

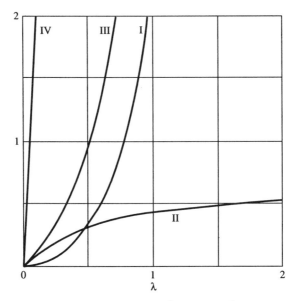

Fig. 8-30   Relation of various parameters to λ: I, $\lambda e^{\lambda^2} \operatorname{erf} \lambda$; II, $\lambda e^{\lambda^2} \operatorname{erfc} \lambda$; III, $\lambda e^{\lambda^2} (1 + \operatorname{erf} \lambda)$; IV, $\lambda e^{\lambda^2} (20 + \operatorname{erf} \lambda)$ [23].

and, with *m* and *n* both zero,

$$x = 1.18$$

$$\theta = \frac{r^2 m\, X}{\alpha} = \frac{(0.01 \text{ cm})^2\ (1.18)}{(2.01 \times 10^{-3} \text{ cm/s}^2)}$$

$$\theta = 5.87 \text{ s}$$

The total cooling time for the polyethylene centerline to reach 90°F is then $(2.40 + 9.61 + 5.87)$ s, or 17.88 s.

If the melt is treated as a case of unsteady-state cooling directly (Fig. 4-7), the calculated cooling time is 13.86 s. This, of course, neglects the thermal effects of the heat of solidification.

REFERENCES

1.   Harper, C.A., *Chemical Engineering*, May 10 (1976).
2.   Frizelle, W.G., and Paulson, D.C., *SPE ANTEC 14*, 405 (1968).
3.   Rubin, I., *Injection Molding Theory and Practice*, Wiley, New York (1972).

4. Thayer, G.B., Mighton, J.W., Dahl, R.B., and Beyer, C.E., in *Processing of Thermoplastic Materials*, E. Bernhardt (ed.), Reinhold, New York (1959), Chap. 5.
5. Schott, N., "Plastics Processing," presentation for Plastics Institute of America (1988).
6. Fenner, R.T., *Principles of Polymer Processing*, Chemical Publishing, New York (1980).
7. Ballman, R.L., Shusman, L., and Toor, H.L., *Ind. Eng. Chem.* **51**, 847 (1959).
8. Ibid., *Modern Plastics* **36**(9), 105 (1959).
9. Ibid., **36**(10), 115 (1959).
10. Barrie, I.T., *Plast. Polymers* **37**, 463 (1969).
11. Ibid., **38**, 47 (1969).
12. Barrie, I.T., *J. Soc. Plast. Eng.* **27**, 64 (1971).
13. Harry, D.H., and Parrot, R.G., *Polymer Eng. Sci.* **10**, 209 (1970).
14. Dealy, J.M., and Wissbrun, K.F., *Melt Rheology and Its Role in Plastics Processing*, Van Nostrand Reinhold, New York (1990).
15. Rose, W., *Nature* **191**, 242 (1961).
16. Janeschitz-Kriegl, H., *Rheol. Acta* **16**, 327 (1977).
17. Van Vijngaarden, Dijksman, J.F., and Wesselling, P.J., *Non-Newtonian Fluids* **11**, 175 (1982).
18. Kamal, M.R., and Kenig, S., *Polymer Eng. Sci.* **12**, 302 (1972).
19. Bird, R.B., Stewart, W.E., and Lightfoot, E.N., *Transport Phenomena*, Wiley, New York (1960).
20. Kenig, S., and Kamal, M.R., *SPE J.* **26**, 50 (1970).
21. Rosato, D.V., and Rosato, D.V., *Injection Molding Handbook*, Van Nostrand Reinhold, New York (1986).
22. Throne, J.L., and Griskey, R.G., *Polymer Eng. Sci.* **15**, 122 (1975).
23. Carslaw, H.S., and Jaeger, J.C., *Heat Conduction in Solids*, Oxford Univ. Press, London (1959).

PROBLEMS

8-1 A young engineer's first attempt at injection-molding a part was not too successful since the part was a short shot, and had obvious shrinkage and considerable sticking. What would you advise in order to better the process?

8-2 Data for a given polymer used in a series of injection-molding experiments are [24]:

C_p = 1880 J/kg K
k = 9.6 $\times$ 10^{-2} W/m K
ρ = 1300 kg/m^3
n = 0.50
K = 4 $\times$ 10^4 N sec^n/m^2

Also, a value for E of 27.8 kcal/g mole is available for the temperature effect in apparent viscosity.

$$\frac{\eta_{APP2}}{\eta_{APP1}} = \exp \frac{E}{R} \frac{T - T_2}{T_1 T_2}$$

where the temperatures are in degrees Kelvin.

Find α and β values for Eq. (8-1) or the modified version in Example 8-1.

Additional data are:

P, N/m^2	T, K	Full time, s
5.52×10^7	483	1.3
8.3×10^7	475	2.0
5.52×10^7	478	4.0
4.14×10^7	478	6.4

8-3 A molten polymer at 295°C is injection-molded into a circular runner (diameter of 0.05 m). The runner wall temperature is kept constant at 90°C. Estimate the solidification profile for the flowing polymer and a temperature distribution across the runner diameter. Polymer properties are:

$k = 6 \times 10^{-4}$ cal/cm^2 s
$C_p = 0.8$ cal/g
$\rho = 1100$ kg/m^3
$n = 0.65$
$K = 2 \times 10^3$ N sec^n/m^2

8-4 An ABS polymer [25] has the following characteristics (at 538 K):

$\rho = 1020$ kg/m^3
$C_p = 2430$ J/kg K
$n = 0.29$
$K = 1.875 \times 104$ N sec^n/m^2

The following pressure–flow rate data are available for a disk-shaped mold (thickness of 0.004 m):

P, N/m^2	Q, m^3/s
4.8	30
4.4	50
4.2	100
4.1	200
4.0	300

Does Eq. (8-4) hold for these data if the diameter of the disk mold is 21 cm? If the diameter is 6.9 cm?

8-5 Fit Eq. (8-1) to the data of Problem 8-4. Assume that T^* is 538 K.

8-6 Use the data of Figs. 8-28 and 8-29 to generate a curve of pressure vs volumetric flow rate.

8-7 A cylindrically shaped object (0.02 m in diameter, 0.10 m long) is molded at 220°C from polystyrene. The mold walls are at 28°C, and the polymer is to be cooled to 40°C at its centerline. How long should the object be retained in the mold?

8-8 Polypropylene (ρ_0 of 500 kg/m³) is molded at 1.17×10^7 N/m² and 550 K. What mold cooling time is required for a slab (0.15 cm) to reach a centerline temperature of 40°C? The mold wall temperature is 26°C. Also, find the slab's average temperature at the end of the cooling cycle.

8-9 A 610 nylon (ρ_0 of 1060 kg/m³) is molded at 1.2×10^7 N/m² and 585 K. If the mold temperature is 36°C, what is the volume shrinkage?

8-10 Estimate the volume shrinkage of an object made from a 50–50 blend of polyethylene and polypropylene. Molding conditions are 1.1×10^7 N/m² and 530 K. The mold temperature is 30°C.

8-11 Suppose the polyethylene of Example 8-5 is molded with nitrogen gas present (i.e., to form a foamed plastic). Estimate the density profile, porosity, and distribution of nitrogen within the molded object. Rheological properties for the polyethylene are $n = 0.6$ and $K = 5 \times 10^3$ N sec^n/m².

8-12 A polymeric material obeys the apparent viscosity–temperature relationship shown in Problem 8-2 (ΔE of 28 kcal/g mole). The rheological values at 463 K are $n = 0.52$ and $K = 4.1 \times 10^4$ N sec^n/m².

   The material is injection-molded in a circular system 0.02 m in diameter and 0.10 m long. If the polymer degrades at 225°C, what is the limit on its injection rate?

   Additional data for the polymer are a density of 1310 kg/m³ and specific heat of 1885 J/k K.

8-13 How will viscous dissipation affect the relation between pressure drop and flow rate in an injection-molding system?

8-14 Consider the polymer in Problems 8-4 and 8-5. If, originally at 538 K, the polymer flows through a system made up of two circular portions (the first 0.015 min diameter, 0.4 m long and the second 0.008 min diameter, 0.05 m long), what will its average temperature be? Take the wall temperature to be 545 K.

8-15 Develop plots of shrinkage vs injection pressure part thickness, mold temperature, and polymer injection temperature. Justify your answers.

8-16 Suppose a system of the type shown in Fig. 8-31 was used in an injection-molding operation for a power law fluid. Develop a relation between lengths and radii for balanced filling.

MINI PROJECT A

Injection-molded parts are sometimes made with infusible particulate matter in the flowing system (glass spheres, carbon particles, mineral fillers, etc.). Develop appropriate relations for the shrinkage behavior of such system as a function of particle size and percent composition.

MINI PROJECT B

Polymer blends are frequently injection-molded. If two polymers form a compatible blend, show how the various equations in Chapter 8 must be altered to describe the process properly (let one polymer be *A*, the other *B*).

MINI PROJECT C

Suppose an injection mold was badly vented (i.e., gases could not easily escape). For this system, develop appropriate relations to show the effect of such gases on the molded part. Apply your relations to the system of Problems 8-4 and 8-5 for varying amounts of gases.

MINI PROJECT D

Develop appropriate equations to describe the molding process shown in Fig. 8-27. By making appropriate assumptions, simplify these results to give process engineers reasonable means to handle such systems. Justify all assumptions.

CHAPTER 9

BLOW MOLDING, ROTATIONAL MOLDING, AND OTHER MOLDING OPERATIONS

9.1 OTHER MOLDING OPERATIONS

A number of molding operations other than injection molding are used in polymer processing. These include blow molding, rotational molding, transfer molding, and compression molding. Some of their advantages and disadvantages, together with those of injection molding, are listed in Table 9-1. Each of these processes will be discussed in the succeeding sections.

9.2 COMPRESSION MOLDING

A very old polymer-processing operation, compression molding was adopted because of the large impact of thermosetting resins, such as phenolics, in the marketplace. In about 1910, thermosets were about 95% of total plastics production. Later, however, increased production of thermoplastic polymers led to the use of other polymer-processing operations. Today, only about 3% of plastics are thermosets.

Basically, compression molding consists of forcing a combination of a resin and a curing, or cross-linking, agent into a mold by means of pressure (Fig. 9-1). Heating and the applied pressure cause the system to cross-link, which results in thermosetting.

Compression-molding operations frequently make use of preheating, which enables the cycle time to be reduced. Such preheating can be brought about by

TABLE 9-1   Advantages and Disadvantages of Molding Operations

Process	Advantages	Disadvantages
Injection molding	Rapid production rates; low per part cost; large complex shapes possible; good dimensional accuracy; good surface finish	High tool and die costs; not for small runs
Blow molding	Low tool and die costs; complex hollow shapes possible; rapid production rates	Limited to hollow shapes; tolerances hard to control
Rotational molding	Low mold cost; large hollow shapes possible; isotropic molded parts	Hollow parts only; very slow production rates
Transfer molding	Rapid production rates; good dimensional accuracy; very intricate parts possible	Expensive molds; large material loss in sprues and runners; size limitations
Compression molding	Low waste; low finishing cost; useful for bulky parts	Poor tolerances; not useful for intricate shapes

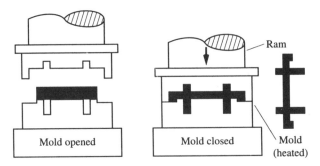

Fig. 9-1   Schematic of compression molding [1].

a variety of means: heated surfaces, infrared lamps, convection, and dielectric heating. The last technique is favored because it quickly and evenly heats (preforms) the molding materials.

Fillers and reinforcements are widely used in compression-molded systems. These materials enhance the molded object in a variety of ways, such as shrinkage reduction, lessened crazing, improved color, and finish. Also, very importantly, the costs of materials is reduced. Widely used fillers include clays and minerals of various types.

Although an old process, compression molding is not well described technically or analytically in the literature. Basically, what is needed is an analysis that combines rheology, heat transfer, chemical reaction (the curing process), and even mass transfer (outgassing) in a dynamic (time-dependent) system. The situation is also complicated by changing temperatures and pressures.

One analysis that has been used involves applying the dimensionless equations of energy and continuity of species to a slab [2]. Calculated reduced temperature curves vs reduced times are shown in Fig. 9-2 [3]. The T^* and t^* terms are defined as

$$T^* = \frac{T - T_o}{T_0 - T_{ad}} \tag{9-1}$$

$$t^* = \frac{xt}{(h')^2} \tag{9-2}$$

$$y^* = \frac{y}{h} \tag{9-3}$$

where T_o is the original temperature, T_{ad} the adiabatic temperature (value if system is adiabatic), x the system's thermal diffusivity, h' the heat-transfer coefficient between the mold and the polymer, h the half-thickness of the slab, and y the distance from the slab centerline.

Figure 9-2 relates the reduced temperature to time for the centerline ($y^* = 0$), a distance halfway between the centerline and wall ($y^* = 0.50$), and a distance three-fourth of the half-thickness ($y^* = 0.75$). Crosses are gel points for the system (see Chapter 6). The dotted curve represents an adiabatic system. Note that the centerline temperature tracks the initial part of the adiabatic curve quite well. This occurs because of a combination of a large heat of reaction and poor polymer heat transfer.

Two other important aspects of compression molding have to do with outgassing and postcuring. During the thermosetting process, gases can and are generated, as they are, for example, for phenolics and ureas. Because of this, the mold is opened slightly at the start of the curing reaction. Failure to outgas properly will blemish the part or actually weaken it.

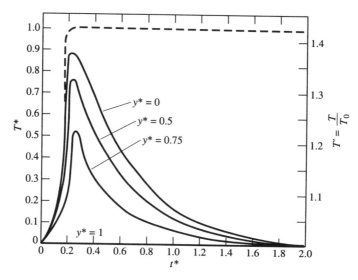

Fig. 9-2   Behavior of temperature in a slab of reacting polyurethane. Centerline is $y^* = 0$. Distances halfway and three-fourths of the half-thickness are $y^* = 0.5$ and $y^* = 0.75$ [3].

Postcuring involves heating the molded object, not only for additional out-gassing but also for improved object properties and dimensional stability.

9.3   TRANSFER MOLDING

Transfer molding (Fig. 9-3) differs from compression molding in that the mix of resin and curing agent is first raised in both temperature and pressure in a

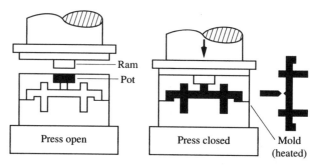

Fig. 9-3   Schematic of transfer molding [1].

holding device called a plasticizing pot or loading well. The slushy mix is then forced into the mold.

This process can also be performed with a screw extruder. There are two ways of doing this. One, the screw-transfer method, uses a plasticating extruder, which not only replaces the preheat and preform portions of the cycle but also loads the pot. Another way of providing the material to be molded is to use a screw injection-molding unit.

Outgassing in transfer molding is accomplished by venting the mold. This is done by using orifices at the parting line of the mold. A typical size is 0.64 cm wide and 0.0025–0.0075 cm deep.

As evident from Table 9-1, there are distinct differences between compression and transfer molding. Table 9-2 illustrates process differences between the two operations.

9.4 BLOW-MOLDING PROCESSES

Blow molding is a polymer-processing operation that produces hollow objects. The large volume of products produced in such operations places blow molding third among polymer-processing categories (exceeded only by extrusion and injection molding).

There are three principal types of blow-molding operations: extrusion blow molding (Fig. 9-4), injection blow molding (Fig. 9-5), and stretch blow molding.

About three-fourths of all blow-molding production is by extrusion blow molding. In this case, a hollow tube of molten or thermally softened polymers, called the parison, is extruded. The parison in this process is usually not supported. After leaving the die at a set parison length, a split cavity mold closes around the parison and crimps one end. A so-called blow pin (opposite the crimped or closed end) is used to inject compressed air, which blows the parison

TABLE 9-2 Process Differences Between Compression and Transfer Molding

Process Variable	Compression Molding	Transfer Molding
Molding temperature	143–232°C	143–182°C
Outgassing	Opening and shutting mold	Venting
Cure time	30–300 s	45–90
Possible size of molding	Limited by press capacity	454 g
Insert use	Limited	Unlimited
Shrinkage	Low	Greater than compression molding

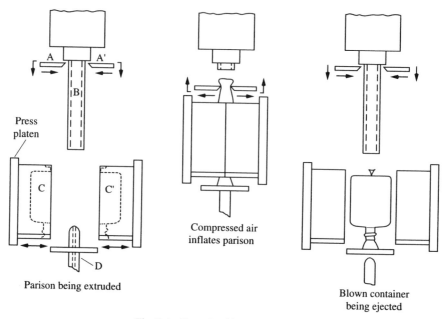

Press platen

Parison being extruded

Compressed air inflates parison

Blown container being ejected

Fig. 9-4   Extrusion blow molding [1].

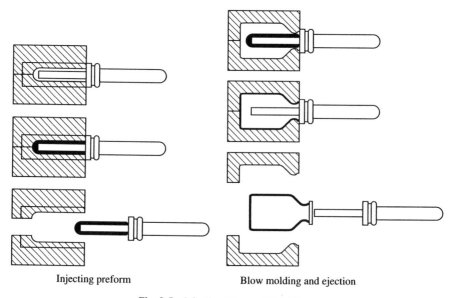

Injecting preform

Blow molding and ejection

Fig. 9-5   Injection blow molding [1].

to fit the mold shape. Contact with the colder mold surfaces causes the polymer to cool and solidify.

The extrusion blow-molding process can be operated either continuously or intermittently. In the continuous process, the parison, in essence, is never-ending. As such, it is necessary to move the mold to and away from the extruded parison (see Fig. 9-6). The intermittent operation, sometimes called short extrusion, uses a reciprocating screw similar to the type used with injection-molding processes (see Fig. 9-7). Intermittent extrusion blow molding is faster than the continuous extrusion operation.

Injection blow molding injects molten or thermally softened polymers into one or more heated preform cavities around a given core pin (Fig. 9-5). The preform mold is then opened. Heated polymer is then moved by using the core pin to the blow-mold station, where it is inflated and then ejected. Figure 9-8 is a schematic of a three-station unit. The injection blow-molding cycle is shown in Fig. 9-9.

Times for each of the cycle portions in Fig. 9-9 are relative. Optimum drying cycle time for the injection cycle is about 1.5 s. This drying cycle time includes opening of the molds, movement of the mold unit, and close time. Basically, plasticated polymer (after an injection delay) is injected into the mold (injection), where it is brought to holding pressure and conditioned. After the injection

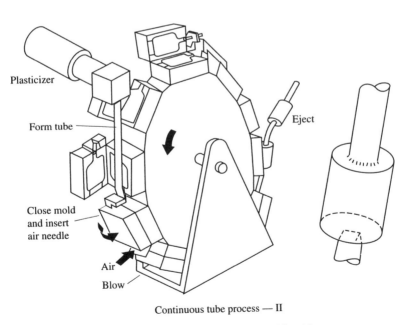

Continuous tube process — II

Fig. 9-6   Continuous extrusion blow molding (4).

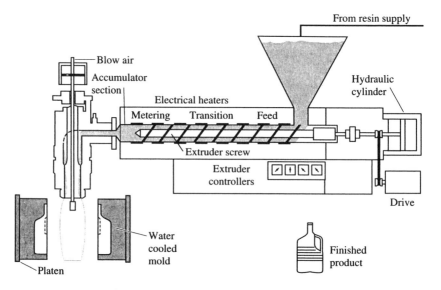

Fig. 9-7   Intermittent extrusion blow molding [4].

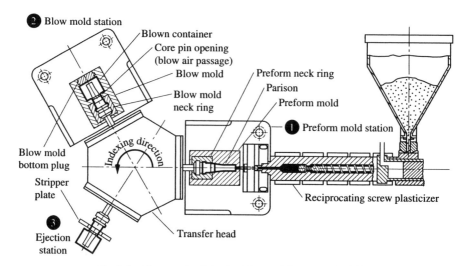

Fig. 9-8   Three-station injection blow-molding operation [1].

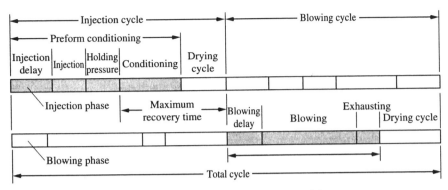

Fig. 9-9   Injection blow-molding cycle [1].

drying cycle, the polymer (after a delay) is blown. This is followed by exhausting and a blowing drying cycle.

Stretch blow molding can be either extrusion- or injection-molded. This process gives a bioriented product that can have significant economic advantages. Figure 9-10 depicts the stretch-injection blow-molding process. In extrusion blow-molding, the stretching takes place after both ends of the parison are physically held. Sometimes, such stretching involves an external gripper.

Table 9-3 shows the advantages and disadvantages of the three principal types of blow-molding operations.

9.5   BASIC PRINCIPLES IN BLOW MOLDING

A technical description of the blow molding of an object would involve both heat transfer and rheology. In this case, the rheology is especially complicated because it involves the movement of the polymer in relation to its exiting the die, to the force of gravity, and to the pressure exerted in the polymer by the gas when the polymer is formed in the mold.

When a molten or thermally softened polymer is forced through a die, its diameter becomes greater than that of the orifice (see Fig. 9-11). This phenomenon, called die swell, jet swell, or extrudate swell, is a complex process. Various researchers [5–9] have attributed it to the following:

1. Velocity rearrangement
2. Elasticity recovery
3. Gravitational and surface-tension effects
4. Normal stresses

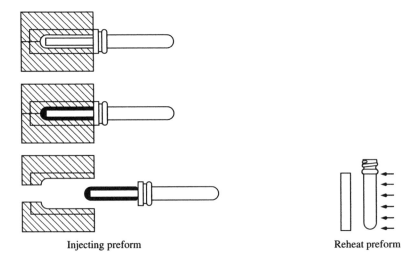

Injecting preform   Reheat preform

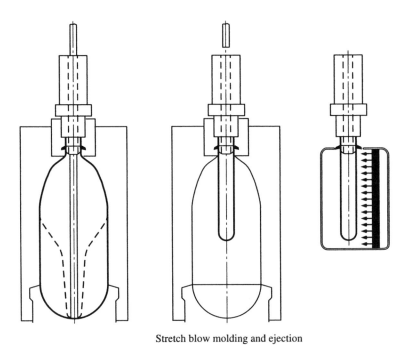

Stretch blow molding and ejection

Fig. 9-10   Stretch-injection blow molding [1].

TABLE 9-3   Advantages and Disadvantages of Blow-Molding Operations

Operation	Advantages	Disadvantages
Extrusion blow molding	High production rates; low tooling costs; wide selection of equipment	Large amount of scrap; uses recycled scrap; limitations on wall thickness; trimming facilities needed
Injection blow molding	No scrap; excellent thickness control; accurate neck finishes; outstanding surfaces, can produce low volume of products	High tooling costs; larger objects not possible
Stretch blow	Economical; improved properties; accurate control of wall thicknesses; reduced weights allowed	

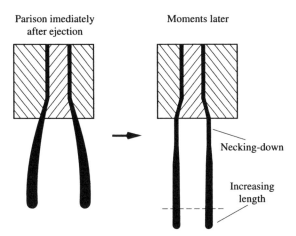

Parison imediately after ejection

Moments later

Necking-down

Increasing length

Fig. 9-11   Formation of blow-molded parison [4].

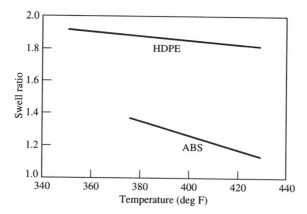

Fig. 9-12   Die swell ratio as a function of temperature for acrylonitrile–butadiene–styrene (ABS) and high-density polyethylene (HDPE) [4].

It appears likely that die swell is a function of at least two, and maybe all four, of the above factors.

Figure 9-12 shows the relationship of die swell to temperature for acrylonitrile–butadiene–styrene (ABS) and high-density polyethylene (HDPE). The effect of the length-to-diameter (L/D) ratio for the die is given in Fig. 9-13.

In Fig. 9-11, the extruded parison, slightly after ejection, is subjected to gravitational force. The results are a necking-down process, as well as increasing length. This portion of the blow-molding process is particularly complex since neither stress nor strain rate is constant [11]. Further, the result is a combination of swell and drawdown (sag).

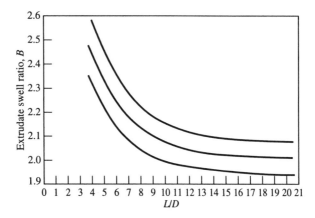

Fig. 9-13   Effect of die length-to-diameter ratio (L/D) on die swell for high-density polyethylene [10].

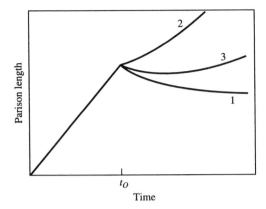

Fig. 9-14 Relation of parison length to time: case 1, swell only; case 2, sag only; case 3, initial recoil, combined swell, and sag [10].

Figure 9-14 is a graph of parison length for three cases at the time that extrusion ceases (t_θ). As can be seen, with an actual parison, length first decreases and then increases at a slower rate than during extrusion (from $t = 0$ to $t = t_\theta$.

Two additional phenomena complicate the blow-molding process. One is pleating (see Fig. 9-15), where the compressive stresses due to the parison weight cause the collapse shown. A rough index [12] of the pleating effect is given by the ratio $t_d/h_0\eta_0$, where t_d is the drop time, η_0 the polymer viscosity at low shear rate, and h_0 the thickness of the parison in the die. The other process

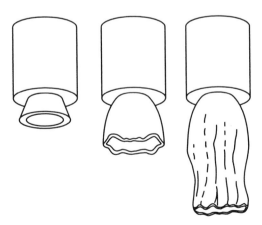

Fig. 9-15 Pleating in blow molding [12].

phenomenon is melt fracture, a situation in which an extruded polymer undergoes surface irregularities or actually fractures.

Figure 9-16 gives an empirical operating diagram for the blow molding of high-density polyethylene. Four parameters—pleating, melt fracture, parison diameter, and object weight—were used to determine the thick line, which represented the overall acceptable operating range for the system.

In the inflation stage, the molten or thermally softened polymer is subjected to the action of the gas pressure. One analysis [13] that involved minimal strain rates showed that a planar extensional viscosity η_p could be correlated with strain rate ($\dot{\epsilon}_{pe}$). The relationship between these quantities was

$$\eta_{pe} = M(\dot{\epsilon}_{pe})^{n-1} \tag{9-4}$$

Ultimately, a relation for inflation time gave

$$t_{\text{inf}} = \frac{1}{2s(CP)^s}\left[\left(\frac{1}{R_{\text{inf}}}\right)^{2s} - \left(\frac{1}{R_o}\right)^{2s}\right] \tag{9-5}$$

where

$$C = \left(\frac{2\pi L}{VM}\right)^s \tag{9-6}$$

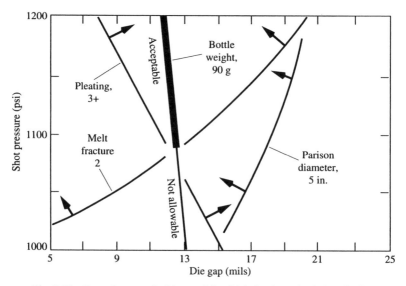

Fig. 9-16   Operating map for blow-molding high-density polyethylene [12].

Here, V represents the volume of the polymer in the parison, L is the parison length, R_{inf} is the inflated radius, R_o is the original radius, and P is the pressure.

The s term results from a rheological analysis of the system [13]. Actually, however, s can be found if an inflation time is measured for a given set of conditions.

Cooling of the expanded parison represents the final step in the cycle. This process is basically unsteady heat transfer (see Chapters 4 and 8). If the object is thick walled as in large containers the time needed for cooling could be a large part of the overall cycle time. In such cases, the process has been accelerated by such procedures as introducing liquid carbon dioxide late in the inflation or using fine crystals of ice that are injected. In either case, the phase change heat transfer has a large effect on the cooling time reducing it by a factor of about four [14].

As with injection molding, process problems can be solved by altering process variables. Table 9-4 presents a chart that relates problems to solutions.

TABLE 9-4   Relations Between Process Problems and Solutions: +, Increase variable; −, decrease variable

Problem	Mold Temp.	Polymer Temp.	Polymer Shear	Blowup Air Pressure	Cycle Time	Venting	Mold Alignment or Close
Rough parison		+					
Gloss	+						
Bubbles			+				
Blow incomplete			−	+			Reduce mold closing speed
Stretched parison	−		+				
Parison blowout				Low initially then increase			Align molds
Handle webbing	+	−					
Thins		−	+				
Weak shoulders		−	−				
Sticking	−				+		
Indented parting line	−			Delay blowup			
Sinks	−					+	
Poor detail	+			+	Increase blowup time	+	

9.6   ROTATIONAL MOLDING

Rotational molding is another polymer-processing operation that is used to produce hollow objects. It is a cyclic and semicontinuous process.

The cycle involves charging a given amount of powder or liquid polymer to
the mold. Figure 9-17 diagrams a four-mold-unit process arrangement. Each unit
is simultaneously rotated along perpendicular axes (i.e., polar and equatorial).
The rotating polymer is spread along the mold surface.

Heating the mold is the next step in the cycle. This can be done by a variety
of means, such as convection oven, direct flame, heat-transfer fluid in a jacket,
or spray.

The polymer, if solid, melts to form a layer of molten material on the mold
surface. This process, which resembles the sintering operations of the metals
industries, involves an action that forces voids out of the polymer melt.

The heating step is a critical part of the cycle. If it is overlong, polymer
degradation can take place. On the other hand, not allowing enough time for
sufficient melting and sintering produces a porous object not suitable for use.

Finally, the mold is cooled. This can be done by using a water spray or
airflow. Proper cooling is also important since the softened object in the mold

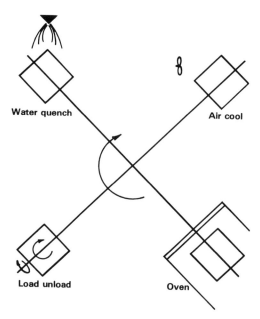

Fig. 9-17   Diagram of a four-station rotational molding operation [15].

can collapse on itself if cooled too fast. Also, in the case of semicrystalline polymers, various dimensionless problems, including warping, can result in severe process problems.

Large hollow products having uniform wall thicknesses can be produced by rotational molding. Objects having capacities as large as 8.52×10^{-2} m^3 with wall thicknesses of 6×10^{-3} m can be molded. Process units for rotational molding operations are not costly.

The molds can have different shapes as needed. A schematic of a typical rotational mold is shown in Fig. 9-18.

Process analysis of rotational molding has been concerned largely with the heat-transfer melting and sintering aspects of the cycle [15]. Further, it has been indicated [15] that, although flow occurs across the inner mold surface, it is not an important effect as long as mold aspect ratio (kept near unity) and mold position (reasonable when contrasted with other such units) are kept under control.

A model [15] based on the schematic shown in Fig. 9-19 was used to analyze the heating of the polymer. The model used an unsteady-state approach with an effective thermal diffusivity ($\alpha_{\text{effective}}$).

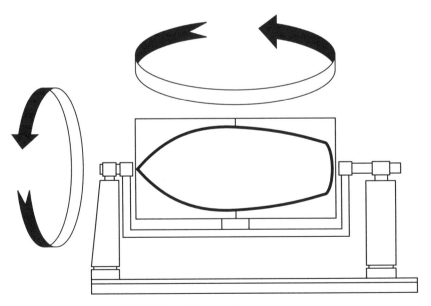

Fig. 9-18   Mold and rotating mechanism [1].

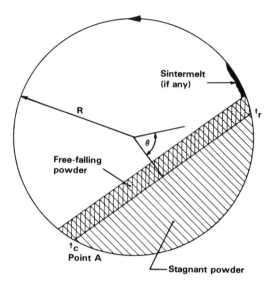

Fig. 9-19   Schematic of flow in rotational molding [15].

This approach gave a δ "thermal penetration thickness," as shown in Eq. (9-7).

$$\delta = (24\alpha_{\text{effective}}t)^{1/2} \tag{9-7}$$

If the surface temperature T_s was linearly related to time ($T_s = at$), the T at a given L was

$$T = at\left[1 - \frac{L}{(8\alpha^{\text{effective}}t)^{1/2}}\right]^3 \tag{9-8}$$

Likewise, if the powder depth was one-half of L_c in Fig. 9-19, the thickness of polymer at or above the sticking temperature was

$$x_1 = \delta\left[1 - \left(\frac{T_x}{T_s(t)}\right)^{1/3}\right] \tag{9-9}$$

where T_x is the sticking temperature and $T_s(t)$ the time-dependent surface temperature.

Finally, the part thickness X_t is

$$x_t = \left(\frac{A_s}{2\pi R}\right)\left(\frac{\rho_{\text{powder}}}{\rho_{\text{polymer}}}\right) \tag{9-10}$$

where the ρ are the powder and polymer densities and A_s is the powder area segment.

Densification [15] of a molded object with significant viscoelastic behavior is governed by Eq. (9-11):

$$\frac{w}{r} = \frac{3r\gamma\tau}{2\eta} \tag{9-11}$$

where w is the thickness of the web that forms between two spheres in the powder, τ is viscoelastic retardation time, γ the surface tension of the molten polymer, and η the polymer low shear (Newtonian) viscosity. When the w given by equation (9-11) is reached, the isothermal sintering would cease.

Liquid rotational molding is more concerned with fluid mechanical behavior. A schematic model [16] of the flow behavior encountered in a liquid system is shown in Fig. 9-20. In the process, the system passes through a series of experiences (Fig. 9-21), including pool withdrawal, cascading flow, rimming flow, and collapsing flow. Figure 9-22 relates these phenomena on the basis of a plot of a Froude number w^2R/g a Reynolds number b^2w/η. Here, w is the angular

Various flow phenomena for pool starting to rotate from quiescent state A.

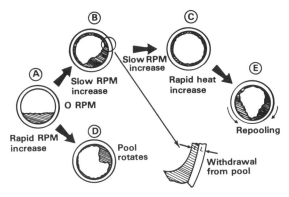

Fig. 9-20   Schematic model of flow in rotational molding [16].

Hydrodynamic regimes in a horizontally rotating cylinder

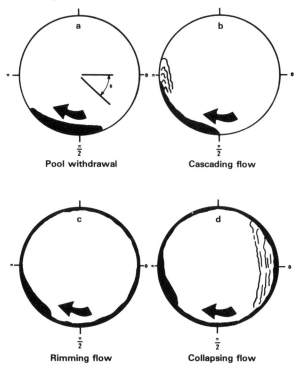

Fig. 9-21   Types of flow regimes in liquid rotational molding [16].

Flow regimes of Deiber and Cerro.
Open circles = SBR.   Solid circles = Type II

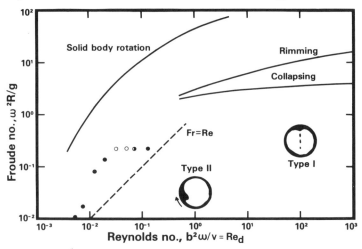

Fig. 9-22   Froude number vs Reynolds number with open circles for transition type I and solid circles for type II [16,17].

velocity, g the acceleration of gravity, b the liquid thickness, and η the polymer liquid's kinematic viscosity. Points are for experimental data [16,17].

REFERENCES

1. ROSATO, D.V., and ROSATO, D.V., *Plastics Processing Data Handbook*, Van Nostrand Reinhold, New York (1990).
2. BIRD, R.B., STEWART, W.E., and LIGHTFOOT, E.N., *Transport Phenomena*, Wiley, New York (1960).
3. BROYER, E., and MACOSKO, C.W., *AIChE J*. **22**, 268 (1970).
4. LEE, N.C., *Plastic Blow Molding Handbook*, Van Nostrand Reinhold, New York (1990).
5. MCKELVEY, J.M., *Polymer Processing*, Wiley, New York (1962).
6. GAVIS, J., and MIDDLEMAN, S., *Phys. Fluids* **4**, 355, 963 (1961).
7. Ibid., *J. Appl. Polymer Sci.* **7**, 493 (1963).
8. METZNER, A.B., *Ind. Eng. Chem.* **50**, 1577 (1958).
9. Ibid., *Trans. Soc. Rheol.* **5**, 133 (1961).
10. DEALY, J.M., and WISSBRUN, K.F., *Melt Rheology and Its Role In Plastics Processing*, Van Nostrand Reinhold, New York (1990).
11. DEALY, J.M., and ORBEY, N., *AIChE J*. **31**, 807 (1985).
12. SCHAUL, J.S., HANNON, M.J., and WISSBURN, K.F., *Trans. Soc. Rheol.* **19**, 351 (1975).
13. DENSON, C.D., *Polymer Eng. Sci.* **13**, 125 (1973).
14. HUNKAR, D.B., *SPE ANTEC* **19**, 448 (1973).
15. RAO, M.A., and THRONE, J.L., *Polymer Eng. Sci* **12**, 237 (1972).
16. THRONE, J.L., and GIANCHANDANI, J., *Polymer Eng. Sci.* **20**, 899 (1980).
17. DIEBER, J.A., and CERRO, R.L., *Ind. Eng. Chem. Fund.* **15**, 102 (1976).

PROBLEMS

9-1   Develop an appropriate relation between the ratio of the closing force at any time to the original closing force ($t = 0$) and time for a compression molding system. Consider the system to be one-dimensional; that is, consider the thickness of the molded object to be the only significant dimension. Plot your results as the force ratio vs the reciprocal of time.

9-2   A reacting system has a rate constant of 0.1 s^{-1} at 475 K (Arrhenius activation energy of 20 kcal/g mole) is compression-molded. The mold wall temperature is at 510 K. Polymer pre-compression temperature is 375 K. Object thickness is 0.03 m (length and width are 0.10 and 0.08 m, respectively). How long a cycle time is needed for a 90% reaction rate?

9-3   Explain in detail how you could use the results shown in Fig. 9-2 for reacting systems other than polyurethane. Show appropriate equations, relations, etc. Justify all assumptions.

9-4   In blow molding, one of the parameters used in the BUR (blowup ratio). The BUR is defined as the ratio of the final radius to the original radius. Would it be possible to obtain a value of unity for BUR for certain temperatures and values of the dimensionless ratio $\pi R_o^3 \Delta P / \mu_{APP} Q$, where R_o is the original radius, ΔP the pressure inside the parison, Q the volumetric flow rate, and μ_{APP} the apparent viscosity of the polymer?

9-5   For a blow-molding operation (0.02 kg/s), with an R_o of 0.05 m and an original thickness of 0.001 m, find a set of operating conditions (final thickness, BUR, etc.). The melt density of the polymer is 900 kg/m^3, and its viscosity can be assumed to be constant at 3×10^4 N s/m^2.

9-6   It has been noted that a rough index of pleating is given by $td/h_0 N_0$. Develop appropriate equations (theoretically, semiempirically, or empirically) that will justify this ratio.

9-7   What drop time would be critical for the polymer of Problem 5-7? What would the drop times be for polyethylene, polypropylene, or polymethyl methacrylate?

9-8   How would curves 1, 2, and 3 of Fig. 9-14 influence the behavior of Fig. 9-16. Justify your answers.

9-9   Find the s value of Eq. (9-4–9-6) for the system of Problem 9-5.

9-10   A polymer with a surface tension of 0.03 N/m and a viscosity of 1.0 N s/m^2 is rotationally molded. When the τ values are low, it is found that w/r is given by $(3\,\tau 2/r\,n)^{1/2}$. Plot curves for both this relation and Eq. (9-11) using assumed values of τ. Do the curves intersect? If not, how can these behavior patterns be reconciled?

9-11   Can appropriate relations be developed from Fig. 9-22 that relate liquid thickness directly to either the Froude or Reynolds number? If so, develop plots of such relations.

9-12   In Fig. 9-22, the experimental data do not seem to directly match any of the lines shown. Why do you think this occurs? Develop your own relationship for the experimental data (i.e., involving Fr and Re).

MINI PROJECT A

Develop an appropriate model for the transfer molding process. Justify all assumptions, and clearly define your terms.

MINI PROJECT B

By building on the relations for blow molding in this chapter, as well as on the other material in the text, develop models for two of the blow-molding processes given.

MINI PROJECT C

An important aspect of rotational molding is the solidification and cooling of the molded object. Using the concepts of this text, develop the appropriate applicable heat-transfer relations needed to describe this process. Assume the mold wall temperature to be constant.

MINI PROJECT D

Develop an equation that describes the behavior of film temperature versus distance from the die for a blown film. Clearly indicate all assumptions.

CHAPTER 10

CALENDERING, THERMOFORMING, AND CASTING

10.1 INTRODUCTION

This chapter summarizes other polymer-processing operations not previously considered. Included are calendering, thermoforming (or sheet forming), and casting.

The first of these, *calendering*, is a continuous process that uses rolls to form a polymer mass into a sheet of uniform thickness (see Fig. 10-1). Calendering, which is analogous to the rolling operations of the metals industries, had its earliest beginnings in the rubber industry and then spread into the manufacture of a variety of polymer-related products, such as linoleum, plastics, and coated papers.

Thermoforming, or *sheet forming*, is a very old polymer-processing operation that involves heating a polymer sheet to its softening temperature and then forming it to a desired shape. The cooled formed sheet represents the product.

There are four classifications of thermoforming (see Fig. 10-2). These are: matched-mold forming, using matched male and female molds; slip forming, using only a male mold; air-blown, with air blowing the sheet into a female mold; and vacuum forming, with a vacuum shaping the sheet around a female mold.

Casting forms a given shape by placing a polymer syrup (partially polymerized system) into a form, shape, or mold in which the syrup is polymerized into a solid mass. The polymers used most often in casting include acrylics, cellulosics, phenolics, and epoxies.

372

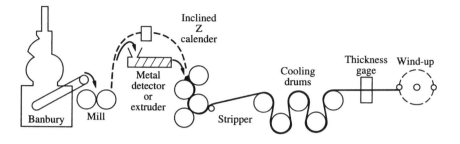

Fig. 10-1   Schematic of calendering operation [1].

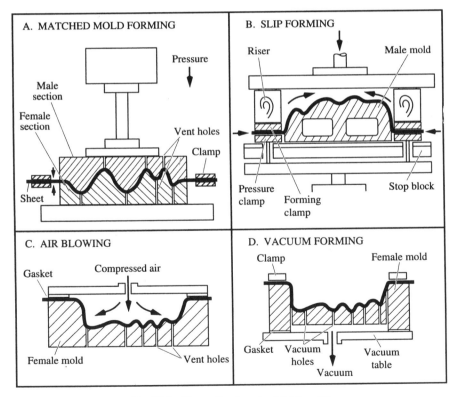

Fig. 10-2   Thermoforming classifications [2].

10.2 CALENDERING

The calendering operation is essentially a process involving laminar flow with heat transfer in a roll system. Typical design outputs include desired roll dimensions, roll speeds, separation of rolls, temperature profiles in the processed material, roll-suspension force, and required power.

Gaskell [4] developed a simplified flow theory that was later modified by McKelvey [3]. The starting point (for the geometry shown in Fig. 10-3) is that

$$\frac{\partial P}{\partial x} = \mu \frac{d^2 V_x}{dy^2} \tag{10-1}$$

Ultimately, equations are obtained for various process parameters, such as P_{max}, maximum pressure; Z, the power; F, force separating rolls; and Q, the volumetric calendering rate). These are given in the following equations:

$$P_{max} = \frac{5\mu U \lambda^3}{H_0} \frac{9R}{8H_0} \tag{10-2}$$

$$F = \frac{3\mu URW}{4H_0} q(\lambda) \tag{10-3}$$

$$Z = 3U^2 \mu \frac{2R}{H_0} f(\lambda) \tag{10-4}$$

$$Q = 2UHW \tag{10-5}$$

$$\frac{V_x}{U} = \frac{2 + 3\lambda^2 \left(1 - \left[\frac{y}{h} \right]^2 \right) - \rho^2 \left(1 - 3 \left[\frac{y}{h} \right]^2 \right)}{2(1 + \rho^2)} \tag{10-6}$$

$$\frac{P}{P_{max}} = \frac{H_0 P}{10\mu U \lambda^3} \sqrt{\frac{32H_0}{9R}} \tag{10-7}$$

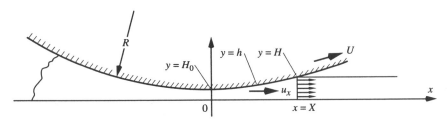

Fig. 10-3   Geometry for calendering equations [3].

where P is pressure, U roll velocity, R roll radius, N roll width, h distance from plane of symmetry to roll surface, y vertical dimension, x horizontal dimension, μ polymer viscosity, and H_0 roll separation; ρ is $x/2RH_0$ and λ is $Q/2UH_0$. The $q(\lambda)$ and $f(\lambda)$ terms, which are complicated functions of λ, are given in Fig. 10-4.

McKelvey [3] compares a calculated pressure profile to experimental values measured by Bergen and Scott [5]. As can be seen (Fig. 10-5) the agreement is quite good. The deviation at values of $\rho < -0.8$ is believed to be due to the non-Newtonian nature of the polymer. In this connection, McKelvey [3] presents an altered set of pressure equations that allow for non-Newtonian behavior. Marshall [1] gives a simplified version that involves an apparent shear rate $2U/H_0$ and a force equation,

$$F = 2\mu_{APP}URW \left(\frac{1}{2H_0} - \frac{1}{2H} \right)$$

Let us now consider calendering examples.

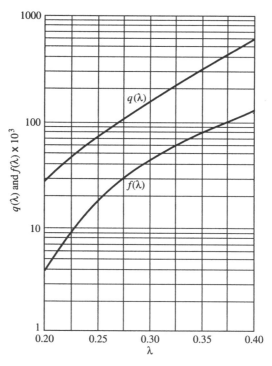

Fig. 10-4   The relationship between the $q(\lambda)$, $f(\lambda)$ factors and λ [3].

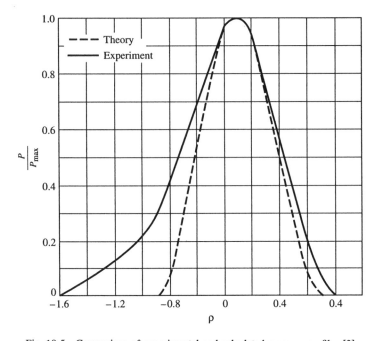

Fig. 10-5   Comparison of experimental and calculated pressure profiles [3].

Example 10-1

A given calendering unit, which has the dimensions $R = 0.1$ m, $W = 1$ m, and $H_0 = 1 \times 10^{-4}$ m, operates with a velocity of 0.4 m/s. The polymer, which has a viscosity of 1000 k/m s, yields a sheet of 2.18×10^{-4} m.

Find P_{max}, Z, F, and Q for the system.

First, determine λ for the system:

$$\lambda = \frac{H}{H_0} - 1$$

which gives a value of 0.3.

The P_{max} occurs at $\rho = -\lambda = -0.3$.

$$P_{max} = \frac{5(1000 \text{ kg/ms})(0.4 \text{ m/ms})(0.3)^3 \dfrac{9(0.1 \text{ m})}{8(0.0001 \text{ m})}}{(0.0001 \text{ m})}$$

$$P_{max} = 1.81 \times 10^7 \frac{\text{N}}{\text{m}^2}$$

$$Z = 3(1 \text{ m})(0.4 \text{ m/s})^2(1000 \text{ kg/m s})\frac{(2(0.1 \text{ m})}{(0.0001 \text{ m})}(0.043)$$

$$Z = 0.92 \text{ kw} = 920 \text{ J/s}$$

$$F = \frac{3(1000 \text{ kg/m s})(0.4 \text{ m/s})(0.1 \text{ m})(1 \text{ m})(0.16)}{4(0.0001 \text{ m})}$$

$$F = 4.8 \times 10^4 \text{N}$$

$$Q = 2(0.4 \text{ m/s})(1.09 \times 10^{-4})(1 \text{ m})$$

$$Q = 8.72 \times 10^{-5}\frac{\text{m}^3}{\text{s}}$$

An estimated temperature rise can be found by assuming that all the required power raises the enthalpy of the polymer (C_p taken to be 2092 J/kg°C and density to be 1000 kg/m^3):

$$mC_p\Delta T = Z$$

$$\Delta T = \frac{Z}{mC_p} = \frac{920 \quad \text{J/s}}{(0.0872 \quad \text{kg/s}) \quad 2092 \quad \text{J/kg°C}}$$

$$\Delta T = 5°C$$

Example 10-2

A plasticized vinyl material is calendered on a 0.203-m-diam system, yielding a sheet 0.381 m wide with a thickness of 4.06×10^{-4} m. The operating speed of the roll is 0.254 m/s at a temperature of 170°C. The separating force is 26,690 N, and the original thickness of the sample is 0.0127 m.

What separating force would be needed for a 0.812-m-diam system producing a 1.5×10^{-4} m sheet with a width of 2.86 m? The calender operates at 0.508 m/s and a temperature of 170°C.

We first find H_0, which can be estimated as three-quarters of the product thickness. Then, using Eq. (10-8), we obtain the apparent viscosity:

$$(\mu_{APP}) = 413.7 \frac{N}{m^2} s$$

This value would have an apparent shear rate V/H_0 of 1670 s^{-1} while, for the larger unit, the V/H_0 is 8900 s^{-1}.

A check of actual rheological data for the system in English units (Fig. 10-6) shows that the line must be shifted to fit the preceding. Further, in the actual calendering case, there will be both an end effect and a time effect.

These can be compensated for by the use of the factor f_s (see Fig. 10-7), which is related to the term N_s:

$$N_s s = 1.73 \frac{H_0}{2R}$$

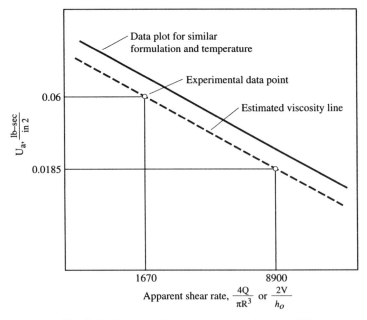

Fig. 10-6   Apparent viscosity vs apparent shear rate [1].

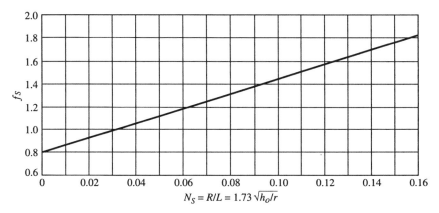

Fig. 10-7   Shear time factor f_s vs shear time number N_s [1].

For the two apparent shear rates, the N_s values are:

Apparent Shear Rate, s^{-1}	N_s
1670	0.095
8900	0.0292

Then, from Fig. 10-7, the f_s values are 1.41 and 0.99.

Finally, the apparent viscosity value is that from Fig. 10-6 connected with the f_s values:

$$\mu_{APP} = 127.6 \ \ N/m^2 \ \ s \ \frac{0.99}{1.41}$$

$$\mu_{APP} = 89.6 \ \ N/m^2 \ \ s$$

Substituting in the force equation gives

$$F_2 = 7.38 \times 10^5 \ \ N$$

10.3   THERMOFORMING

As noted earlier, there are four main techniques for thermoforming. Each of these, however, has a number of modifications, with the largest being variants of vacuum forming.

The various types of vacuum forming include:

380   *Calendering, Thermoforming, and Casting*

1. Straight vacuum forming (Fig. 10-8) can be done with either male or female molds, as well as with a double clamping frame. It is used for rigid materials.
2. Vacuum snap-back forming (Fig. 10-9) is used for rubbery materials. The male mold can be moved into position as in part A of the figure or used with vacuum as in B.
3. Vacuum drape forming (Fig. 10-10) is used for deep draws. At the shaping temperature, the sheet is pulled over the male mold.
4. Plug-assist vacuum forming (Fig. 10-11) is used to overcome thinning problems. Because the plug drops, this technique is also known as drop forming.

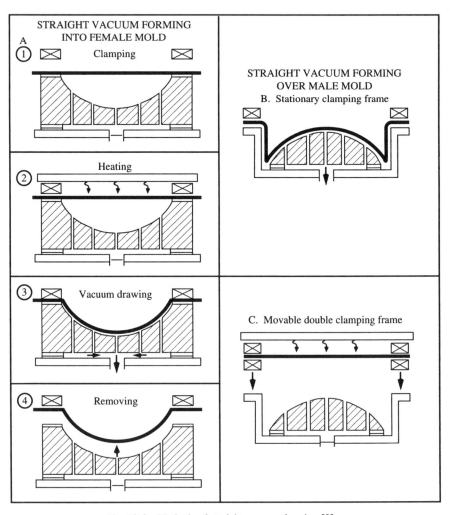

Fig. 10-8   Methods of straight vacuum forming [2].

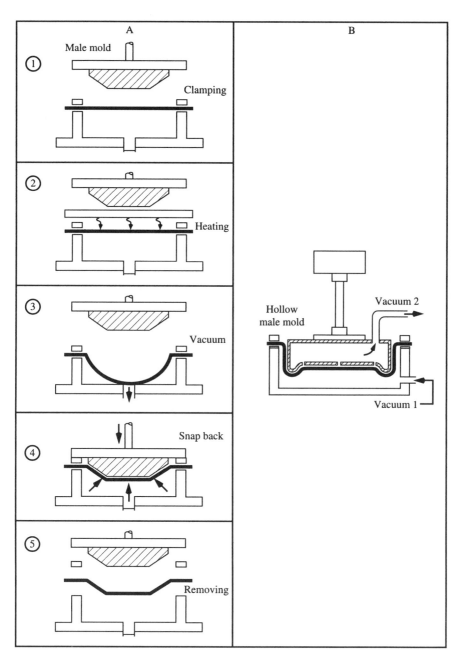

Fig. 10-9   Schematic of snap-back vacuum forming [2].

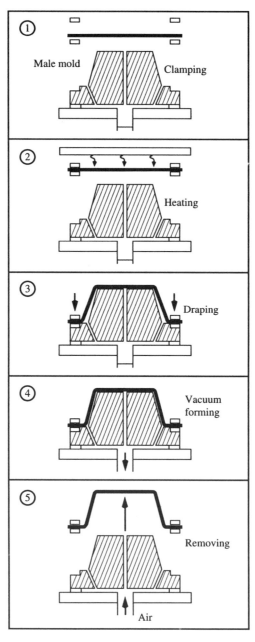

Fig. 10-10 Vacuum drape forming [2].

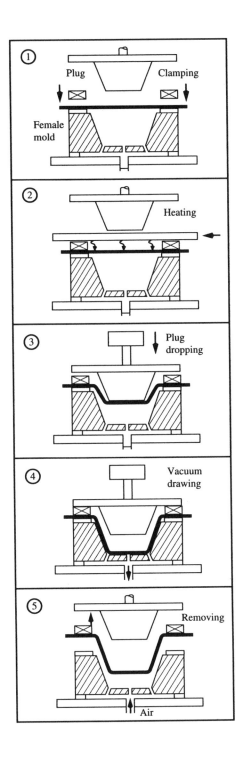

① Plug   Clamping

Female mold

② Heating

③ Plug dropping

④ Vacuum drawing

⑤ Removing

Air

383

5. Vacuum air-slip forming (Fig. 10-12) is used to reduce thinning in deep-drawn items. The polymer is prestretched in a bubble shape prior to forming. When the bubble is evacuated, it drops around the mold.
6. Air-cushioning vacuum operation (Fig. 10-13) combines features of the plug-assist and air-slip processes. Essentially, the softened sheet is prestretched by air.

The two principal elements in thermoforming, as with all processing operations, are the transfer of energy and the application of stress (rheology).

Heat transfer is extremely important in the thermoforming operation since the time required to reach the softening temperature can be 50%–80% of the total cycle time. Obviously, the less time needed to reach the forming temperature, the less the overall cycle time.

Heating can be done by conduction, convection, radiation, or some combination of these. The usual mode of heat transfer in thermoforming is radiation. Radiation heat transfer in polymeric systems was discussed in Chapter 4.

As pointed out previously, such radiation heat transfer will be different for opaque and semitransparent polymer systems. In the opaque case, the incident radiation absorbed by the polymer is converted to surface heat. In the semitransparent system, the thermal radiation passes through the entire slab. Because of this, semitransparent systems heat at much lower rates.

An important parameter for both systems is the eveness index discussed in Chapter 4. These indexes for each of the cases are

$$I = \frac{T_s - T_i}{T_c - T_i}$$

for the opaque case, where T_s is surface temperature, T_i the initial slab temperature, and T_c the centerline temperature, and

$$E1 = \frac{T_{max} - T_{min}}{T_{max} - T_o}$$

for the semitransparent case where T_{max}, T_{min} are, respectively, the maximum and minimum temperatures in the slab and T_o is the original temperature.

Examples 4-11 and 4-12 illustrate the use of the eveness index concept, as well as radiation heat transfer to polymer sheets.

Figure 10-14 charts the thermal history and time required to bring a polystyrene sheet to its optimum forming temperature. Note that the top and bottom of the sheet experience different thermal paths.

The principal rheological aspect of thermoforming is the stretching of the softened polymer sheet. As such, the material is subjected to orientation. Such orientation can affect the mechanical behavior of the finished product.

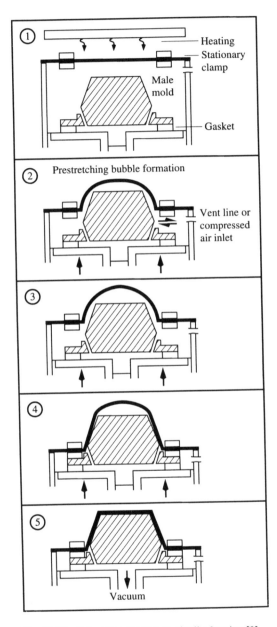

Fig. 10-12   Schematic of vacuum air-slip forming [2].

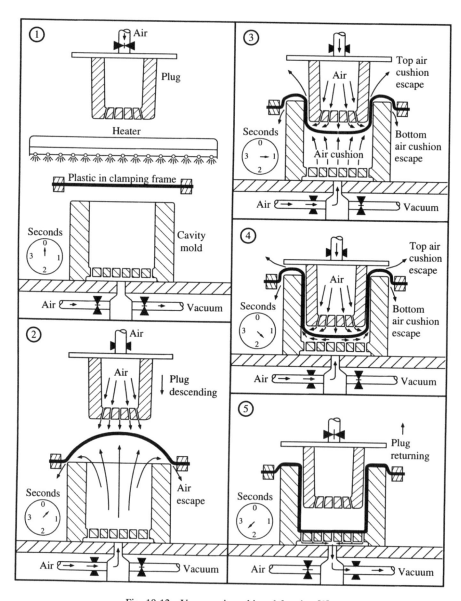

Fig. 10-13   Vacuum air-cushioned forming [2].

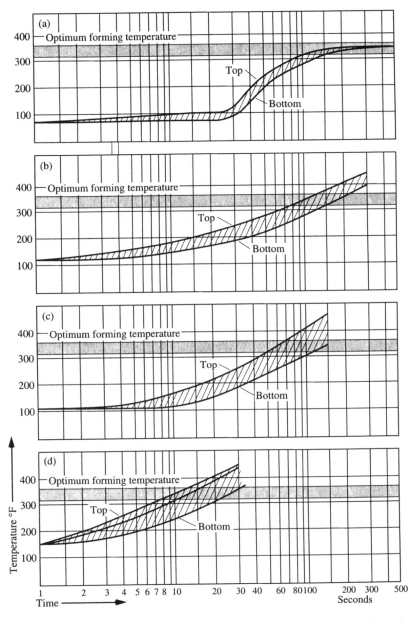

Fig. 10-14   Heating of 0.08-in.-thick polystyrene sheet [1]. The heaters and distances from the sheet air: (a) Pyrex, 8 in.; (b) glass fiber, $4\frac{1}{2}$ inches; (c) Chromalox (700°F), $3\frac{1}{2}$ in.; and (d) Chromalox (100°F), $3\frac{1}{2}$ in.

10.4 CASTING

There are a number of applications of casting, including:

(1) The use of special molds for phenol-formaldehyde polymers
(2) Forming specific shapes, rods, and sheets of acrylic and allyl polymers
(3) Production of tooling using cellulosic hot melts
(4) Encapsulation and potting
(5) Slush molding
(6) Film casting

Phenol-formaldehyde polymers are usually cast in lead molds. The polymer is poured while it is liquid, that is, before extensive cross-linking has occurred. The liquid has the consistency of a syrup. Filled molds are cured at a temperature controlled somewhere in the 60–80°C range. The cure time can range from hours to days.

The final products of cast phenol-formaldehyde polymers have a number or exceptional properties, including high tensile and compressive strengths, good electrical insulating capabilities, and excellent adhesive qualities. Also, they can be polished and machined. Finally, the presence of very small water droplets in the material gives the surface a superb appearance.

Casting of acrylic polymers differs from that of phenol-formaldehyde polymers in that considerably more polymerization must take place. Further, the time, temperature, and catalyst must be carefully controlled. Benzoyl peroxide is usually employed as a catalyst. However, other materials can be used, such as diacyl peroxides, aldehyde peroxides, ketone peroxides, alkyl hydroperoxides, alkyl peresters, and alkyl acid peresters.

Catalyzed monomer or prepolymer used for a cast object must be handled carefully to avoid formation of bubbles. This involves a slow, controlled addition to the mold, followed by the proper allowance of time for the escape of any accelerated air bubbles.

The molds used in casting acrylics must be highly polished (either glass or metal can be used). Molded objects include rods, cylinders, and special shapes.

Tooling production involves a mix of ethyl cellulose or cellulose acetate-butyrate with a wax. Plasticized cellulose acetate-butyrate can also be used. These mixes are poured into molds as hot liquid melts. This process differs from the others previously described in that no additional polymerization takes place. Objects produced in this manner include jigs, fixtures, blocks, hammer forms, dies, and punches. Molds are usually simple and are made of plaster or wood.

Two of the principal processes connected with the electronic industries are encapsulation and potting. Encapsulation coats a unit with a plastic, whereas potting is total penetration of all voids in a unit by a polymer. Typical polymers used in these processes include phenolics, epoxies, polyurethanes, and unsaturated polyesters. In the case of potting, preheating the object and mold above 100°C is a required first step in order to expel moisture. Liquid polymers are poured into the cooled object-mold system and then cured for at least 1 h. The curing operation typically involves the use of vacuum, pressure, or centrifugal action to guarantee permeation of the material into the device's voids.

Slush molding uses plastisols (dispersions of thermoplastic polymers in liquid plasticizers; 50–100 parts plasticizer to 100 parts polymer). A hollow mold filled with the plastisol is carried on a conveyor belt through an oven (temperature 500–600°F). Excess liquid is then poured out. The mold goes next to a second oven, where the curing is finished. This technique, used for toys, produces a product with fine details.

Finally, films can be cast on highly polished drums or on belts or bands that are driven by drums (see Fig. 10-15). A solution flows onto the highly polished surface, where the solvent is then evaporated, at first to give a film with some consistency and then to remove even more solvent. A better surface can be obtained on the drum. However, the longer drying times needed require slower speeds and produce thinner films. The belt method is preferred since production rates are higher and thicker films can be formed.

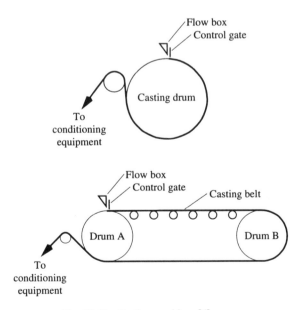

Fig. 10-15 Casting machines [6].

REFERENCES

1. MARSHALL, D.I., in *Processing of Thermoplastic Materials*, E.C. Bernhardt (ed.), Rheinhold, New York (1959), Chap. 6.
2. PLATZER, N., in *Processing of Thermoplastic Materials*, E.C. Bernhardt (ed.), Rheinhold, New York (1959), Chap. 8
3. MCKELVEY, J.M., *Polymer Processing*, Wiley, New York (1962), Chap. 9.
4. GASKELL, R.E., *J. Appl. Mech.* **17**, 334 (1950).
5. BERGEN, J.T., and SCOTT, G.W., *J. Appl. Mech.* **18**, 101 (1951).
6. WINDING, C.C., and HIATT, G.D., *Polymeric Materials*, McGraw-Hill, New York (1961).

PROBLEMS

10-1   A calendering system (R of 0.17 m, H_0 of 0.006 m, roll speed of 0.1 m/s) processes a material whose viscosity is 1×10^4 N s/m^2. For this system, find the sheet thickness, maximum pressure, and roll-separating force.

10-2   For the calendering system of Example 10-1 and a non-Newtonian fluid (n of 0.5 and K of $10^3 \dfrac{\text{N s}^n}{\text{m}^2}$, find the maximum pressure.

10-3   Develop an equation for maximum shearing stress if the calendering system rolls are of equal size but different roll speeds.

10-4   For the system of Problem 10-1, find the power delivered to the rolls and the adiabatic temperature rise of the material.

10-5   A polymer is calendered at 100°C in a system that has the following dimensions: $R = 0.1$ m, width of 0.4 m, H_0 of 0.004 m. If the polymer's rheological parameters are $n = 0.4$ and $K = 3 \times 10^5$ N s^n/m^2, find and plot the relation between the parameters $F/2RW$ and U/H_0.

10-6   Find the maximum pressure, roll-separating force, and required power for a calendering system (R of 0.1 m, W of 1 m, H_0 of 0.012 m, U of 0.4 m/s) processing a material whose viscosity is 1000 N s/m^2.

10-7   If the roll-separating force is balanced by the weight of the calendering system (1100 lbf), what is the sheet thickness (material viscosity of 1200 N s/m^2)? The system's dimension are $R = 0.15$ m, $W = 1.7$ m. Roll speed is 0.11 m/s.

10-8   Analyze the data of Fig. 10-14 using the material of Chapter 4. Assume that the glass fiber heater is at 700°F.

10-9   Sagging of a plastic sheet is a common problem in sheet-forming operations. A set of polyethylenes *A, B, C, D* have the following relative percent sagging (all compared to *A*) in a sheet-forming process: *A*, 100%; *B*, 135%; *C*, 150%; *D*, 300%. The same polyethylenes have the following melt index* (g/10 min) values: *A*, 0.6; *B*, 1.0; *C*, 7.1; *D* 3.0. (The melt index is a dead weight piston-type device used to obtain a measure of a polymer's flow behavior.)

Based on the foregoing and other material in the text, explain the sagging behavior.

For polyethylene at 190°C, the approximate pressure is 3×10^5 N/m². The capillary for the flow is 0.021 cm in diameter and 0.80 cm in length.

10-10   Figure 10-16 gives data for plug-assisted sheet forming of polystyrene. Analyze the overall process to the best possible degree of engineering reliability.

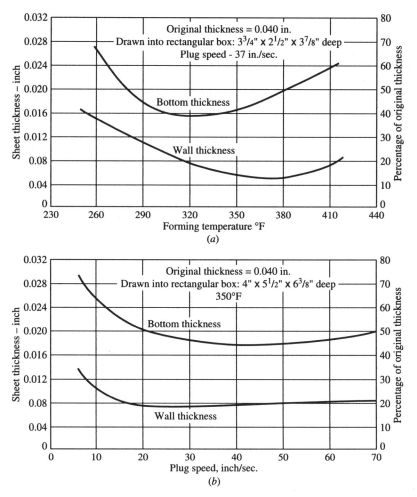

Fig. 10-16   Plug-assisted sheet forming for polystyrene. Graph (a) is thickness versus temperature. Graph (b) is thickness versus plug speed [2].

10-11   A plastic sheet 0.003 m thick, with thermal diffusivity of $2 \times 10^{-7} \frac{m^2}{sec}$, moves through a 0.6-m-long radiant oven at 0.10 m/s. The sheet is 0.5 m wide, has a density of 1450 kg/m^3, and is to be heated at a rate of 275 kJ/kg. What should the temperature of the heating elements be? (Assume that the area of the heaters is 0.019 m^2 and that their emissivity is 0.9.

MINI PROJECT A

Develop appropriate relationships for the calendering of a coextruded sheet of two polymers (i.e., one side is polymer A, the other polymer B; overall sheet thickness is $t_A + t_B$).

MINI PROJECT B

Develop a model to describe the vacuum air-slip forming (Fig. 10-12). Use the concepts of Chapter 2 to help in this endeavor. Illustrate how you would apply this for a sheet made out of polymethyl methacrylate.

MINI PROJECT C

Develop the calendering system equations for a case in which the rolls have unequal diameters.

MINI PROJECT D

Develop a heat-transfer analysis for radiant heating of a coextruded sheet of plastic material (example thickness of 0.5 cm of polymethyl methacrylate and 0.75 cm of acrylonitrile–butadiene–styrene, or ABS). Assume that the heating flux to each side is the same.

CHAPTER 11

FIBER-SPINNING PROCESSES

11.1 SYNTHETIC FIBERS

A fiber can be defined as a material whose length is at least 100 times its diameter. Synthetic fibers are organic or inorganic materials that have been processed (spun) into the fiber form (thread lines). The synthetic fiber can be categorized as either staple (short) or continuous filament (very long).

Organic synthetic fibers can be formed from polymers in a variety of ways, depending on the nature of the material. In each case, however, the fiber-forming process (spinning) is similar (see Fig. 11-1) in that a polymer melt or solution is forced or extruded through a small orifice, called a spinneret or jet, which shapes it into the fiber form. Next, the material undergoes a phase transformation to the solid form, after which it is wound up, or ''taken up'', on a bobbin (spool or pirn). The processed or spun yarn can then pass through other operations so that it is brought to some final performance level.

The principal forms of spinning are listed in Table 11-1, together with the phase-transformation process as well as the principal commercial fibers formed by each technique. Generally, the selection of a particular type of spinning process is related to the material being spun. For example, nylons are semicrystalline in the solid state and have a definite melting point. Whereas the nature of solid nylon makes it difficult to put it into a solution, it does not prevent a melt from being formed. Hence, nylon is a melt-spun fiber. On the other hand, polyacrylonitrile is an amorphous solid and, as such, is dissolved more readily than a semicrystalline polymer. Here, the result is that polyacrylonitrile is either dry-

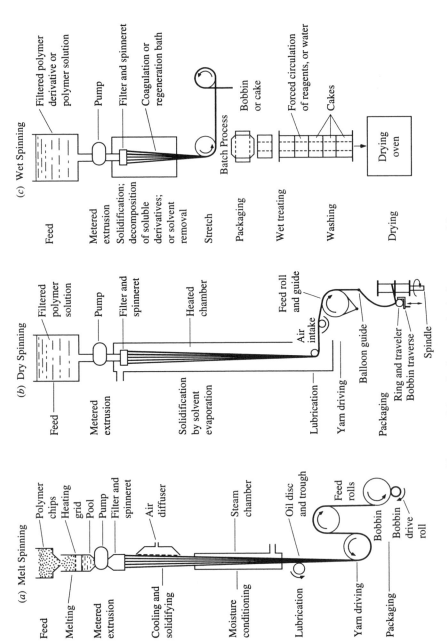

(a) Melt Spinning

Feed — Polymer chips

Melting — Heating grid, Pool

Metered extrusion — Pump, Filter and spinneret

Cooling and solidifying — Air diffuser

Moisture conditioning — Steam chamber

Lubrication — Oil disc and trough

Yarn driving — Feed rolls

Packaging — Bobbin, Bobbin drive roll

(b) Dry Spinning

Feed — Filtered polymer solution

Metered extrusion — Pump, Filter and spinneret

Solidification by solvent evaporation — Heated chamber, Air intake

Lubrication — Feed roll and guide

Yarn driving

Packaging — Balloon guide, Ring and traveler, Bobbin traverse, Spindle

(c) Wet Spinning

Feed — Filtered polymer derivative or polymer solution

Metered extrusion — Pump, Filter and spinneret

Solidification; decomposition of soluble derivatives; or solvent removal — Coagulation or regeneration bath

Stretch — Batch Process

Packaging — Bobbin or cake

Wet treating — Forced circulation of reagents, or water, Cakes

Washing

Drying — Drying oven

Fig. 11-1 Schematic diagrams of spinning processes [1].

394

TABLE 11-1   Spinning Types

Spinning Process	Type of Phase Transformation	Materials Spun
Melt	Solidification from molten mass	Nylon, polyester, polypropylene, glass
Dry	Evaporation of solvent with solidification	Cellulose acetate, cellulose triacetate, polyacrylonitrile (Orlon[a])
Wet	Countercurrent diffusion in a bath with solidification	Polyacrylonitrile (Acrilan[b])
Reaction	Material reacts and solidifies	

[a] Trademark: E. I. duPont DeNemours & Co., Inc.
[b] Trademark: Monsanto Corporation.

spun as in Orlon or wet-spun as in Acrilan (which are, respectively, trademarks of E. I. duPont deNemours and Company, Inc., and Monsanto Corporation).

11.2   FIBER REQUIREMENTS AND SPINNABILITY

Synthetic fibers must have certain characteristics in order to be useful materials. They should, for example, have a high thermal softening point, to permit processing or ironing; a high initial modulus of elasticity or stiffness; and reasonable tensile strength over a wide temperature range. Tensile strength is expressed as grams/denier. The denier is a measure of the size of a fiber and is defined as the weight in grams of 9000 m of fiber length. This means that denier is proportional to both the density and the cross-sectional area of the fiber. Hence, the larger the denier, the greater the cross-sectional area of the fiber.

Approximate property ranges for synthetic fibers are:

1. Tensile strengths should be about 5 g/denier for textile applications to 7–8 g/denier for industrial applications.
2. Initial modulus should be 30–60 g/denier for textile applications and 50–80 g/denier for industrial applications.
3. Elongation at break should not be less than 10% for textile use and 8%–15% for industrial use.
4. Temperatures at which creep or softening occurs should be no lower than 215°C for textile applications or 250°C for industrial applications.
5. Fibers should have good abrasion resistance, moisture resistance for textile applications, and chemical resistance for industrial use.

An important basis of all fiber-spinning operations is degree of *spinnability* or, as the Germans put it, *Spinnbarkeit*. A polymer is considered spinnable if it is: (1) chemically and thermally stable under spinning conditions, (2) capable of yielding continuous fluid or semisolid fibers, and (3) easily transformed into solid polymer. Although spinnability is a necessary property for fiber formation, it is not sufficient since many spinnable liquids, such as oils, are not fiber-forming. There are also spinnability limits for fiber-forming materials. For example, if the viscosity of the spinning fluid is too low, that is, a melt temperature is too high or a solution too dilute, the fluid will be extruded in individual drops. On the other hand, the extreme of high viscosity can lead to cohesive fracture of the spinning fiber.

The usual measure of spinnability is the maximum thread length x^*, which appears to be a complex function of velocity and viscosity (as shown in Fig. 11-2 for a series of cellulose acetate solutions). This behavior has also been demonstrated by the work of Ueda and Katoaka [1B], who found that, for molten polyethylene, $x^* = (\eta V)^{0.86}$.

Basically, the overall limits of spinnability are set by two end conditions. The first of these is cohesive brittle fracture caused by applied tensile force (see Fig. 11-3) and the second by spontaneous development of surface waves, with resultant breakup into drops (see Fig. 11-4).

In the first case,

$$W_{el} = \frac{(P_{xx})^2}{2E} \approx K \tag{11-1}$$

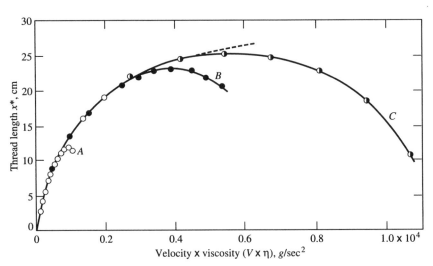

Fig. 11-2   Thread length vs velocity $\times$ viscosity [1A].

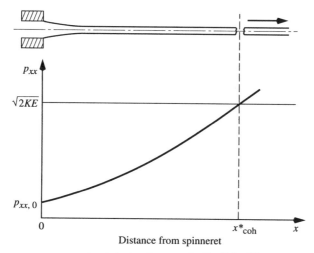

Fig. 11-3   Cohesive fracture of a liquid jet [6].

where

W_{el} = critical elastic energy density
E = Young's modulus
P_{xx} = critical tensile strength
K = cohesive energy density

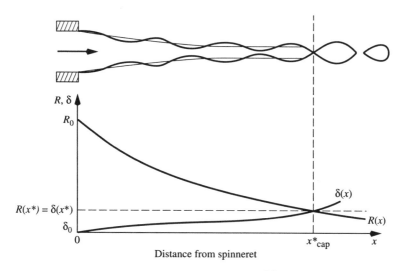

Fig. 11-4   Capillary breakup [6].

The W_{el} is reached when e_{xx} (the deformation extension rate) is not less than some critical value.

$$\dot{e}_{xx} = \frac{2K}{E\tau^2} \tag{11-2}$$

where τ = relaxation time.

If, then, an exponential velocity distribution in the χ or axial direction is assumed, we obtain an approximate expression for the maximum thread length due to cohesive failure:

$$\chi^* \text{ cohesive} = \frac{1}{\dfrac{d\ln V}{d\chi}} \ln\left[\left(\frac{2K}{E}\right)^{1/2} \left(\frac{1}{3V_0 \dfrac{d\ln V}{d\chi} \tau}\right) \right] \tag{11-3}$$

where V_0 is the initial thread line velocity.

In the second case for wave development and breakup into drops, the axially symmetrical distortion of the fiber radius R is

$$B(t) = B_0 \exp(ct)\cos\left(\frac{2\Pi\chi}{\lambda}\right) \tag{11-4}$$

where

B = axially symmetrical distortion of radius
c = growth factor
λ = wavelength

The breakup condition occurs when

$$B(\chi^*) = R(\chi^*) \tag{11-5}$$

and,

$$\chi^*_{\text{capillary}} \approx \frac{2}{\dfrac{d\ln V}{dx}} \left[\ln\left(\frac{R_0}{B_0}\right) - \frac{\sigma}{\left(3\eta V_0 \dfrac{d\ln V}{dx}\right)} \right] \tag{11-6}$$

where

σ = surface tension
R_0 = initial thread line radius

If both mechanisms are combined, the results shown in Fig. 11-5 are obtained where curve 1 is capillary breakup, curve 2 cohesive failure and curve *2a* combined cohesive failure and capillary waves.

11.3   TYPES OF SPINNING; ROLE OF THE TRANSPORT PHENOMENA

There are basically four types of fiber spinning:

1.   Melt spinning
2.   Dry spinning
3.   Wet spinning
4.   Reaction spinning

The distinction between them is made by the technique used for solidification. Melt spinning consists of extruding a molten polymer and into an appropriate medium (gas or liquid), where it is solidified by the transfer of heat. Dry spinning involves the extrusion of a polymer solution into a heated gas, where the solvent is removed and the fiber solidified. Wet spinning represents the extrusion of a polymer solution into a liquid chemical bath. The subsequent solidification takes place by mass transfer. In reaction spinning, a prepolymer (partially reacted material) is extruded into a heated fluid medium, where solidification takes place by chemical reaction.

It is apparent that the transport phenomena (momentum transfer or fluid flow, energy transfer or heat transfer, mass transfer, and chemical reaction) play a role

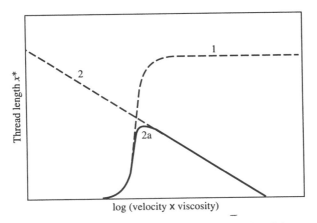

Fig. 11-5   Combined mechanisms of thread line failure [6].

in fiber spinning; the relation to the various spinning processes is shown in Table 11-2. The notation *primary* means that the transfer process is important, whereas *secondary* means that the process can occur but is not usually of prime importance in the fiber-spinning process.

Figure 11-1 gives schematic representation (1) of three of the processes. Generally, reaction spinning will resemble either dry or wet spinning.

11.4   MELT SPINNING

Melt spinning is one of the most widespread of all the fiber-processing operations. Typical products include polyesters, polypropylene, and nylons. The process involves either melting polymer pellets in, or pumping molten polymer to, an appropriate spinning unit, which carefully meters the flow of the polymer. The molten polymer passes through a sand or screen pack and then through a spinneret or jet, where the fibers are formed and extruded. After the extrusion, the filaments are quenched and cooled in some fluid medium (air, other gases, or even water) and eventually taken up on an appropriate mechanical device. In some cases, the fiber is then processed further, that is, drawn, separately. In others, the process is continuous in that the polymer is spun and drawn on one overall apparatus.

If the overall process of melt spinning is analyzed, three distinct regions or zones become apparent. These are (see Fig. 11-6):

1. The experience or behavior of the melt prior to extrusion
2. The region after extrusion until solidification
3. The behavior from solidification until take-up

Each zone will be discussed in detail in the following sections.

TABLE 11-2   Relation of Transport Processes to Fiber-Spinning Types

Fiber-Spinning Type	Momentum Transfer (Fluid Flow)	Energy Transfer (Heat Transfer)	Mass Transfer	Chemical Reaction
Melt	Primary	Primary	Secondary	Secondary
Dry	Primary	Primary	Primary	—
Wet	Primary	Secondary	Primary	—
Reaction	Primary	Primary	Primary	Primary

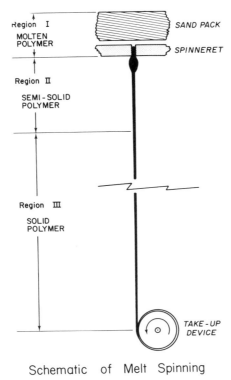

Schematic of Melt Spinning

Fig. 11-6   Melt spinning zones.

Flow Through Sand or Screen Packs

In this region of flow, the molten polymer is pumped first through a sand or screen pack and thence to a set of spinneret holes, from which it is extruded.

Originally, the purpose of the sand or screen pack was to act as a filter prior to spinning. Industrial experience shows, however, that other functions of the sand pack are apparently far more important than the filtration aspect.

This importance has been demonstrated in a variety of ways. First of all, the porosities of the sand or screen pack are quite critical. It has been shown that these can have a definite effect on final fiber properties. Beyond this, even the placement of a given porosity medium (sand or screen) in a layered pack can exert a strong influence on the product properties. Such behavior shows that the rheological history of the melt in the pack is a factor in setting yarn properties.

In addition to the effect of the pack itself, there is the pattern of behavior experienced in spinning operations. At times in industry, high pack pressure occurs without a related change in polymer degree of polymerization, inhomogeneity content, or spinning rate. The occurrence of these patterns also coincides with improved fiber properties. The high pack pressures encountered reflect a change in the polymer's rheological behavior.

In order to develop some possible mechanisms, consider the flow situation in a pack system. First of all, a molten polymer is not only non-Newtonian but also viscoelastic. The flow of a polymer melt through a porous media with a large number of tortuous capillaries of varying cross section can lead to complex flow and shear patterns [8–10].

For example, the fluid can be alternately stressed and relaxed because of its viscoelastic nature. Such treatment could have an ultimate effect on fiber properties.

At this stage, consider the flow behavior of the melt in porous media. Experimental and theoretical studies have shown that a modified Blake–Kozeny [9,10] equation could be used to describe such flow with the application of the power law:

$$\tau = K\left(\frac{d\gamma}{dt}\right)^n \tag{11-7}$$

where τ = shear stress, $d\gamma/dt$ = strain rate or shear rate, and K,n are empirical constants for a power law fluid.

The appropriate Blake–Kozeny equations were:

$$v_0 = \frac{n\epsilon}{3n + 1}\left(\frac{D_p}{3(1 - \epsilon)}\right)^{(n+1)/n}\left(\frac{6\Delta P}{25KL}\right)^{1/n} \tag{11-8}$$

$$v_0 = \left(\frac{k}{H}\frac{\Delta P}{L}\right)^{1/n} \tag{11-9}$$

$$H = \frac{K}{12}\left(9 + \frac{3}{n}\right)^n (150K\epsilon)^{\frac{1-n}{2}} \tag{11-10}$$

$$f^* = \frac{\Delta P D_p \epsilon^3 \rho}{LG_0^2(1 - \epsilon)} \tag{11-11}$$

$$N_{Re}^* = \frac{D_P G_0^{2-n} \rho^{n-1}}{150H(1 - \epsilon)} \tag{11-12}$$

where v_0 = superficial velocity, ϵ = porosity, D_P = particle diameter, ΔP = pressure drop, k = permeability, H = non-Newtonian bed factor, f^* = modified friction factor, G_0 = superficial mass velocity, and N_{Re}^* = modified Reynolds number.

Application of Eqs. (11-12–11-16) yielded the relation

$$f^* = \frac{1}{N_{Re}^*} \tag{11-13}$$

Molten polymer experimental data were found to fit the relation (see Fig. 11-7) for the N_{Re}^* range from 5×10^{-10} to 10^{-8}. For this range, little effect of viscoelasticity appeared. However, Marshall and Metzner [11], working with polymer solutions, found that (see Fig. 11-8), in a higher N_{Re}^* range, viscoelastic effects did occur. Since the flow rates encountered in industrial operations would be considerably higher than those of the experimental molten polymer studies [9,10], viscoelastic effects could very well play an important role.

Beyond the rheological experiences in the small capillaries of the pack, there are other possible effects to consider. Such flow of a polymer melt can lead to

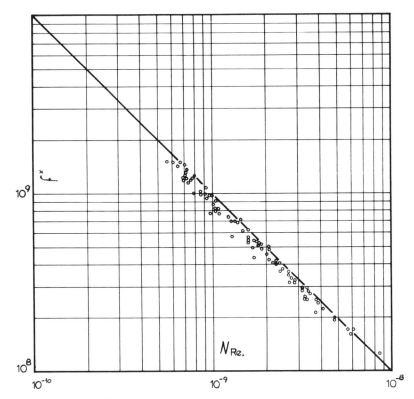

Fig. 11-7 $f^* - N_{Re}^*$ Relation for flow of molten polymers through porous media [10].

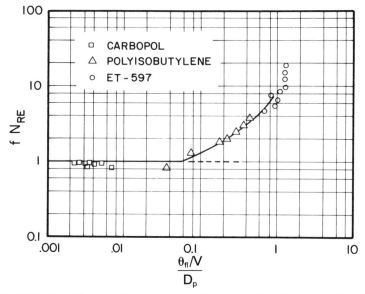

Fig. 11-8 $f^* - N_{Re}^*$ versus Deborah number showing possible viscoelastic effects [11].

molecular orientation. Furthermore, the effect of the tortuous, nonconstant cross sections will be to vary that orientation experience considerably. It has been proposed [12,13] that molecular fractionation could take place across a flowing stream of polymer.

The possibility of rheological effects (shear, relaxation), molecular orientation, and molecular fractionation taking place are further reinforced by industrial observation of the melt flow between the pack and the spinneret capillaries. Here, it has been found that the space between the pack and the spinneret capillaries is a critical factor in determining fiber properties. The sensitivity of these properties to this distance strongly indicates the possible effects of relaxation, changes in molecular orientation, or fractionation, all of which could influence fiber properties.

Extrusion Through Capillaries (Spinnerettes)

The flow of the polymer melt through the spinneret capillaries themselves constitutes still another portion of the pre-extrusion experience of the polymer. The complexity of this aspect can be demonstrated by considering the phenomenon of melt fracture. When a molten polymer is extruded, it can exhibit surface irregularities that culminate in the extrudate being twisted or actually broken (see Fig. 11-9). Three main theories have been proposed to try to explain this

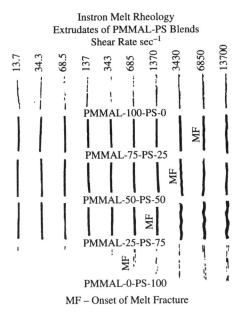

Instron Melt Rheology
Extrudates of PMMAL-PS Blends
Shear Rate sec^{-1}

MF – Onset of Melt Fracture

Fig. 11-9   Melt fracture effects.

behavior relating to flow entering the capillary, leaving the capillary, and in the capillary itself.

For example, Clegg [14] found, by visual observation of transparent dyes, that extrudate distortion was related to inlet flow behavior to the capillaries. Carefully tapering the inlet to the capillary improved the situation with respect to such distortion. In essence, the result indicated that the extrudate's coiled position in the reservoir before the capillary was retained through a viscoelastic "memory."

Other investigators have shown that extrudate irregularities also appear to occur as a result of capillary flow behavior. For example, Howells and Benbow [15], who extruded molten polymer directly from tubes without an inlet region, found that irregularities occurred. Others have attributed the phenomenon to boundary slip [16]. Attempts have also been made to try to explain the phenomenon on the basis of stability theory [17] and the compressibility of the polymer melt [17].

In any case, it can be seen that the experiences of the melt in the inlet region and capillary itself can result in downstream occurrences. Before leaving the capillary situation, it is probably well to examine the applicable flow equations and corrections that apply.

It has been shown [18,19] that, for steady, isothermal, laminar flow, the relationship between τ_{wall} and the term $8V/D$ (where V is the average fluid velocity) must be independent of tube diameter as long as:

1. Shear stress–shear rate relations are not time-dependent.
2. Slippage does not occur at the tube wall.

In general, both effects are not found to any extent in the flow of polymer melts through a capillary.

A thorough discussion of capillary flow, together with end effects, kinetic energy, and other aspects, is given in Chapter 3.

The Extruded Semimolten Fiber

In this region, several important factors are taking place. First of all, the extruded melt experiences the phenomenon known as die swell. In this process, the extruded polymer melt expands its diameter beyond the limits of the spinning capillary. The diameter increase can be as much as 200% to 300% (see Fig. 9-12).

The complex changes taking place in the die swell phenomenon could have sizable effects on the development of melt-spun fiber properties.

In the region from extrusion to solidification, the fiber undergoes both momentum- and energy-transfer situations. After the die swell momentum experience, the fiber is subjected to the application of a number of forces. The applicable force balance [30] is

$$F_{ext} + F_{grav} = F_{rheo} + F_{in} + F_{surf} + F_{aero} \tag{11-14}$$

where F_{ext} is the external tensile force (take-up force), F_{grav} the gravitational force, F_{rheo} the force due to the rheological nature of the fluid, F_{in} the force of inertia, F_{surf} the surface-tension force, and F_{aero} the air-drag (skin-friction) force.

Analysis of this force balance in the region of solidification has been somewhat limited. One set of investigators [31] solved a modified form,

$$F_{rheo} = F_{ext} + F_{grav} - F_{in} \tag{11-15}$$

using for the rheological term a combination of the equation

$$_{zz} = 3\eta \left(\frac{\partial V}{\partial z} \right) \tag{11-16}$$

where η is a viscosity, P_{zz} a tensile stress in the z direction, and $(\partial V/\partial z)$ the velocity gradient in the z direction; they also used equations for viscosity temperature dependence and a z-direction temperature distribution. The results of the analysis were found to be only qualitatively correct.

Likewise, another set of workers [32] used an even simpler form of the force balance:

$$F_{\text{rheo}} = F_{\text{ext}} = \text{const} \tag{11-17}$$

Their results were even less satisfactory although they qualitatively seemed to indicate the trend of V, the velocity in the z direction.

It should be pointed out that all these treatments deal essentially with a gross balance of forces and the behavior of V, the average fluid velocity. There apparently are no treatments of the velocity profile (velocity with radius) within the spinning fiber. In spite of this, a qualitative understanding of the situation can be developed.

As a starting point, consider the behavior of the fluid after extrusion. Immediately after extrusion, the fluid experiences the die swell phenomenon, where the velocity profile flattens. Ultimately, when the fiber solidifies, the velocity profile will be flat (i.e., plug flow). In between, a velocity profile will possibly be first formed and then distorted by solidification at the fiber exterior. Even if the profile becomes fully developed, however, it will not have a parabolic shape but rather will have a blunted form because of the polymer's non-Newtonian fluid behavior. In essence, then, the fiber in the post–extrusion–solidification region will have a velocity profile that can be closely approximated by assuming plug flow (i.e., a constant V across the fiber cross section).

This velocity profile behavior greatly simplifies the treatment of the heat-transfer portion of the solidification process. From the energy equation (4-2), the applicable equation is

$$\rho c_p V \frac{\partial T}{\partial z} = k \left[\frac{1}{r} \frac{\partial}{\partial r} \left(r \frac{\partial T}{\partial r} \right) \right] + A_0 \tag{11-18}$$

where it is assumed that:

1. The fiber radius is constant.
2. Fiber properties ρ, c_p, and k are constant.
3. $v_z = V$ (i.e., approximate flat velocity profile).
4. Convection in the z (axial) direction greatly exceeds conduction:

$$\left(\rho c_p V \frac{\partial T}{\partial z} >>> k \frac{\partial^2 T}{\partial z^2} \right)$$

The A_0 term in Eq. (38) is a heat-generation term, in this case the heat of fusion for the solidifying polymer. Also, the V is the average fiber velocity relative to the quenching gas flow. The applicable boundary conditions are:

$$T = T_0 \text{ at } z = 0, \text{ all } r \tag{11-19}$$

$$T \text{ is finite at } z = z, r = 0 \tag{11-20}$$

$$\left(\frac{\partial T}{\partial r}\right)_{r=R} = \frac{h'(T_R - T_g)}{k} \tag{11-21}$$

where R is the fiber radius, T_g the ambient gas temperature, and h' the heat–transfer coefficient between the ambient gas and the fiber surface. The foregoing equation has not been solved in the open literature. In spite of this, some understanding can be attained by considering a simplified form of Eq. (38) that does not include the A_0 term:

$$\rho c_p V \frac{\partial T}{\partial z} = k\left[\frac{1}{r}\frac{\partial}{\partial r}\left(r \frac{\partial T}{\partial r}\right)\right] \tag{11-22}$$

A solution applied to typical melt-spinning conditions, indicates that the ambient air temperature would be reached not far from the spinneret face. Of course, in actuality, the requirement of matching the polymer's heat of fusion will greatly delay the fiber's reaching ambient temperature.

A number of investigators [33–36] have used a somewhat different approach than Eq. (39). Their technique has been essentially to ignore A_0 and to take a coordinate system moving with the fiber velocity. Then,

$$\rho c_p \left(\frac{\partial z}{\partial t}\right)\left(\frac{\partial T}{\partial z}\right) = k\left[\frac{1}{r}\frac{\partial}{\partial r}\left(r \frac{\partial T}{\partial r}\right)\right] \tag{11-23}$$

and

$$\rho c_p \frac{\partial T}{\partial t} = k\left[\frac{1}{r}\frac{\partial}{\partial r}\left(r \frac{\partial T}{\partial r}\right)\right] \tag{11-24}$$

The boundary conditions used by most investigators were the same as Eqs. (38a–c). Solutions of Eq. 11-24 were directed mainly toward determining average fiber temperature rather than the radial temperature profile. Also, many of the solutions did not in any way take account of the solidification process occurring in the fiber.

Basically, this mean that most of the investigators were essentially determining a heat-transfer coefficient h' that was related either to the Reynolds and Prandtl numbers or to the Reynolds number itself. The result was that such coefficients differed by as much as a factor 8 [41,42]. The reason for this can be appreciated. First, all the investigators computing h' did not consider the freezing or solidification aspect of the process. Next, the average temperatures were determined in some cases by means that were not very precise.

In this latter regard, a considerable improvement was made in measuring fiber surface temperatures by infrared techniques [36] or with a polarizing interferometer [34]. However, even in these cases, the investigators still utilized Eq. (11-24), with its associated boundary conditions. An example was the work of Wilhelm [36], some of whose data are presented in Fig. 11-10. The curves are calculated from Eq. (11-24), and the points represent experimentally observed data. At first glance, the fit appears to be acceptable. It should be pointed out, however, that the heat-transfer coefficients used in the boundary condition (11-21) were, in essence, computed from experimental observations. In essence, then, the curves actually represent a fit of the data themselves rather than a separate theoretical solution. The trend of surface temperature change is, of course, valid. Furthermore, the plots all show a plateau (constant temperature) over a range of distances. These plateaus represent the final solidification of the fiber. For the cases studied, these occurred 8–20 cm from the spinneret face.

While all the foregoing contributions are worthwhile, they would be more

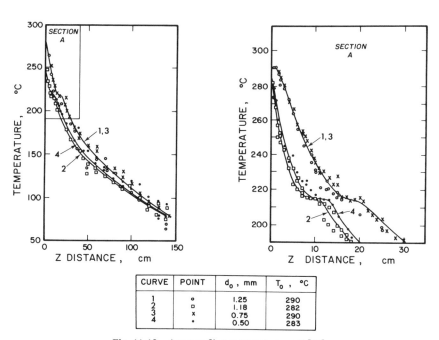

CURVE	POINT	d_0 , mm	T_0 , °C
1	o	1.25	290
2	□	1.18	282
3	x	0.75	290
4	•	0.50	283

Fig. 11-10   Average fiber temperature vs Z [41].

significant if heat-transfer studies included the solidification process and also developed information regarding the temperature profile within the fiber. Two sets of investigators have considered the heat transfer from this viewpoint. Morrison [37] approached this by writing separate forms of Eq. 11-18 for the liquid and solid phases, as well as a third equation,

$$-k_{\text{liquid}} \frac{\partial T_{\text{liquid}}}{\partial r}/r^* + k_{\text{solid}} \frac{\partial T_{\text{solid}}}{\partial r}/r^* - L\rho_{\text{solid}} \frac{dr^*}{d\left(\dfrac{z}{V}\right)} = 0 \qquad (11\text{-}25)$$

where k_{liquid} and k_{solid} are liquid and solid thermal conductivities, respectively; ρ_{solid} is the solid density; r^* is the radius for the solid/liquid interface; and L is the heat of fusion. All three equations were solved numerically to give both radial and axial effects on point temperature. Figure 11-11 is a typical set of data.

Although these cannot be confirmed, they are of interest in giving a representation of the solidification process with the associated temperature profile. The fiber is transformed from a liquid to a solid at an Ω value of about 0.832. Since Ω is defined as distance from the spinneret z times thermal diffusivity α divided by the fiber velocity V times the fiber radius squared (R^2), it appears possible to estimate the z required for solidification. When this is done, the estimated z is found to be several thousand centimeters. This finding is obviously not correct and indicates that the data of Fig. 11-11 cannot be used directly to calculate the axial distance required for complete solidification.

There is, however, another use for data of the type of Fig. 11-11. That use is to estimate the reduced radius at which the solid/liquid interface is positioned for various z values. All that is needed in such as case is some rough measure of complete solidification, such as a quench point generally measured as a plant parameter. The rough measure of z is then taken as the value for the Ω for solidification ($\Omega = 0.832$). Then, for example, for the case in which the liquid/solid interface is at an r/R of 0.5 ($\Omega = 0.510$), the following will hold:

$$\frac{\Omega_e(\text{complete solidification})}{\Omega_e(\text{liquid/solid at } r/R = 0.5)} = \frac{z(\text{complete solidification})}{z(\text{liquid/solid at } r/R = 0.5)} \qquad (11\text{-}26)$$

if it is assumed that the ratio α/VR^2 does not appreciably change. For the case cited,

$$z(\text{solidification at } r/R = 0.5) = \frac{(0.510)}{(0.832)} z(\text{complete solidification}) \qquad (11\text{-}27)$$

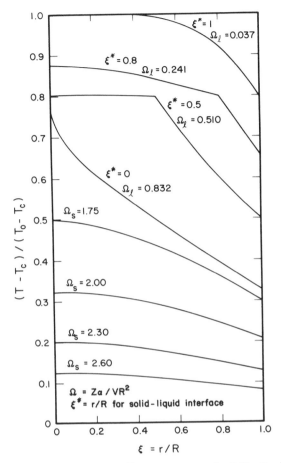

Fig. 11-11   Calculated point fiber temperature and solidification [37].

The same estimate can be made for other r/R points of solidification. Such data, together with the temperature profiles, could be useful in determining fiber properties.

Attention should also be directed to the work of Fox and Wanger [38], who combined infrared experimental studies of surface temperatures with a refinement of Morrison's numerical solution. They obtained very precise fits of their numerical and experimental data for surface temperatures. This overall result lends confidence for the use of such approaches in the study of temperature and solidification behavior in the region after extrusion and prior to solidification.

The complete solidification of the fiber simplifies considerably the analysis for this region. Basically, in a rheological sense, the fiber is a solid being subjected to an elongating force. Although the mechanical force balance cannot be handled completely quantitatively, it is still not as complex as the two-phase situation taking place during solidification.

In terms of the heat-transfer situation, the solidification establishes plug flow (all point velocities = V), which confirms the earlier assumption. Furthermore, the completely solid polymer allows the A_0 term (for the polymer heat of fusion) to be dropped. The result is that equation (11-22) can be used to describe the heat-transfer situation. In fact, the analysis cited earlier shows that the solid fiber rapidly attains the ambient temperature of the quenching gas.

Furthermore, the relatively uncomplicated situation in the melt-spinning process when the fiber has solidified actually permits meaningful analysis of the development of fiber properties for this region.

11.5 DRY SPINNING

Mechanisms

The principal feature of dry spinning is the formation of solid fibers by the evaporation of a solvent. In dry-spinning operations, the polymer to be spun is first dissolved in an appropriate solvent, filtered, and then charged to the spinning device. The spinning solution (usually called a *dope*) is at a temperature higher than room temperature and at a pressure above atmospheric pressure before extrusion. As a result, when the dope is extruded, it undergoes a process similar to flash vaporization [43–48] except that, in this case, solvent vapor is rapidly evolved from the extruded material. Observations [43–49] made close to the spinning jet orifices confirm that large amounts of vapor are present. After this initial surge of vapor evolution, the process seems to resemble conventional drying.

Figure 11-12 presents typical experimental curves of solvent content vs distance from the spinning jet [43–45]. These curves are for polymethyl methacrylate spun from a benzene solution. Similar curves [49] have been found for such industrially important fibers as cellulose triacetate.

As can be seen, the solvent concentration curves show a sharp decline followed by a plateau and then a much more gradual curve. The initial sharp decrease corresponds to the flash vaporization that occurs after extrusion. The later more gradual curve is the same as that of a normal drying curve. The nature of the plateau will be considered later. However, it should be mentioned that such plateaus are also found in industrial data.

Basically, then, it can be concluded that the evolution of solvent vapor takes place in three stages: (1) a rapid initial process caused by extruding a solution

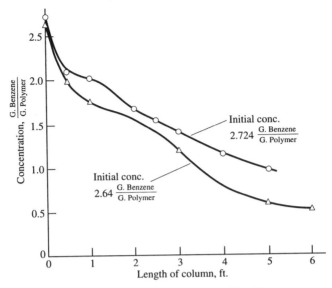

Fig. 11-12   Concentration data vs Z [43–45].

from a higher ambient pressure into a lower ambient pressure, (2) a later process in which standard (diffusion-controlled) drying takes place, and (3) an intermediate or transition process, that is, the plateau.

Momentum Transfer

Many of the effects noted earlier, such as screen or sand pack behavior and entrance effects in spinning jets, are not as pronounced in dry spinning. The result is that pre-extrusion effects associated with momentum transport usually do not relate to fiber-spinning performance or properties. One exception is the effect of the shape of the entry into the spinning jet or orifice. It has been found that fiber properties could be correlated with the shape of these entries.

An experimental study by Brazinsky et al. [49A] investigated the flow behavior of dry-spun fibers. Specifically, they determined spin line radius, spin line velocity, and flow rate as functions of axial distance. Their results are shown in Figs. 11-13–11-15.

In considering the actual aspects of momentum transport in the postextrusion dry-spinning process, two factors should be discussed. The first of these is the shape of the velocity profile in the spinning fiber. When the fiber is first extruded, the velocity profile is flat. From this point, it ultimately develops into an appropriate profile. Since the fluid is non-Newtonian, this profile will not be parabolic but will be blunted. Finally, as the fiber solidifies, the solid portion

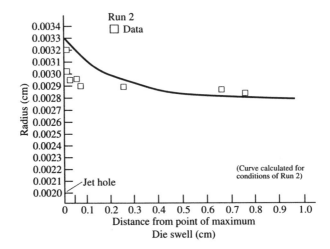

Fig. 11-13   Spin line radius as a function of axial position.

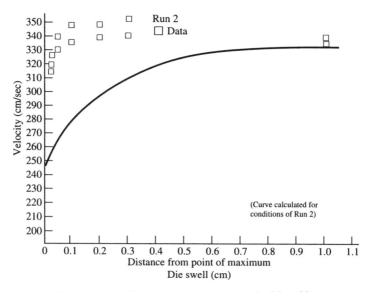

Fig. 11-14   Spin line velocity as a function of axial position.

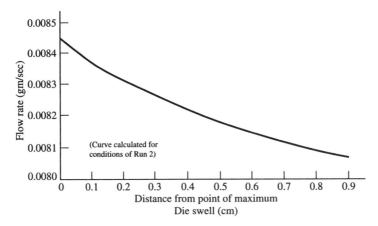

Fig. 11-15   Flow rate as a function of axial position.

will have a uniform velocity until, finally, the entirely solid fiber will have plug flow (i.e., $U_z = V$, where V = average velocity).

The change in velocity profile from a flat to a blunted profile and then rapidly to a flat or plug flow condition means that the assumption of a flat velocity profile ($U_z = V$) in the spinning fiber is reasonable. Furthermore, this assumption will greatly simplify consideration of the heat- and mass-transfer aspects of the dry-spinning process.

The second factor relating to momentum transport is the nature of the average velocity V. Generally, in dry-spinning operations, heated air or another gas is passed concurrently or countercurrently in parallel flow to the spinning fiber. This means that the V value used in the heat- and mass-transfer analysis must be relative to the gas velocity.

Heat Transfer

Consider Eq. (4-2), the energy equation presented earlier. If it is assumed that:

1. The fiber radius is constant.
2. Fiber properties ρ, c_p, k are constant.
3. $v_z = V$ (i.e., approximate flat velocity profile).
4. Convection in the z (axial) direction greatly exceeds conduction:

$$\left(\rho c_p V \frac{\partial T}{\partial z} \ggg k \frac{\partial^2 T}{\partial z^2} \right)$$

Then,

$$\rho c_p V \frac{\partial T}{\partial z} = k \left[\frac{1}{r} \frac{\partial}{\partial r} \left(r \frac{\partial T}{\partial r} \right) \right] + A_0 \tag{11-28}$$

where the terms are as defined earlier.

In the equation, the only term relating to the mass-transferred (i.e., solvent-vaporized) is A_0, the energy required for solvent evaporation. If small z segments are taken, the equation can be solved since A_0 will be essentially constant over small z values. For such a case, the boundary conditions are:

$$T = T_0 \text{ at } z = 0, \text{ all } r$$

$$T \text{ is finite at } z = z, r = 0$$

$$T = T_s \text{ at } z = z, r = R$$

where T_0 = extrusion temperature, T_s = fiber surface temperature, and R = fiber radius. The solution is

$$T = T_s + \frac{A_0}{4k} [R^2 - r^2]$$

$$- \frac{A_0 R^2}{4k} \sum \left\{ \left[\frac{2}{(\alpha_n)^3 J_1(\alpha_n)} - \frac{2}{\alpha_n [J_1(\alpha_n)]^2} - \frac{4 J_2(\alpha_n)}{(\alpha_n)^2 [J_1(\alpha_n)]^2} \right] \beta \right\}$$

$$- \frac{\bar{T} - T_\infty}{T_\infty} \sum \frac{2}{(\alpha_n)^3 J_1(\alpha_n)} J_0 \left(\alpha_n \frac{r}{R} \right) \exp - \left(\frac{\alpha_n^2 k \pi z}{c_p Q} \right)$$

where α_n = the solution of the equation $J_0(x) = 0$, J_0 = the Bessel function of the first kind and order zero, J_1 = the Bessel function of the first kind and order 1, J_2 = the Bessel function of the first kind and order 2, and Q = average volumetric flow rate.

Also,

$$\beta = J_0 \left(\alpha_n \frac{r}{R} \right) \exp \left(\frac{-\pi \alpha_n^2 k z}{c_p Q} \right) \tag{11-30}$$

The overall solution rapidly reduces to

$$T = T_s + \frac{A_0}{4k} (R^2 - r^2) \tag{11-31}$$

within a very short z distance.

This equation should describe the temperature behavior for the dry-spun fiber. There are difficulties in using it, however, since the fiber surface temperature could not be measured [50]. The reason is that dry-spun fibers are at too low a temperature to use infrared techniques and the surface too wet to use a contact system.

The problem of the surface temperature was bypassed in the following manner: By an energy balance over a fiber segment, the following equation results:

$$hA(T_\infty - T_s) - w\lambda = Qc_p(\overline{T} - \overline{T}_1) \tag{11-32}$$

where h = the heat-transfer coefficient from ambient fluid to the fiber, T_∞ = the temperature of the ambient fluid, w = the weight of solvent evaporated in a given fiber segment, λ = the heat of vaporization of solvent, $\overline{T}$ = the average temperature of fiber segment, $\overline{T}_1$ = the average temperature of the preceding segment, and A = the surface area.

Terms such as T_∞, A, Q, c_p, and w can be found experimentally (i.e., Fig. 11-12); c_p and h values are derived from the literature. The h values can be taken from the work of Mueller [51], who studied heat transfer to long thin wires in parallel airstreams. $\overline{T}_1$ can be closely approximated in the first z segment since the temperature profile is flat at $z = 0$. This, however, still leaves $\overline{T}$ and T_s to be determined. If Eq. (56) can be averaged over the cross section to give an expression for $\overline{T}$,

$$\overline{T} - T_s = - \left(\frac{1}{6\pi k}L\right)w\lambda \tag{11-33}$$

Solving Eqs. (57) and (58) simultaneously eliminates T_s and allows $\overline{T}$ to be determined.

A plot of average fiber temperatures calculated in this manner for polymethyl methacrylate fibers is shown in Fig. 11-16. The stars on the plot are average fiber temperatures measured experimentally with a miniature calorimeter.

As can be seen, there is some agreement at shorter column lengths. However, at the longer lengths, the experimental values are considerably lower than the calculated values. This lack of agreement is probably due to heat losses from the calorimeter, which would be accentuated at the higher fiber temperatures of the longer column lengths. It is interesting to note, however, that the shapes of the experimental and the calculated curves are similar, even though displaced.

Curves similar in shape to the one shown in Fig. 11-16 have also been calculated for industrially spun fibers [49].

In general, there is a decrease of temperature well below the initial extrusion temperature, followed by a gradual rise until a plateau is reached. This initial

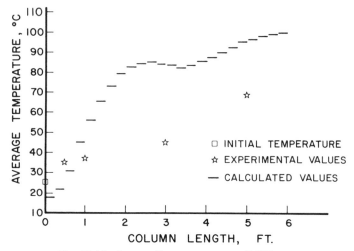

Fig. 11-16   Average fiber temperature vs Z [50].

temperature is readily explained in view of the flash vaporization of solvent that takes place upon extrusion. The large amount of solvent liberated brings about a severe cooling effect. Rough measurements made in the vicinity of the jet or orifice confirm this result.

Figure 11-16 shows that the spun fiber will reach an average temperature close to the ambient temperature within a relatively short distance (2 ft) from the spin jet face. Industrial data show that fiber average temperature is also ambient at approximately the same distance [49].

The fairly rapid attainment of ambient temperature should not be too surprising in light of the calculated result when solvent evaporation does not take place. For such a case, the applicable equation is

$$\rho c_p V \frac{\partial t}{\partial z} = k\left[\frac{1}{r} \frac{\partial}{\partial r}\left(r\, \frac{\partial T}{\partial r}\right)\right] \tag{11-34}$$

When the equation is solved, the result shows that ambient temperature should be reached within 1 cm of the spin jet face. With solvent evaporation, the resultant cooling depresses the fiber temperature and thereby makes the distance necessary to attain ambient temperature greater than it would be without solvent evaporation.

Mass Transfer

The foregoing discussion of momentum and heat transfer sets the stage for the consideration of mass transfer. First, the approximate flat velocity means that average velocity relative to the ambient gas stream can be used in describing mass transfer. Next, heat-transfer studies show that the small fiber diameter offers negligible resistance to heat transfer. The result is that the solvent can easily be evaporated but still has to diffuse through the fiber at some given rate.

In essence, then, the rapid heat transfer and the nearly flat or flat velocity profile mean that the equation of continuity of species [Eq. (5-2)] can properly describe the situation. If that equation is subjected to the following conditions:

1. $v_z = V$
2. Steady-state $\partial c/\partial t = 0$.
3. No chemical reaction ($R_A = 0$).
4. Diffusing solvent is a vapor.
5. No diffusion of ambient gas into the fiber.
6. Constant density and diffusivity.
7. Axial convective mass transfer exceeded axial mass transfer by diffusion:

$$\left[v_z \left(\frac{\partial c}{\partial z} \right) \gg D_{AB} \left(\frac{\partial^2 c}{\partial z^2} \right) \right]$$

8. No mass transport in the θ direction:

$$\left(\frac{D_{AB}}{r^2} \frac{\partial^2 c}{\partial \theta^2} = 0 \right)$$

then, the resultant equation is

$$\frac{\partial c}{\partial z} = \frac{D_{AB}}{V} \left[\frac{1}{r} \frac{\partial}{\partial r} \left(r \frac{\partial c}{\partial r} \right) \right] \tag{11-35}$$

Now, the diffusivity of the solvent within the fiber is far less than in air or any ambient gas (about 10^{-6} cm^2/s in the fiber compared to about 10^{-1} cm^2/s in gas). This means that the principal diffusional resistance is within the fiber and that the solvent, on reaching the fiber surface, will rapidly diffuse away into the ambient gas. This result establishes a boundary condition for the solution of Eq. (35), namely that $c = 0$ at $r = R$ (fiber surface) for all z (actually $c \cong 0$). Taking this condition together with

$$C = C_0 \text{ at } z = 0 \text{ for all } r$$

$$C \text{ is finite at } r = 0 \text{ for all } z$$

Eq. (60) is solved to yield

$$C = C_0 \sum \left[\frac{2}{\alpha_n J_1(\alpha_n)} \right] \left[J_0 \frac{\alpha_n r}{R} \right] \exp \left[-\left(\frac{\alpha_n}{R} \right)^2 \left(\frac{D_{AB} z}{V} \right) \right] \qquad (11\text{-}36)$$

where J_0, J_1 and α_n are as defined earlier and C_0 is the initial concentration.

It should be noted that the work of Sano and Nishikawa [52] obtained a somewhat similar result empirically. In treating the residual solvent concentration data empirically, they found that the logarithm of solvent concentration correlated with column length divided by the product of fiber density and cross-sectional area. This is essentially a simplification of Eq. (11-36).

Ohzawa et al. [53,54] also considered the solution of Eq. (60). However, they did not solve for concentration as a function of r and z but, instead, averaged the solution of Eq. (11-35). This gave them average concentration vs z distance. The average concentration curve they obtained did not match the shapes either of Fig. 11-11 or of industrial data [49]. Other dry-spinning mass-transfer studies include those of Yoshida and Hyodo [55], who studied the use of superheated solvent vapor as a drying atmosphere and Tokahashi and Watanabe [56], who made rough experimental measurements of solvent content in spun fibers.

The series in Eq. (36) converges rapidly and, generally, the first four terms are sufficient. Concentration profiles were calculated by using Eq. (36) and various diffusivity values. A number of different diffusitivities were considered since previous work [57] on diffusion coefficients in acrylic polymers indicated that values ranged from 1 to 5×10^{-5} cm²/s.

Figure 11-17 shows computed curves of average fiber concentrations, together with experimental data. As can be seen, no one calculated curve fit the experimental data. Actually, a combination of the 2×10^{-6} cm²/s and 1×10^{-6} cm²/s curves would seem to fit the experimental data. The lower diffusivity would hold at the longer column lengths. There is a physical precedent for this since the fiber is more like a solid at the longer lengths and, hence, should have a lower diffusivity.

A more rigorous approach yielded somewhat better results. Equation (11-35) was used with a new boundary replacing the former one of zero concentration at $r = R$. The new condition was that surface concentration ($r = R$; $z = z$) was a function of z. In addition, changes in fiber diameter and volumetric flow rate (see Fig. 11-18) with z were considered. Equation (35) was then solved by numerical analysis [43–45]. The resultant calculated data are compared to experimental data in Figure 11-19. The calculated data are presented as point concentrations for r/R values of 0, 1/2, and 1 (centerline of fiber, half the distance between centerline and fiber surface, and fiber surface) for two different diffusivities. Note that the calculated average concentrations exactly matched

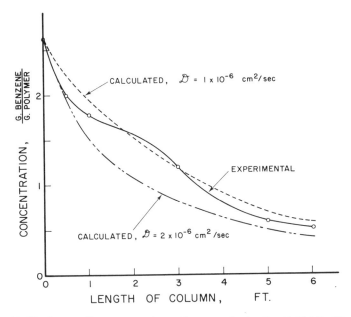

Fig. 11-17   Average fiber concentration vs Z computed using Eq. 11-61 [43–45].

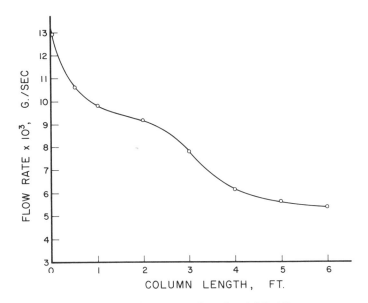

Fig. 11-18   Flow rate vs column length [43–45].

the experimental curve for all diffusivities ranging from 6×10^{-6} cm^2/s to 1×10^{-5} cm^2/s.

The revised approach produced a much more satisfactory solution because not only could one diffusivity be used for the entire column but also because calculated average concentrations exactly matched the experimental curve.

Certain features of Fig. 11-19 merit additional consideration. The calculated curve for $r/R = 0$ and $r/R = 1/2$ show a steady, gradual decline whereas those for the fiber surface ($r/R = 1$) show a rapid initial decline, an increase, and then a decline, with eventual leveling off. It would be expected that the initial rapid effusion of solvent would come mainly from or near the surface of the fiber (rapid initial decline: 0–1 ft). Then, as the rapid solvent evaporation subsided, there would be an actual buildup of solvent at and near the surface since the slower diffusional process predominated (increase from 1 to 2 ft). Next the concentration decreases because of the large solvent concentration gradient for the fiber surface and region immediately adjacent to it in the fiber (2–4 ft). Finally, the decrease in overall solvent content would slow the entire mass-transfer process (4–6 ft). The lower concentration position of the $r/R = 1$ curves is due to a much larger diffusivity of the solvent in air (10^{-1} cm^2/s).

11.6 WET SPINNING

Momentum and Heat Transfer

Wet-spun fibers are formed by extruding a highly viscous polymer solution through a spinneret into an appropriate liquid bath, where it is solidified. The solidification is brought about by a diffusional interchange between the extruded polymer filaments and the bath. In this process, called *coagulation*, one or more components from the bath diffuse into the fiber while, in turn, solvent leaves the forming filaments. The net result is a solid fiber. In some fiber systems, such as viscose rayon, there is also a chemical reaction superimposed on the diffusional process [60–62]. In others, the situation is strictly diffusional. Only the diffusional process will be considered in this treatment.

Wet spinning represents a somewhat different case from melt or dry spinning. For example, temperature effects and heat transfer play a much smaller role than in the other two principal spinning types. On the other hand, pre-extrusion effects may be more important for wet spinning than for dry spinning. Unfortunately, data related to such effects are even sparser for wet spinning than for melt spinning.

One investigator who has reported some data on pre-extrusion effects in wet spinning is Paul [63]. He pointed out, for example, that, in a typical wet-spinning process, the residence time in a spinning capillary is an order of magnitude of

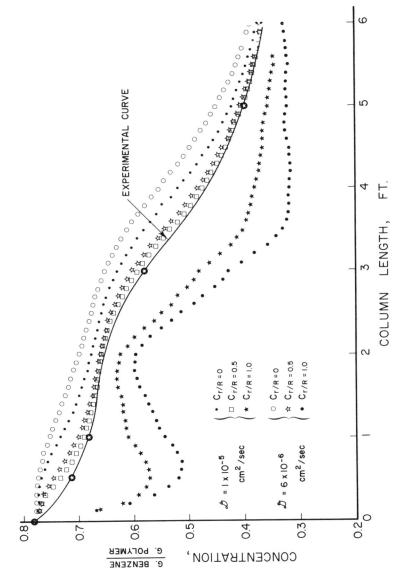

Fig. 11-19 Average and point fiber concentration vs Z computed by Numerical Analysis [43–45].

one or two less than the longest relaxation time in a calculated spectrum of relaxation time. This points out the potential importance of elastic effects on the wet-spinning process in the postextrusion stage.

In addition, Paul studied the behavior of jet swell with wet-spinning solution. Specifically, he measured the ratio D_f/D (where D_f is the diameter of filaments attained if diffusional exchange does not alter solids level or density and if D is the diameter of the spinneret) vs $4Q/\pi R^3$. These data, given in Fig. 11-20 for spinnerets ranging from 3.5 to 11.0 mils, show that the D_f/D ratio steadily increases with $4Q/\pi R^3$ and reaches a maximum near 10^5 s^{-1}. Paul postulates that D_f, since it affects the freestream velocity V_f, influences tension and ultimately spinnability and fiber breakage.

The assumption made earlier for melt and dry spinning, namely, that the velocity profile across a fiber or filament can be represented by plug flow appears also applicable to wet spinning.

Once again, the reasoning is that the non-Newtonian polymer solution, with its blunted profile, will move toward a solid where plug flow definitely holds. This assumption is, of course, quite useful if the heat- or mass-transfer situation with respect to a filament is treated.

The process of wet spinning introduces still another aspect of momentum transport, namely, that of the boundary layer flow induced by the moving fiber. This aspect has been considered by Griffith [64], who used a model similar to

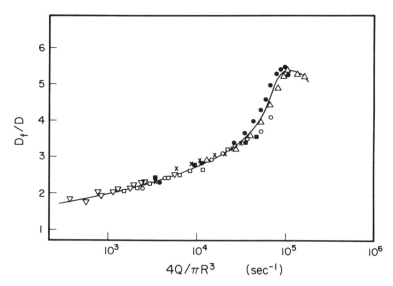

Fig. 11-20 D/D_f vs $4Q/\pi R^3$ [63].

Fig. 11-21. Based on the model, he set up equations of change as shown:

$$\frac{\partial(rU)}{\partial x} + \frac{\partial(rV)}{\partial r} = 0 \tag{11-37}$$

$$Ur\frac{\partial U}{\partial x} + Vr\frac{\partial U}{\partial r} = \frac{1}{N_{Re}}\frac{\partial}{\partial r}\left(r\frac{\partial U}{\partial r}\right) \tag{11-38}$$

where U and V are x and r velocities divided by the fiber velocity U_f, and N_{Re} is $aU_f\rho/\mu$. Velocity profile data are plotted in Fig. 11-22, where it can be seen that the velocity induced in the liquid bath ranges from the fiber velocity at the fiber surface to a zero value somewhere from 30 to 100 radii distances from the fiber surface. In addition, Griffith developed a rough criterion for transition from laminar to turbulent flow around the fiber based on $N_{Re\delta} = \delta U_f\rho/\mu$. Also, he found that

$$N_{Re\delta} \cong 4\sqrt{N_{Re}} \tag{11-39}$$

This meant that, initially, flows around the fiber were laminar (with low values of x) and the became turbulent over a large final section of the bath. Such fluid mechanical behavior would obviously play a strong role in the countercurrent diffusion found in wet spinning.

Griffith [64] also considered the heat-transfer situation for wet spinning by means of the equation

$$Ur\frac{\partial T}{\partial x} + Vr\frac{\partial T}{\partial r} = \frac{1}{Pe_H}\frac{\partial}{\partial r}\left(r\frac{\partial T}{\partial r}\right) \tag{11-40}$$

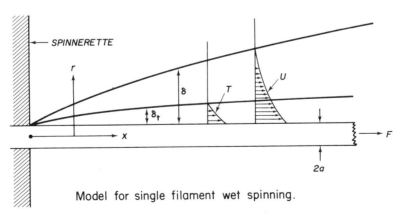

Model for single filament wet spinning.

Fig. 11-21 Wet-spinning filament model [64].

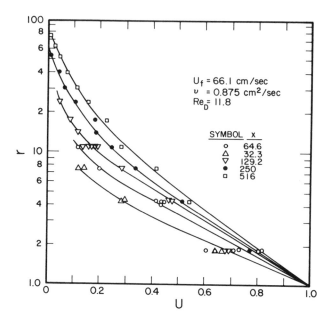

Fig. 11-22   Velocity profile data for wet spinning [64].

where $Pe_H = aU_f/k$. The solution of Eq. (65) showed little change in the fiber temperature for most wet-spinning cases. This is, of course, only a theoretical solution and, in essence, represents at best an approximation.

Mass Transfer

Paul [65] has pointed out that the coagulation process in wet spinning is quite complex since it involves:

1. A ternary system with possible coupling effects
2. A polymer component
3. A phase change for the polymer
4. Thermal effects, such as heat of mixing or a latent heat associated with the phase change
5. Boundaries and volume of sample change.

Simplified approaches could be taken [65] that neglected items 4 and 5 above and considered the system as a pseudobinary (diffusion of solvent into a water bath, and vice versa). Paul used these and the assumption that any phase change was instantaneous when a critical composition was reached to try to treat the mass transfer theoretically. One additional restraint was necessary, namely, the situation with respect to the fluxes of the solvent and water.

Paul [65] considered three such cases: an equal-flux model, a constant-flux ratio model, and a variable-flux model. These models were used together with a vector form of Fick's law:

$$\underline{n}_3 = -\rho D \, \nabla \, w_3 + w_3(\underline{n}_1 + \underline{n}_3) \tag{11-41}$$

where $\underline{n}_1$ and $\underline{n}_3$ are mass fluxes of solvent and water, w_3 is the water weight fraction ($w_1 + w_3 = 1$), ρ is the liquid density, and D is a pseudobinary diffusivity.

In the equal-flux case, a solution derived by Jost [66] was used. The result was compared indirectly to experimental data for coagulation boundary movement for gelled rods of polymer solution in coagulant baths. The results are compared in Figs. (11-23 and 11-24). Although the data are not exactly the same, they do show similar trends, namely, that the boundary position is independent of R and proportional to $t^{1/2}$. The approach here does not account for system weight changes. It does, however, represent a useful qualitative approach to coagulation. Tazikawa [67] has, in fact, used such a model for interpreting coagulation.

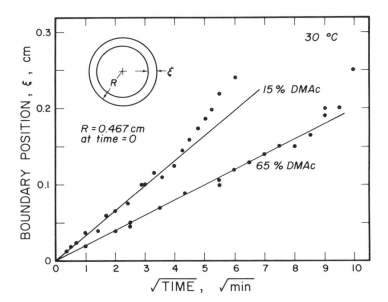

Fig. 11-23   Rate of boundary motion during wet spinning coagulation [65].

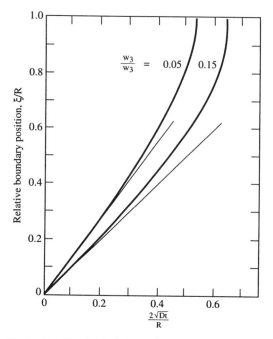

Fig. 11-24   Model calculation of boundary motion [65].

The constant-flux ratio model gave two differential equations:

$$\frac{\partial f}{\partial t} = D_u \frac{\partial^2 f}{\partial x^2} \qquad (11\text{-}42)$$

and

$$\frac{\partial u}{\partial t} = D_c \frac{\partial^2 u}{\partial x^2} \qquad (11\text{-}43)$$

Equation (42) was for the fiber, where $x > \xi$, and Eq. (43) for the region $0 < x < \xi$. In the preceding equations, $f = kw_3/(1 + kw_3)$, $u = \ln(1 + kw_3)$, $k = [(n_1 + n_3)/n_3]$, D_u is the diffusion coefficient in uncoagulated material, D_c is the diffusion coefficient in the coagulated material, and x is the distance measured from the bath/polymer solution interface.

The solutions to Eqs. (11-42) and (11-43) are, respectively, Eqs. (11-44) and (11-45):

$$f = f^*[1 - \text{erf}(x - \rho)]/(D_u/t)^{1/2} \tag{11-44}$$

$$u = u_0 - (u_0 - u^*)\left[\text{erf}\frac{x}{2(D_c t)^{1/2}}\right]/\text{erf}(B_c) \tag{11-45}$$

The asterisk indicates conditions at $x = \xi$ and the zero subscript conditions at $x = 0$. B_c is a constant equal to $\xi/2(D_c t)^{1/2}$.

The model is not compared directly to coagulation data but rather is compared indirectly by means of density of liquid in coagulated solution as a function of weight fraction of solvent in the bath and the diffusivities. The density data are fairly represented by the constant-flux ratio model. However, the temperature behavior of D_u is anomalous. In spite of this, Paul [65] feels that the model gives a reasonable physical picture of coagulation.

The variable-flux model was found to be the least satisfactory of the models. This is probably due to the failure to consider liquid density variations. Again, however, such consideration would make the problem a very difficult mathematical proposition.

An additional piece of experimental data determined by Paul [65] is also of interest. In Fig. 11-25, the respective ratios of weight of solvent removed (M_{1t}) and water added (M_{3t}) to the gel weight M_0 are plotted as functions of solvent concentration in the bath and temperature. The plot shows that transfer rates increase with temperature (i.e., since diffusivities increase), relative rates of solvent and water differ (because of differing diffusivities), and increased bath content of solvent decreases mass transport generally.

Booth [68] made experimental coagulation studies of wet-spun single filaments. This contrasts with Paul's work [65], which experimentally studied only gelled rods of polymer solution and not spun fibers per se. The data of Booth were of several forms. First, the ratio for the uncoagulated radius was obtained as a function of t/R_0^2, where t = time and R_0 = radius of fiber (see Fig. 11-26). The data, after a short initial curvature, are linear as with ξ vs $t^{1/2}$ in the case of Paul's data (Fig. 11-23). Booth also determined diffusion velocities ($\xi^2/4t$) for a number of spun fibers with varying bath contact times. The diffusion velocity values were found to decrease with increasing solvent bath content. In addition, the slope of diffusion velocity with coagulation bath content

$$\left(\lim_{\xi\to 0}\frac{d\left(\frac{\xi^2}{4t}\right)}{dC_{cB}}\right)$$

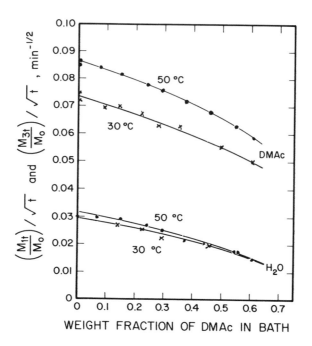

Fig. 11-25   Mass-transfer rates during coagulation [65].

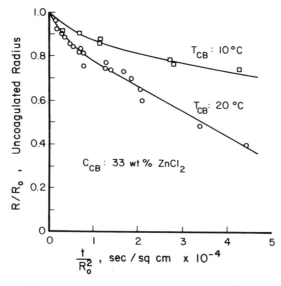

Fig. 11-26   Ratio for uncoagulated radius vs t/R_0^2 [68].

was related to the weight percent solids in the polymer solution and temperature. In both cases, the slope was found to increase with increasing values of percent solids and temperature, respectively.

Another investigation relevant to the transfer of mass in the dry-spinning operation was the work of Rotte et al. [68A]. These researchers studied the transfer of mass to a moving continuous cylinder by an electrochemical technique involving a moving wire that was drawn through a bath. Although this was not a wet-spinning operation, it nonetheless represents an analogous situation.

The results of their work are expressed in terms of Sherwood (Sh_H) and Peclet (Pe_H) numbers, as shown in Figs. 11-27 and 11-28. The numbers are defined as follows:

$$Sh_H = \frac{k_H H}{D_{AB}} \qquad (11\text{-}46)$$

$$Pe_H = \frac{U_s H}{D_{AB}} \qquad (11\text{-}47)$$

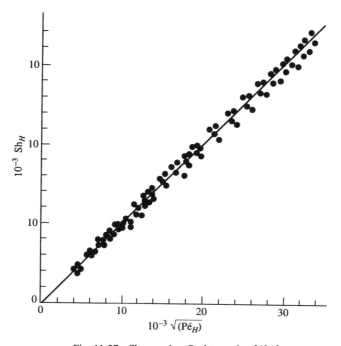

Fig. 11-27   Sherwood vs Peclet number [68A].

where

D_{AB} = diffusivity, m²/s
H = height of liquid in the vessel, m
k_H = average mass-transfer coefficient, m/s
U_s = wire velocity, m/s

These plots can be used to find appropriate mass-transfer coefficients for wet-spinning operations.

11.7 REACTION SPINNING

Synthetic fiber formation basically involves bringing about a change in phase of a material between the spinneret and take-up device. In most cases, this change is brought about by transfer of energy, mass, or momentum or a combination of these processes. Examples are the widely used operations of wet, dry, and melt spinning.

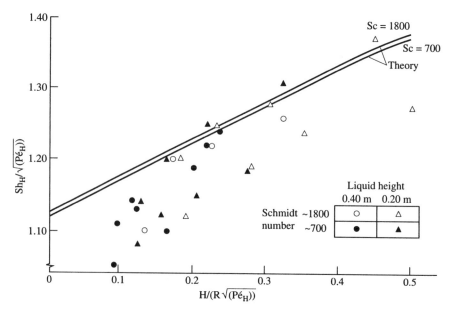

Fig. 11-28   Sherwood vs Peclet number [68A].

Chemical reaction, though present, is not generally an important factor in fiber formation. This is in direct contrast to nature, where the chief mechanism in forming fibers by spiders, silkworms, and ant lions is chemical reaction [77]. All these creatures, in essence, eject or spin a prepolymer, which then reacts and forms the final polymeric material. Attempts have been made from time to time to emulate the reaction-spun fibers formed by the spider, silkworm, and ant lion. For example, in the early days of World War II, spiderweb fibers were used to construct the cross hairs for bomb sights. There was, at that time, an effort to reproduce these fibers by laboratory techniques.

Later, reaction spinning again received attention as a means of forming fibers from a manageable prepolymer that could be taken to a polymer with desirable characteristics but that, in itself, could not be processed. A specific example is the production of a cross-linked polymer fiber. Such a material could be spun as a linear polymer and then cross-linked (reacted) during the spinning process.

Published research in the area of reaction spinning has been somewhat limited. The principal papers to date are the work of Pohl [78] and Fok et al. [79]. Mention should also be made of the work done on postspinning treatment of fibers. This type of process, widely used for high-temperature fibers [80–83], consists of subjecting dry- or wet-spun fibers of a given polymer to time and temperature to bring about a reaction that takes the material to another polymer compound. This process is used because the final high-temperature polymer is either infusible or insoluble or both.

Published studies [78,79] of reaction spinning, as well as those in nature (e.g., the spider), resemble dry spinning most clearly. In essence, the process could be likened to dry spinning with a chemical reaction included.

The similarity of dry spinning makes for a like momentum-transport case. For example, pre-extrusion effects, such as sand or screen pack behavior, entrance effects in spinning jets are not as pronounced as for the melt-spinning process. The net result is that pre-extrusion experiences do not appear to have a sizable effect on fiber-spinning performance or ultimate properties.

Again, during the postextrusion process, the shape of the velocity profile should follow the situation found in dry spinning. On extrusion, the velocity profile is flat and ultimately begins to develop to an appropriate profile. However, the non-Newtonian nature of the fluid will tend to blunt the final profile. Furthermore, as the fiber solidifies, the solid portion will have a uniform velocity until, finally, the entirely solid fiber will be in plug flow (constant V across the flow, where V is the average velocity).

The change in velocity profile from a flat to a blunted velocity profile and then to a flat or plug flow condition means that the assumption of a flat profile (V = constant across the fiber) is a reasonable assumption.

Heat and Mass Transfer

As mentioned earlier, the literature on reaction spinning is not extensive. In fact, there are only two items in the open literature, the work of Pohl [78] and that of Fok [79]. Pohl [78] directed his efforts to measuring some intrinsic viscosities (molecular weights) and tenacities for laboratory-scale spun fibers of polymethane. In addition, a rough indication was given of how the viscosity changed during the process. Pohl also showed that a certain viscosity level (molecular weight) was necessary before a fiber could actually be formed.

The other work, that of Fok et al. [79], was undertaken to study reaction spinning in a more quantitative sense. This research involved the reaction spinning of prepolymer syrups of polymethyl methacrylate to form fibers.

Before considering the data of Fok and co-workers [79], the basic equation for heat and mass transfer will be presented. The equation of energy (the heat-transfer equation) is similar to that for dry spinning:

$$\rho c_p V \frac{\partial T}{\partial z} = k\left[\frac{1}{r}\frac{\partial}{\partial r}\left(r\frac{\partial T}{\partial r}\right)\right] + A_0 \tag{11-48}$$

where it is assumed that

1. The fiber radius is constant.
2. Fiber properties ρ, c_p, and k are constant.
3. $v_z = V$ (i.e., approximate flat velocity profile).
4. Convection in the z (axial) direction greatly exceeds conduction:

$$\left(\rho c_p V \frac{\partial T}{\partial z} >>> k\frac{\partial^2 T}{\partial z^2}\right).$$

It should be understood that the heat-generation term A_0 is more complex than that found in dry spinning since it includes not only energy required for evaporation of either a solvent or monomer but also the heat of reaction associated with the chemical transformation of monomer to polymer. Hence, for reaction spinning,

$$A_0 = \Delta H_{vap} + \Delta H_{reaction}(k'M^n) \tag{11-49}$$

where ΔH_{vap} is the heat of vaporization, $\Delta H_{reaction}$ is the heat of reaction, k' is the specific reaction rate constant, M is the concentration of monomer, and n is the reaction order.

Furthermore, from the Arrhenius relation,

$$k' = A \exp\left(-\frac{E}{RT}\right) \tag{11-50}$$

so that

$$A_0 = \Delta H_{vap} + (\Delta H_{reaction})(A)(M^n)\left[\exp\left(-\frac{E}{RT}\right)\right]$$
(11-51)

where A is the frequency factor, E the activation energy, R the gas constant, and T temperature. The net result is to complicate considerably the energy equation, which now must have an exponential in temperature.

In the case of mass transfer, the equation of continuity of species resembles the case for dry spinning except that an R_A term for chemical conversion is included:

$$\frac{\partial c}{\partial z} = \frac{D_{AB}}{V}\left[\frac{1}{r}\frac{\partial}{\partial r}\left(r\frac{\partial c}{\partial r}\right)\right] + R_A$$
(11-52)

where it is assumed that

1. $v_z = V$
2. Steady-state $\partial c/\partial t = 0$
3. A chemical reaction ($R_A \neq 0$).
4. Diffusing solvent is a vapor.
5. No diffusion of ambient gas into the fiber.
6. Constant density and diffusivity.
7. Axial convective mass transfer exceeded axial mass transfer by diffusion:

$$\left[v_z\left(\frac{\partial c}{\partial z}\right) \gg D_{AB}\left(\frac{\partial^2 c}{\partial z^2}\right)\right]$$

8. No mass transport in the θ direction:

$$\left(\frac{D_{AB}}{r^2}\frac{\partial^2 c}{\partial \theta^2} = 0\right)$$

The R_A term is defined as

$$R_A = k'M^n$$
(11-53)

where k', M, and n are as defined above. If the R_A is written in the form of the Arrhenius relation for k', then,

$$R_A = A\left[\exp\left(-\frac{E_A}{RT}\right)\right](M^n)$$
(11-54)

This means that the mass-transfer equation is even more closely locked to the heat-transfer relationship. Furthermore, the introduction of the exponential term in temperature considerably complicates the mass-transfer case.

The foregoing equations in heat and mass transfer are extremely difficult to handle and, as such, have not as yet been solved analytically. The situation is, in fact, overly complicated, even from the standpoint of numerical analysis. In spite of these considerations, the combination of Eqs. (48) and (52) can assist in analyzing the reaction-spinning situation.

With the foregoing background, it is appropriate to consider the published experimental data of Fok and co-workers [79]. These are shown in Figs. 11-29–11-31. Figure 11-31 shows the average molecular weight of the fiber as a function of column height (i.e., z distance) for two different temperatures of the air passed countercurrently to the spinning fibers. The extrusion rate in Fig. 11-29 is 0.485 g/min.

As can be seen, the fiber molecular weight increases with z (i.e., residence time) and temperature. Since loss of methyl methacrylate monomer can increase average molecular weight as well as reaction itself, there is some question as to whether reaction or diffusion of monomer out of the fiber is the predominating process.

Several factors can be used to reach a conclusion. First of all, consider Fig. 11-19, the typical plots of concentration of solvent vs column height for dry spinning. Approximately 50% of the solvent is lost in the first foot of the col-

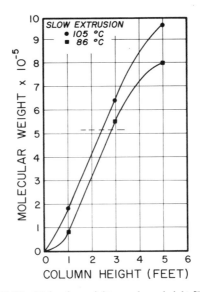

Fig. 11-29 Molecular weight vs column height [79].

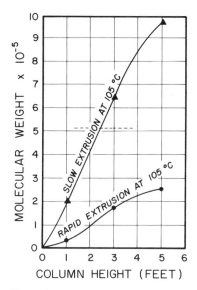

Fig. 11-30   Molecular weight vs column height [79].

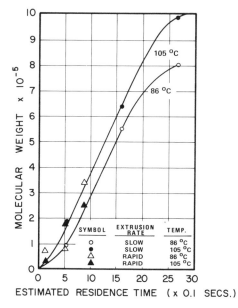

Fig. 11-31   Molecular weight vs residence time [79].

umn. This would mean that a diffusion loss of monomer would be most evident in the first foot of the column. Figure 11-29, however, shows no major effect on molecular weight in this region. This would indicate that monomer diffusion did not play a major role in the reaction-spinning process. Additionally, Eqs. (52) and (54), the mass-transfer representation, show that temperature will have an exponential effect on the chemical-reaction term. The magnitude of an exponential type of function would indicate the predominance of the chemical-reaction aspect. Finally, the curves in Fig. 11-29 bear an amazing resemblance to curves of polymethyl methacrylate polymerization data [84–88].

The experimental data plotted in Fig. 11-30 show in more depth the effect of residence time. All the points are for one column temperature (105°C) but for two different extrusion rates (0.485 and 1.476 g/min). Here again, the predominance of chemical reaction is reinforced by the residence time effect.

Figure 11-31 combines all the data as average molecular weight vs an estimated residence time at constant temperature. The estimated residence time was calculated by assuming a constant extrudate density and cross-sectional area. Both these quantities actually changed during the extrudate's travel. However, the assumptions made should at least indicate trends.

All the data shown in Fig. 11-31 fall on a definite curve regardless of extrusion rate. Again, the curve shapes match polymethyl methacrylate polymerization curves. And, again, this would seem to indicate that chemical reaction was the prevailing rate process.

REFERENCES

1. RILEY, J.L., *Polymer Processes*, C. E. Schildknect (ed.), Interscience, New York (1956), Chapter XVIII.
1A. OSHIMA, Y., MAEDA, H., and T. KAWAI, *J. Chem. Soc. Japan* **60**, 311, (1957).
1B. UEDA, S. and KATOAKA, T., *Chem. High Polymers (Tokyo)* **17**, 534 (1960).
2. NITSCHMANN, H., and SCHRODE, J., *Helv. Chim. Acta* **31**, 297 (1948).
3. HIRAI, J., *J. Chem Soc. Japan,* **75**, 1019, 1022, 1024 (1954).
4. HIRAI, N., *Rheol. Acta* **1**, 213 (1958).
5. ZIABICKI, A., TAKSERMAN-KROZER, R., *Roczniki Chem.* **38** 465, 1221 (1964).
6. Ibid., *Kolloid-Z.* **199**, 9 (1964).
7. THIELE, H., and LAMP, H., *Kolloid-Z.* **173**, 63 (1960).
7A. NAKAGAWA, T., *Bull. Chem. Soc. Japan* **25**, 93 (1952).
8. GREGORY, D.R., Ph.D. Thesis, Virginia Polytechnic Inst. and State Univ., Blacksburg VA (1965).
9. GRISKEY, R.G., and D.R. GREGORY, *AIChE J.* **13**, 122 (1967).
10. GRISKEY, R.G., GREGORY, D.R., and SISKOVIC, N., *AIChE J.* **17**, 281 (1971).
11. MARSHALL, R.J., and METZNER, A.B., *Ind. Eng. Chem. Fund.* **6**, 393 (1967).

12. SCHRIEBER, H.P., STOREY, S.H., and BAGLEY, E.G., *Trans. Soc. Rheol.* **10**, 275 (1966).
13. SCHREIBER, H.P., and STOREY, S.H., *J. Polymer Sci.* **B3**, 723 (1965).
14. CLEGG, P.L., *Trans. Plast. Inst.* **26**, 151 (1958).
15. HOWELLS, E.R., and BENBOW, J.J., *Trans. Plast. Inst.* **30**, 240 (1962).
16. MAXWELL, B., and GALT, J.C., *J. Polymer Sci.* **62**, S50 (1962).
17. PEARSON, J.R.A., and PETRIE, C.J.S., *Proc. 4th Intern. Cong. Rheol.*, Paper 118 (1963).
18. RABINOWITSCH, B., *Z. Physik. Chem.*, **A145**, 1 (1929).
19. MOONEY, M. *J. Rheol.* **2**, 210 (1931).
20. BAGLEY, E.B., *J. Appl. Phys.* **28**, 624 (1957).
21. GERRARD, J.E., and PHILIPPOFF, W., *Proceedings of the 4th International Rheological Congress*, Paper 51 (1963).
22. MACOSKO, C., and COX, H., Paper presented at AIChE National Meeting, Minneapolis, MN, Sept. 1972.
23. GRISKEY, R.G., SALTUK, I., and SISKOVIC, N., *Polymer Eng. Sci.* **12**, 402 (1972).
24. MCKELVEY, J.M., *Polymer Processing*, Wiley, New York (1962) p. 91.
25. MCINTOSH, D.L., Doctoral Dissertation, Washington Univ., St. Louis, MO (1960).
26. GAVIS, J., and MIDDLEMAN, S., *Phys. Fluids* **4**, 355, 963 (1961).
27. Ibid., *J. Appl. Polymer Sci.* **7**, 493 (1963).
28. METZNER, A.B., *Ind. Eng. Chem.* **50**, 1577 (1958).
29. Ibid., *Trans. Soc. Rheol.* **5**, 133 (1961).
30. ZIABICKI, A., *Kolloid-Z.* **175**, 14 (1961).
31. ZIABICKI, A., CYBULSKI, A., and GROMADOWSKI, J., unpublished report (1964).
32. KASE, S., and MATSUN, T., *J. Polymer Sci.* **3A**, 2541 (1965).
33. ANDREWS, E.H., *Brit. J. Appl. Phys.* **10**, 39 (1959).
34. BARNETT, T.R., *Appl. Polymer Symposia* **6**, 51 (1967).
35. COPLEY, M., and CHAMBERLAIN, N.H., *Appl. Polymer Symposia* **6**, 27 (1967).
36. WILHELM, G., *Kolloid-Z.* **208**, 97 (1966).
37. MORRISON, M.E., *AIChE J.* **16**, 57 (1970).
38. FOX, V.G., and WANGER, W.H., Paper presented at AIChE Annual Meeting, Chicago, IL, Dec. 1970.
39. HOWELLS, E.R., and BENBOW, J.J., *Trans. Plast. Inst.* **30**, 240 (1962).
40. SANO, Y., and NISHIKAWA, S., *Kagaku Kogaku* **28**, 275 (1964).
41. Ibid., **30**, 135 (1966).
42. ZIABICKI, A., *Appl. Polymer Symposia* **6**, 1 (1967).
43. GRISKEY, R.G., and FOK, S.Y., *Proceedings, Material Engineering Science Division Biennial Conference*, American Institute of Chemical Engineers (1970), p. 332.
44. FOK, S.Y., Ph.D. Dissertation, Virginia Polytechnic Inst. and State Univ., Blacksburg, VA (1965).
45. GRISKEY, R.G., and FOK, S.Y., *J. Appl. Polymer Sci.* **11**, 2417 (1967).
46. CORBIERE, J., *Man Made Fibers*, H. Mark, S. M. Atlas, and E. Cernia, (eds.), Interscience, New York (1967).
47. COULSON, J.M., and RICHARDSON, J.F., *Chemical Engineering*, Pergamon, London (1968), p. 321.

48. McCabe, W.L., and Smith, J.C., *Unit Operations of Chemical Engineering*, McGraw-Hill, New York (1967), p. 540.
49. Griskey, R.G., Celanese Fibers Corp., unpublished data.
49A. Brazinsky, I., Williams, A., and LaNieve, H.L., *33rd Annual Technical Conference SPE*, Vol. XXI (1975), p. 182.
50. Griskey, R.G., and Fok, S.Y., *Appl. Sci. Res.* **16**, 141 (1966).
51. Mueller, A.C., *Trans. Am. Inst. Chem. Engrs.* **38**, 613 (1942).
52. Sano, Y., and Nishikawa, S., *Kagaku Kogaku* **30**, 245 (1966).
52A. Spruiell, J.E., and White, J.L., *33rd Annual Technical Conference SPE*, Vol. XXI (1975), p. 188.
52B. Matsuo, T., and Kase, S., *Sen-i-Gakkashi* **24**, 512, (1968).
53. Ohzawa, Y., Nagano, Y., and Matsuo, T., *J. Appl. Polymer Sci.* **13**, 257 (1969).
54. Ibid., *Proceedings of the Fifth International Congress on Rheology*, University Park Press, Baltimore, MD (1970), p. 393.
55. Yoshida, T., and Hyodo, T., *Ind. Eng. Chem.* **2**, 52 (1963).
56. Takahashi, M., and Watanable, M., *Sen-i-Gakkashi,* **16**, 458 (1960).
57. Ryskin, G.Y., *Zh. Tekh. Fiz.* **25**, 458 (1955).
58. Geil, P.H., *Polymer Single Crystals*, Interscience, New York (1963), p. 29.
59. Sippel, A., *Z. Elektrochem.* **50**, 152 (1944).
60. Sisson, W.A., *Textile Res. J.* **30**, 153 (1960).
61. Vermaas, D., *Textile Res. J.* **32**, 353 (1962).
62. Hermans, J.J., *J. Colloid Sci.* **2**, 387 (1947).
63. Paul, D.R., *J. Appl. Polymer Sci.* **12**, 2273 (1968).
64. Griffith, R.M., *Ind. Eng. Chem. Fund* **3**, 245 (1964).
65. Paul, D.R., *J. Appl. Polymer Sci.* **12**, 383 (1968).
66. Jost, W., *Diffusion in Solids, Liquids and Gases*, Academic Press, New York (1960), pp. 31–46.
67. Tazikawa, A., *Sen-i-Gakkashi* **17**, 397 (1961).
68. Booth, J.R., *Appl. Polymer Symposia* **6**, 99 (1967).
68A. Rotte, J.W., Tummers, G., and Dekker, J.L., *Chem. Eng. Sci.* **24**, 1009 (1969).
69. Grobe, V., Mann, G., and Durve, G., *Faserforch. Textiletechn.* **17**, 142 (1966).
70. Grobe, V., and Heyer, H.J., *Faserborch. Textiletechn.* **18**, 577 (1967).
71. Hill, A.V., *Proc. Roy. Soc. (London)* **B104**, 41 (1929).
72. Tazikawa, A., *Sen-i-Gakkashi* **16**, 842 (1960).
73. Ibid., 950 (1960).
74. Murakami, E., *Sen-i-Gakkashi* **20**, 448 (1964).
75. Epstein, M., and Bringar, W.C., *Appl. Polymer Symposia* **6**, 99 (1967).
76. Paul, D.R., *Proceedings, Material Engineering Science Division, Biennial Conference*, American Institute of Chemical Engineers (1970), p. 325.
76A. Ziabicki, A., *Fundamentals of Fibre Formation*, Wiley, New York, (1976), pp. 333, 344.
77. Meyer, K.H., *Natural and Synthetic High Polymers*, Interscience, New York (1942), p. 452.
78. Pohl, H.A., *Textile Res. J.* **28**, 473 (1958).
79. Fok, S.Y., Mickles, R.A., and Griskey, R.G., *Textile Res. J.* **36**, 131 (1966).

80. TEMIN, S.C., *Appl. Polymer Symposia* **9**, 3 (1969).
81. FRAZER, A.H., and FITZERALD, W.P., *J. Polymer Sci.* **C19**, 95 (1967).
82. YODA, N., IKEDA, K., KURIBARA, M., TOHYAMA, S., and NAKANSHI, R., *J. Polymer Sci.*, **A1, 5**, 2359 (1967).
83. HIRSCH, S.S., and HOLSTEN, J.R., *Appl. Polymer Symposia* **9**, 187 (1969).
84. NAYLOR, M.A., and BILLMEYER, F.W., Jr., *J. Am. Chem. Soc.* **75**, 2181 (1953).
85. NORRISH, R.G.W., and BROOKMAN, E.F., *Proc. Roy. Soc. (London)*, **A171**, 147 (1939).
86. NORTH, A.M., and REED, G., *Trans. Faraday Soc.* **57**, 589 (1961).
87. SCHULZ, G.V., and BLASCHKE, F., *Z. Physik Chem.* **B151**, 75 (1942).
88. SCHULZ, G.V., and HARBORTH, G., *Makromol. Chem.* **1**, 106 (1947).
89. ZIABICKI, A., *Fundamentals of Fibre Formation*, Wiley, New York (1976), pp. 362, 364, 367, 370, 407, 415, 416, 427.
90. MARK, H.F., ATLAS, S.M., and CERNIN, E., *Man Made Fibers*, Vol. I, Interscience, New York (1967), p. 153.
91. HICKS, E.M. JR., RYAN, J.R. JR., TAYLOR, R.B. Jr., and TICHNOR, R. L., *Textile Res. J.* **30**, 675 (1960).
92. BROSSLER (Chemstrand), U.S. Patent 2,979,882.
93. MIDDLEMAN S., *Fundamentals of Polymer Processing*, McGraw-Hill, New York (1977).
94. WANGER, W.H. Jr., Ph.D. Thesis, Univ. of Denver, Denver, CO (1969).
95. HILL, J.A., and CUCOLO, J.A., *J. Appl. Polymer Sci.* **18**, 2569 (1974).
96. TAKIZAWA, A., *Sen-i Gakkashi* **16**, 842 (1960).
97. BOUNDY, R.H., and BOYER, R.T., *Styrene—Its Polymers, Copolymers and Derivatives*, Reinhold, New York (1952).

PROBLEMS

11-1   An alternative equation [93] for capillary breakup length is

$$\frac{X^*}{2R_0} = \left[\ln\left(\frac{R_0}{C_0}\right)\right] We^{1/2}\left(1 + 3\,\frac{We^{1/2}}{Re}\right)$$

where C_0 is $(2.2 \times 10^{-6}\,R_0)$, We the Weber number, Re the Reynolds number, and X^* the capillary breakup length.

Assume that the material to be spun has a viscosity of 100 N s/m² and a surface tension of 0.1 N/m. Plot the behavior of X^* with the product of viscosity times velocity, and compare the results to Figs. 11-2, 11-4, and 11-5.

11-2   In Fig. 11-32, the $f^*N^*_{Re}$ relation is plotted against the Deborah number $(\theta V/D_p)$ for molten polyethylene. Compare these data to those of Fig. 11-8. If there are differences, explain why these occurred. Also, if the correlations are different, explain how they might be reconciled in another format.

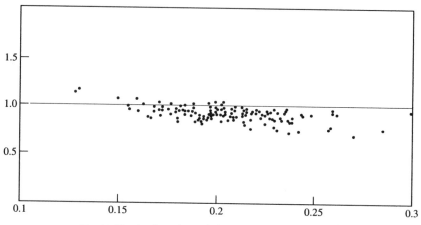

Fig. 11-32 $f*N_{Re}^*$ vs $(\theta V/D\rho)$ for molten polyethylene [10].

11-3   The shear stress on the surface of a moving cylinder can be written as

$$\tau^1 = \frac{1}{2} \rho V^2 C_f$$

where C_f is a skin-friction coefficient. Given that

$$c_f = \frac{0.7}{Re^{0.8}}$$

Then, for a spun molten nylon 6 (with μ_0 of 200 N s/m², flow rate of 0.05 g/s), estimate the behavior of the take-up stress with take-up velocity.

11-4   Wanger [38,94] studied the melt spinning of polypropylene fibers. His data have been plotted [93] as shown in Fig. 11-33 (the melt-spinning temperature of the polymer was 260°C; the surrounding air was 25°C).
   Analyze on qualitative and phenomenological bases the data of Fig. 11-33. Some particular items to consider are: (1) the temperature and z at which the polymer, solidified, (2) the behavior of the filament diameter, and (3) the heat transfer taking place.

11-5   Analyze the heat transfer for the data of Problem 11-4 quantitatively (i.e., get some numbers).

11-6   Hill and Cucolo [95] studied the melt spinning of polyethylene terephthalate. A plot of their data [93] is shown in Fig. 11-34. Repeat the analyses of Problems 11-4 and 11-5 for their data.

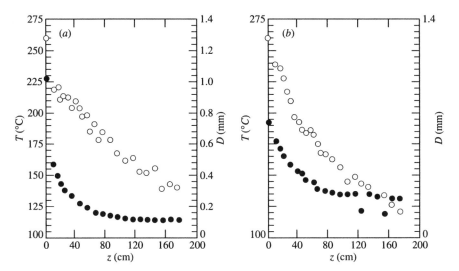

Fig. 11-33 Surface temperature (open circles) and filament diameter (solid circles) vs z, distance from spinneret. Polymer spun is polypropylene, initially at 260°C (air at 25°C). Mass flow rate and spin speed are: (a) 2 g/min, 200 m/min and (b) 4 g/min, 100 m/min [38,93,94].

11-7   Determine the solidification profiles (i.e., r/R at which polymer becomes solid) as a function of z for the data of Figs. 11-10, 11-33, and 11-34. (Hint: Use the data of Morrison [37].)

11-8   Reevaluate the mass transfer for the data shown in Figs. 11-12 and 11-17–11-19 using the concept of changes in solvent diffusivity because of solidification, concentration, temperature, etc. In other words, assume that a changing diffusivity (with radical and axial position) was the cause of the mass-transfer behavior.

11-9   Suppose the mass flux at the fiber surface was governed by the relation.

$$N_A = k_A A(C_{\text{surface}} - C_{\text{air}})$$

where N_A is the solvent mass flux, k_A the mass-transfer coefficient, A the surface area; the C are the solvent concentration at the surface and in the air.

   If so, how would the data considered in the preceding example be affected (i.e., analyze these data using the above concept; note that only short axial lengths should be considered since concentration is at the fiber surface).

   Mass-transfer coefficients can be obtained by using an analogy with Mueller's [51] heat-transfer relation

$$Nu = 0.516 Pr^{0.3} Re^{0.43}$$

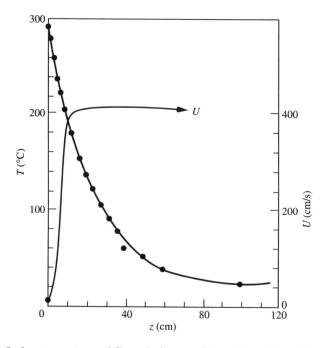

Fig. 11-34   Surface temperature and fiber velocity vs z, distance from spinneret. Mass flow rate is 6.3×10^{-3} g/s. Initial melt temperature is 290°C (air at 22°C). Spinneret diameter is 0.0031 cm [93,95].

where Nu, Pr, and Re are, respectively, the Nusselt, Prandtl, and Reynolds numbers.

11-10   Sano and Nishikawa [52] obtained the experimental data shown [76A] in Fig. 11-35. Based on their work, they concluded that dry-spinning mass transfer was a diffusion controlled process.

   Is this assumption correct? What value(s) of diffusivity did they use?

11-11   Compare the coagulation results of Booth [68] and Paul [65] given in Chapter 11 (i.e., Figs. 11-23–11-26). Explain any similarities or dissimilarities. Why do you think this behavior took place?

11-12   Takizawa [96] has theoretically solved the countercurrent diffusion problem with a moving boundary and with both diffusion coefficients having the same value (see Fig. 11-36). How well does this theoretical analysis fit the work of Booth [68] and Paul [65]? Explain any similarities or dissimilarities.

11-13   Polymers that are dry- or wet-spun generally can be spun by the other process. Would you suppose that there would be any differences in the finished product

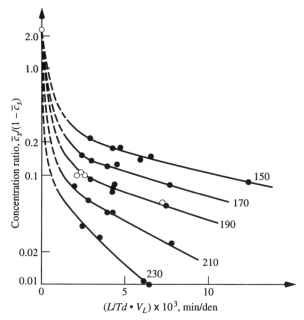

Fig. 11-35   Relative solvent concentration vs $L/V_L^2 R_L^2$ for dry spinning of polyacrylonitrile in dimethylformamide; each curve at temperature in °C with an extrusion temperature of 110°C; L, overall spinning length of 6.45 m; air velocity of 2 m/s; V_L and R_L are taken-up velocity and radius [52].

(i.e., cross-sectional shape, interior texture, etc.)? Justify your answers. Why would you select one of the processes over the other?

11-14   Figure 11-37 gives molecular weight, rate, and temperature data for the polymerization of styrene. Also available is a value of k_p for styrene polymerization at 50°C and 209 L s/mole. Using these data and Chapter 11, develop curves of the type shown in Figs. 11-29–11-31.

MINI PROJECT A

Essentially, all the basic treatments of heat and mass transfer discussed in Chapter 11 or the preceding problem section involved theoretical, analog, or single-filament experiment data.

In actuality, fiber-spinning processes involve multifilament spinning. Pick one of the treatments of Chapter 11 and extend it to the multifilament case.

Clearly indicate and justify all assumptions. Be sure to present your approach in a manner that a process engineer could use.

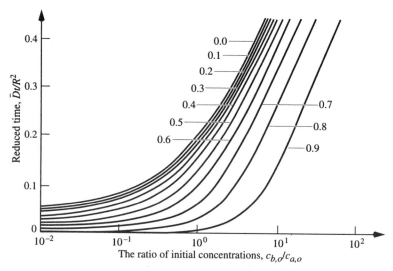

Fig. 11-36   Theoretical results for moving boundary of reduced time vs concentration ratio [96].

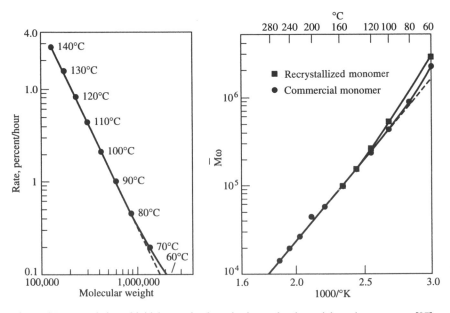

Fig. 11-37   Interrelation of initial rate of polymerization molecular weight and temperature [97].

MINI PROJECT B

Many fibers are spun from noncircular spinnerets (i.e., jets or capillaries) to obtain special cross-sectional shapes. The resultant cross sections are different from the die cross section. Further, these differences are accentuated by the distance from the spinneret.

Develop a model or models for this process. Apply it to the following systems, and predict the change in shape of the fiber cross section:

1. Melt-spun through a y-shaped spinneret.
2. Melt-spun through a star-shaped spinneret.
3. Dry-spun through a square spinneret.
4. Dry-spun through a rectangular spinneret.

MINI PROJECT C

Chapter 11 indicates that it might be possible to develop a radial solidification profile in a melt-spun fiber.

Develop a model (using both available information and your ingenuity) that will describe the development of solidification in a dry-spun fiber.

MINI PROJECT D

A number of the melt-spun fibers (the nylons and polyesters) are polycondensation polymers. In such polymers, there is an equilibrium:

$$P_1 + P_2 \rightleftharpoons P_3 + \text{by-product molecule}$$

where P_1 and P_2 are shorter polymer chains that combine to form a longer chain (P_3) and a small by-product molecule (such as water).

Develop the appropriate equations needed to model a melt-spinning process that would use the behavior to increase the fiber's molecular weight during spinning.

What spinning variables (spin speed, cooling gas temperature, etc.) would you expect to be important?

MINI PROJECT E

Fibers are sometimes spun from polymer blends to produce bicomponent fibers that have special properties.

Transform and/or alter the melt-spinning relation and equations of Chapter 11 for this case. Use these to obtain practical or empirical forms that can be used for processing.

Also, what special properties do you expect bicomponent fibers to have? Justify your answers.

CHAPTER 12

THE INTERRELATION OF POLYMER PROCESSING, POLYMER STRUCTURE, AND POLYMER PROPERTIES

12.1 INTRODUCTION

Polymers are materials whose mechanical, electrical, chemical and thermal properties depend on their structural characteristics. Processing operations can and do change these characteristics. This means that the end-product polymer as formed and shaped is a different material than the starting resin. Furthermore, the polymer engineer has the ability to alter processing operations in order to optimize product properties by controlling structural characteristics.

Some of the principal mechanical properties of polymers are shown on the idealized stress-strain diagram of Fig. 12-1. These include:

1. *Ultimate tensile strength*: The stress at which the specimen breaks or fractures.
2. *Elongation at break*: The ultimate elongation or the strain at which the material breaks.
3. *Elongation at yield*: The strain at which the substance undergoes a permanent deformation (i.e., no longer has elastic behavior).
4. *Initial or modulus of elasticity*: The linear slope of the stress-strain curve. This represents the elastic behavior of the specimen.
5. *Impact strength*: Also known as toughness or resilience, this is the *area* under the stress-strain curve.

Other important mechanical properties are surface hardness and creep or cold flow behavior. The latter represents the tendency of a polymer to react with a

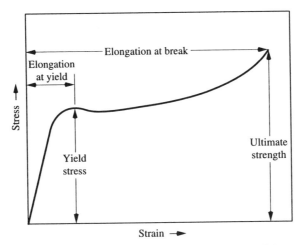

Fig. 12-1   Idealized polymer stress-strain diagram [1].

continued deformation after a long period of time has elapsed since the application of a stress.

Chemical properties include swelling (or solubility) resistance, moisture resistance, acid resistance, and alkali resistance.

The electrical properties most often considered are electrical resistance (or, conversely, conductance) and the polymer's dielectric behavior (dielectric constant is the ratio of the capacitance of a given volume of polymer relative to that of the same volume of air).

In Chapter 7 (Table 7-1) the relations between polymer properties and polymer structural characteristics were presented. Figures 12-2 and 12-3 list the same sort of data but for specific commercial polymers. Note that some additional items of importance (i.e., processibility, price, and flammability) are included in the tabulation. These data can help to provide us with a qualitative picture of the effect of processing conditions on a given polymer property by allowing us to relate changes in structural characteristics to processing conditions. For example, if we increase orientation during processing, we find that many mechanical properties, such as ultimate tensile strength, modulus of elasticity, and resistance to creep or cold flow, increase.

Conversely, an increase in cross-linking ultimately decreases impact strength and elongation at break, whereas an increase in polymer chain flexibility decreases the polymer's modulus of elasticity.

As has been noted, all polymer-processing operations involve time, temperature, and rheological interactions. In the succeeding sections, we will consider the effect of these parameters on polymer structure characteristics. These, in turn, can be used to predict property development in a given system.

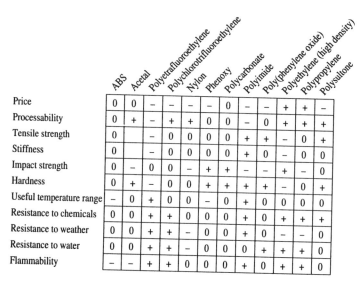

Fig. 12-2   Properties for various polymers [3].

12.2 PROCESSING AND MOLECULAR WEIGHT, BRANCHING, AND CROSS-LINKING

Polymer molecular weights can be altered during processing. One such change can be a decrease in molecular weight caused by a degradation reaction. Such reactions can be a hydrolytic mechanism or can result from oxidation.

Many polymers are very susceptible to the action of oxygen, which brings about a degradation reaction. These reactions are chain-type processes consisting of an initiation step, which produces an active organic or peroxide radical, and a propagation step, which terminates in products that have peroxide, hydroperoxide, or carbon–carbon bonds.

The most susceptible points of attack are bonds adjacent to double bonds, ether linkages, or those on a tertiary carbon.

Antioxidants are either materials that themselves are readily oxidized or that form a stable product by combining with a polymer that is oxidizing. Chemical groupings used as antioxidants include phosphorous compounds, cadmium and nickel compounds, aromatic amines, products of amines or aminophenols with a variety of compounds (ketones, aldehydes), and phenols.

Particular antioxidants include butylated hydroxy-toluol (BHT) phenyl-B-napthyl amine (PBNA), and polymerized trimethyl-dihydroquinoline.

Polymers are also susceptible to attack by ozone. Certain materials (dialkyl-*p*-phenyldiamines) can react directly with either free ozone or ozone-polymer

Fig. 12-3 Properties for various polymers [3].

Property	Poly (methyl methacrylate)	Modacrylic	Cellulose acetate	Cellulose acetate-butyrate	Cellulose propionate	Ethyl cellulose	Chlorinate polyether	Epoxy polyether	Epoxy (cast)	FEP Fluorocarbon	Ionomer	Melamine-formaldehyde (cellulose filled)	Phenol-formaldehyde (cellulose filled)	Polybutylene	Polyethylene (low density)	Poly(phenylene oxide)	Polystyrene	Poly (vinyl chloride)(Modified)	Poly (vinyl chloride)	Poly (vinylidene fluoride)	Poly (vinylidene chloride) (copolymer)	SAN	Silicone	Urea-formaldehyde (cellulose filled)
Price	o	o	o	o	o	o	o	o	–	o	o	+	o	+	o	+	+	o	–	o	–	+	–	o
Processability	+	+	+	+	+	+	+	+	+	+	o	o	o	+	+	o	+	+	+	+	+	–	+	+
Tensile strength	+	o	o	o	o	o	+	–	–	–	o	o	–	+	+	–	–	o	o	o	o	–	+	+
Stiffness	+	+	o	o	–	+	o	o	+	–	o	+	–	+	o	+	–	–	–	–	–	–	+	+
Impact strength	o	o	o	o	o	o	–	+	+	–	+	–	+	–	–	+	+	–	–	o	o	–	–	–
Hardness	+	+	+	+	+	o	+	–	–	+	+	+	+	+	+	–	+	+	+	+	–	–	+	+
Useful temperature range	o	–	o	o	–	o	+	+	o	o	o	o	o	o	o	–	o	+	o	o	o	–	–	–
Resistance to chemicals	–	o	–	–	–	–	+	+	+	o	o	–	+	–	+	+	–	o	o	+	+	o	o	–
Resistance to weather	+	o	–	–	–	o	o	+	+	o	o	o	+	+	–	o	o	+	–	+	+	+	o	o
Resistance to water	o	–	–	–	+	+	o	o	+	o	o	+	+	o	o	o	o	o	o	–	+	o	+	o
Flammability	–	o	–	–	–	o	+	+	+	–	o	–	–	–	–	o	o	o	–	–	o	o	o	o

products. Others (various waxes) diffuse to the polymer surface and form a surface barrier to ozone diffusion.

Polymer molecular weights can also be increased during processing. This situation occurs in condensation polymers where, effectively, a chemical equilibrium exists as shown in Eq. (12-1):

$$P_1 + P_2 \rightleftarrows P_3 + \text{by-product molecule} \qquad (12\text{-}1)$$

where P_1 and P_2 are shorter polymer chains that combine to form P_3, a longer chain, together with a by-product molecule.

If the small by-product diffuses to the polymer surface and is swept away, the equilibrium of Eq. (12-1) shifts to the right. The net result is an increase in polymer molecular weight, with the attendant effect on properties shown in Table 12-1. Incidentally, such a polycondensation process can take place even in the solid state [4,5].

Cross-linking can be brought about purposefully in processing operations such as compression molding or reaction injection molding. The ultimate degree of cross-linking will then affect the properties shown in Table 12-1 as indicated.

Branching is not usually affected by processing. Further, as Table 12-1 shows, the effect of branching on polymer properties is unusually unclear.

12.3   ORIENTATION

Orientation is a structural characteristic of polymers that can be greatly affected by processing operations. Basically, application of a tensile stress will cause the polymer chains to orient. Such stress can be brought about in a variety of ways, such as mechanical means, flows, and temperature changes.

Injection Molding

Sometimes, the orientation effects are brought about by a combination of causes. In injection molding, for example (see Fig. 12-4), an orientation profile results from flow and heat transfer. The portion of the polymer nearest the wall at the mold is quickly frozen and, thus, is not oriented. Likewise, the material in the region near the center will have a low orientation. The layer nearest the solidified skin at the wall will have the highest orientation. The values will decrease from this zone to the center.

The effects of a number of injection-molding process conditions (gate size, melt temperature, part thickness, mold temperature, ram forward time) on the maximum birefringence value of polystyrene are shown in Figs. 12-5–12-9. Examination of these plots shows that maximum orientation values occur for

TABLE 12-1 **Relation of Polymer Properties to Structure**

Property	D.P.	Branching	Cross Linking	Polar Structures	Chain Flexibility	Crystallinity	Size of Spherulites	Orientation	Molecular Weight Dist.
Ultimate tensile strength	+	?	?	+	?	+	?	+	?
Modulus of elasticity	+	?	+	?	–	+	? Smaller +	+	? Narrow +
Impact strength	+	?	–	?	+	–	Smaller –	+	Narrow +
Elongation at break	+	?	–	?	+	–	?	?	+
Range of reversible extensibility	+	?	–	?	+	?	?	?	?
Surface hardness	+	?	+	+	–	+	?	+	?
Temperature resistance	+	?	+	+	–	+	? Large +	+	High D.P. +
Electrical resistance	?	?	?	–	–	?	?	?	?
Dielectric constant	?	?	?	+	+	–	?	?	?
Swelling resistance	+	?	+	?	–	+	? Smaller –	+	High D.P. +
Creep or cold flow resistance	+	?	+	+	–	+	?	+	?
Moisture resistance	+	+	+	?	–	+	?	+	?
Alkali resistance	?	?	+	?	?	+	?	+	?
Acid resistance	?	?	+	?	?	+	?	+	?
Adhesive power	?	+	–	+	+	–	?	?	? Broad +

Adapted from Mark (2).

453

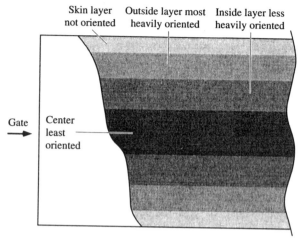

Fig. 12-4   Effect of injection molding on plastic orientation distribution [6].

larger gates, thinner molded objects, lower melt temperatures, lower mold temperatures, and longer forward ram times. It has also been found that higher
pressures and slower mold-fill rates favor orientation.

Those effects favoring higher temperatures in the molded object, such as high
melt temperatures, high mold temperatures, and thicker objects, all work to make
the object warmer. This higher temperature causes the polymer to relax and
hence decreases orientation. Those parameters giving greater shearing stresses

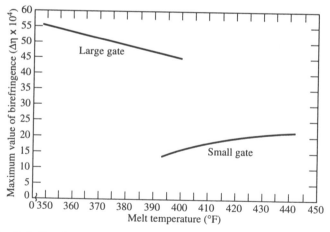

Fig. 12-5   Relation of orientation to melt temperature for different gate sizes [7].

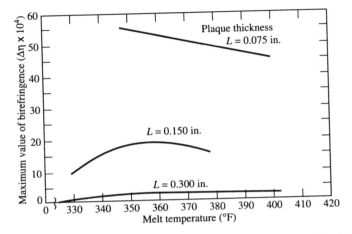

Fig. 12-6 Relation of orientation to melt temperature for various thicknesses of molded parts [7].

(high pressure, slow fill rate) increase the orientation. The larger gate sizes promote slower mold fills and thus favor orientation.

Increased orientation should produce the effects predicted in Table 12-1. This, in fact, is the case as shown for ultimate tensile strength and impact strength (Figs. 12-10 and 12-11).

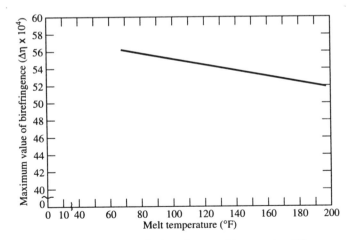

Fig. 12-7 Relation of orientation to mold temperature [7].

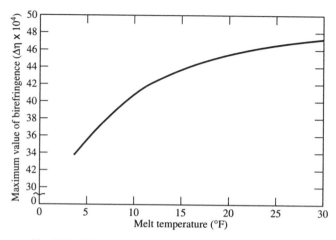

Fig. 12-8   Relation of orientation to ram forward time [7].

Other Polymer-Processing Operations

Orientation effects are also important in other polymer-processing operations. One in particular is the spinning of fibers. Here the solidifying and ultimately solid fibers are subjected to tensile stresses that bring about significant orientation.

A theoretical analysis [9] of the orientation taking place in dry or wet spinning (see Chapter 11) indicated that four cases or types could occur:

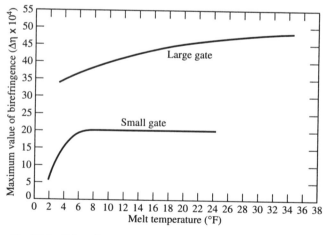

Fig. 12-9   Orientation vs ram forward time for different gate sizes [7].

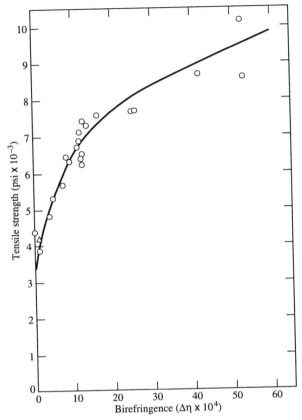

Fig. 12-10   Tensile strength vs orientation (parallel to injection direction) for general-purpose polystyrene [8].

1. Orientation of the first type takes place in the spinneret capillary.
2. Orientation of the second type takes place perpendicular to the fiber centerline axis in the unsolidified fiber.
3. Orientation of the third type takes place in the solidifying skin of the fiber parallel to the fiber axis.
4. Orientation of the fourth type takes place in the direction of the fiber axis inside the forming, but not yet solidified, fiber.

The most important type of orientation indicated was type 3, which takes place in the solidified skin.

Although the foregoing was developed for dry- or wet-spun fibers, the importance of type 3 orientation also holds for melt-spun fibers. It therefore be-

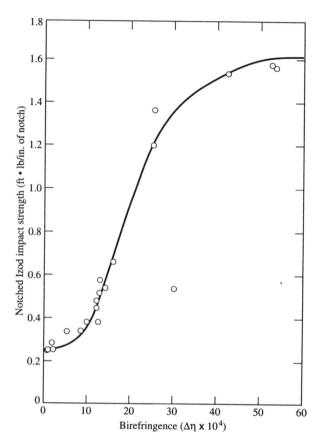

Fig. 12-11   Impact strength vs orientation (parallel to injection direction) for general-purpose polystyrene [8].

comes apparent that fiber orientation is dependent not only on the mechanical tensile stress exerted on the fiber thread line but also on the solidification process. In this regard, see Fig. 12-12, which shows the cross-filament variation of birefringence with the cooling gas flow as shown. These data show a higher level of orientation where the fiber is more rapidly cooled (and hence solidified).

Figures 12-13 and 12-14 show the behavior of orientation with increasing take-up velocity and spin line stress (f_c, the function along the fiber axis, is the important characteristic). As expected, orientation increases with increasing take-up velocity and spin-line stress.

The net effect of orientation on tensile strength is shown in Fig. 12-15. Here, again, the prediction of Table 12-1 (increasing orientation increases tensile strength) is verified.

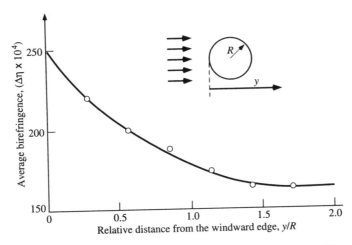

Fig. 12-12   Cross-filament variation of birefringence (orientation) [10].

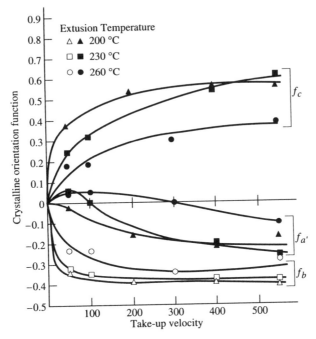

Fig. 12-13   Orientation function vs take-up velocity for polypropylene fiber [11].

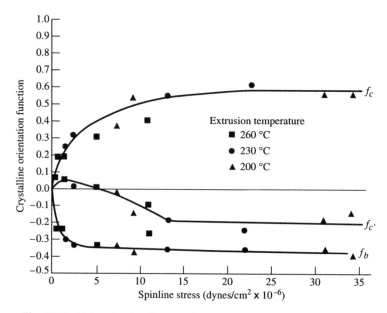

Fig. 12-14   Orientation function vs spinline stress for polypropylene fiber [11].

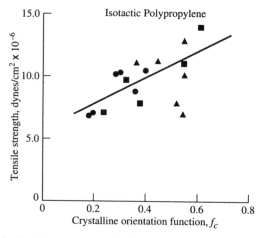

Fig. 12-15   Tensile strength vs orientation function for polypropylene fiber [11].

Further, in a qualitative sense, the spinning factors affecting orientation are temperature history, spinning rate, fiber diameter, and polymer nature.

In dry spinning, the overall effect of increasing orientation is shown in Fig. 12-16, which diagrams the changes in tensile strength and percent elongation at break. Fiber orientation in Fig. 12-16 increases with increasing column length. In this case, the fiber was in free fall without a take-up.

Wet-spinning orientation takes place first in the solidfying skin, then later in the solid fiber itself. The effect of process variables on orientation can be found from the data in Table 12-2.

According to Table 12-1, the indicated mechanical behavior is a function of orientation. Hence, increasing spinning velocity and residence time in the bath increase orientation. Likewise, lower bath temperatures and smaller spinneret radii increase orientation.

Orientation processes also take place in blow-molding operations. A particularly significant aspect is biaxial stretch blow molding, in which the polymer preform is stretched both in the hoop and axial directions.

Two ratios involving preform-to-product dimensions are utilized to describe biaxial stretch molding. The first of these is the hoop ratio H:

$$H = \frac{D_1}{D_2} \tag{12-2}$$

where D_1 is the largest inside diameter of the product and D_2 is the preform diameter.

The second ratio is the axial ratio A:

$$A = \frac{L_1}{L_2} \tag{12-3}$$

where L_1 is the inside length of the product and L_2 is the inside length of the preform.

Another term is the product of the two ratios known as the blowup ratio (BUR):

$$BUR = HA = \frac{D_1}{D_2} \frac{L_1}{L_2} \tag{12-4}$$

The BUR for a container holding the contents under pressure is 10 or higher. Hoop ratios (4–7) are usually more important than axial ratios (1.4–2.6).

As in the other cases discussed previously, increased orientation results in improved properties. In Table 12-3, barrier properties are given for oriented and nonoriented blow-molded polyvinyl chloride (PVC) objects. As can be seen,

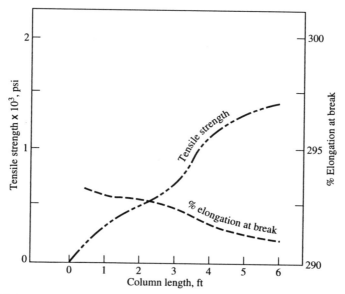

Fig. 12-16   Property variation with increasing orientation (column length) for dry-spun polymethyl methacrylate [12].

orientation significantly increases the resistance to penetration of both gases and liquids.

Another processing operation in which orientation plays an important role is thermoforming. Biaxial stretching that takes place can be correlated using the expression

$$t_f = \frac{t_i}{\left(\dfrac{L_f - L_i + 1}{L_i}\right)\left(\dfrac{W_f - W_i + 1}{w_i}\right)} \tag{12-5}$$

TABLE 12-2   Effect of Wet Spinning Conditions on Mechanical Properties

Variable	Tensile Strength	Elongation at Break
Spinning velocity	+	−
Time spent in spinning bath	+	−
Bath temperature	−	+
Spinneret orifice radius	−	+

Source: Ref. 13.

where t_i and t_f are initial and final sheet thicknesses, L_f and L_i final and initial lengths, and W_i and W_f initial and final widths.

The effect of percent stretch (orientation) on impact strength is shown in Figs. 12-17 and 12-18. As predicted (Table 12-1), increasing orientation increases impact strength.

Extruded products (sheet, film, pipe, etc.) involve the use of take-off and windup equipment. The former are usually sets of rolls that move the product

TABLE 12-3   Barrier Properties

Penetrant	Nonoriented PVC	Oriented PVC
Oxygen	12.2	9.0
Carbon dioxide	35.7	16.5
Water	1.5	0.7

Note: Oxygen and carbon dioxide units are $cm^3/mL/24$ h/atm/100 in^2; water is $g/ml/24$ h/atm/100 in^2.
Source: Ref. 14.

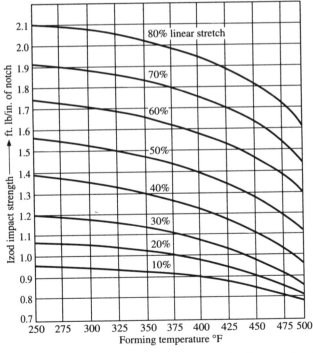

Fig. 12-17   Effect of biaxial stretching on impact strength of polystyrene sheets [15].

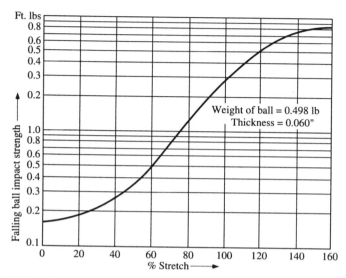

Fig. 12-18   Effect on stretching on falling ball impact strength of acrylic sheet [15].

along the postextrusion process. Windup devices use a constant tension or torque (or some combination of these) to package the extruded products. In either case, the polymeric material is obviously subjected to orientation, which has an effect on the properties of the product.

Finally, compression molding produces orientation profiles in a molded object. Figure 12-19 shows a birefringence vs position for samples of poly(ether-imide) (PEI) and poly(phenylene ether) (PPE). As can be seen, more rapid cooling (quenching) produces a more definitive profile in the polymer.

12.4   CRYSTALLINITY IN PROCESSING OPERATIONS

Polymer-processing operations can alter, change, and ultimately fix the structural characteristic of crystallinity in the product. As a semicrystalline polymer solidifies and cools, it also develops its crystallinity. The formation of this crystallinity depends on both the number of nuclei present and the growth of these nuclei. Lower temperatures produce more nuclei (Fig. 12-20), whereas higher temperatures favor growth. Further, nonpolymeric materials, such as fillers and coloring agents, present in the melt act as nucleating sites. In essence then, we can say, from a qualitative standpoint, that temperature (i.e., fixing the nuclei, influencing nuclei growth) and time (since crystal growth is a kinetic process) set the polymer's crystalline structure.

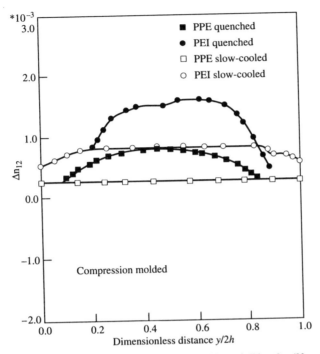

Fig. 12-19   Birefringence profile in compression-molded objects ($y/2h = 0$; $y/2h = 1.0$ are walls) [16].

This combination of factors is graphed in Fig. 12-21, which shows an optimum peak that results from the interaction of temperature and time. An empirical temperature for maximum growth [19] is taken to be 8/9 of the polymer's melting temperature in absolute units, that is, degrees Kelvin or Rankine.

There are, however, other aspects that can influence the polymer's morphology. One of these is the orientation of the polymer chains. Another is the stress applied to the polymer.

Polymer crystallinity is a particularly important property in melt-spun fibers. In such cases, crystallinity first appears in the solidifying skin formed on the outside of the fiber. The combination of molecular orientation, applied stress, temperature, and time influences the polymer's crystallinity.

A simulation of the effects in melt spinning on crystallization is shown in Fig. 12-22, where temperature-time plots for crystallization are given for sheared, quiescent melts as well as for melt spinning itself. Changes in the crystallinity along the spinning fiber thread line are shown in Fig. 12-23 for high-density polyethylene.

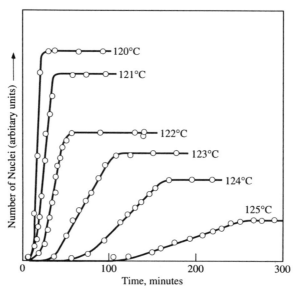

Fig. 12-20   Nucleation rate of decamethylene terephthalate as a function of temperature [17].

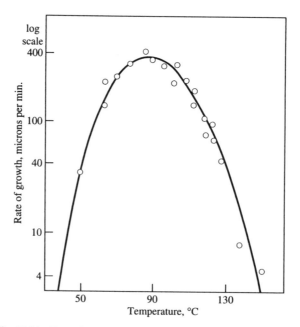

Fig. 12-21   Rate of crystallization for polyoxymethylene (acetal) [17].

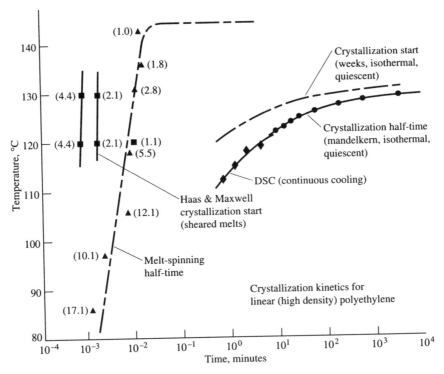

Fig. 12-22 Relations among time, temperature, and stress for crystallization [11].

If the ordinate of Fig. 12-21 is taken to be a temperature-dependent crystal-lization rate constant $k(T)$, then, for a nonisothermal case (as in melt spinning), the degree of crystallinity v is approximately as shown below:

$$v = \int_0^{t_L} K[T(t)]\, dt \qquad (12\text{-}6)$$

The $K[T(t)]$ term indicates that the crystallization rate constant is for a noniso-thermal condition; t_L is the time needed to reach the take-up device.

Analyses [13] of spinning data for polyethylene [21,22] and polypropylene [21,22] using a simplified heat-transfer approach showed that predicted changes in degree of crystallinity with single variable changes, such as extrusion tem-perature, cooling air temperature, heat transfer, and take-up velocity, matched density changes in product fibers. When more than one variable was changed,

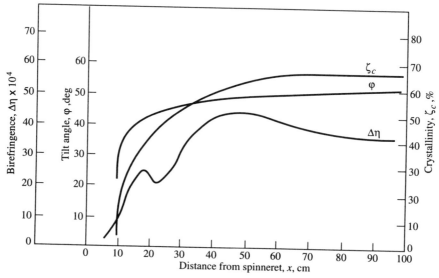

Fig. 12-23   Behavior of structural characteristics along spin thread line [20].

however, agreement between estimated degrees of crystallinity and densities was poor.

Crystallization also takes place in injection-molded objects. The resultant morphology reflects the temperature, stress, and orientation existing in the system. At the mold wall where the polymer is first cooled, there is also a large shear difference. These factors combine to produce crystalline material that is perpendicular to the mold wall and the flow direction (see Fig. 12-24). In the next layer, the crystalline aggregates are perpendicular to the mold surface. Finally, unoriented crystalline aggregates predominate in the molded object's central region.

Crystallinity can also be altered for injection-molded objects by a postmolding process of heating known as annealing. Figure 12-25 illustrates the effect of annealing (400°F for 2 h) on objects molded at various temperatures. As can be seen, the annealing process raises the polymer's crystallinity to the value associated with the 400°F mold temperature.

12.5 CHAIN FLEXIBILITY, REINFORCEMENTS, AND BLENDS

Polymer chain flexibility can also be changed during processing. This basically results from the action of smaller molecules taken up in the polymer. These molecules, which make the polymer more flexible and less rigid, are termed

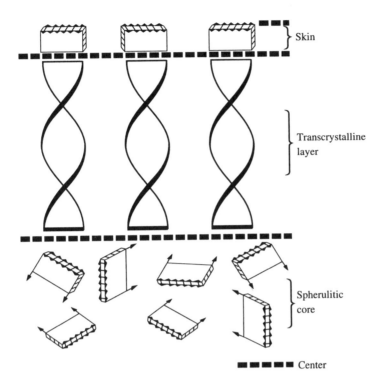

Fig. 12-24   Representation of crystalline regions in a molded polyacetal object [23,24].

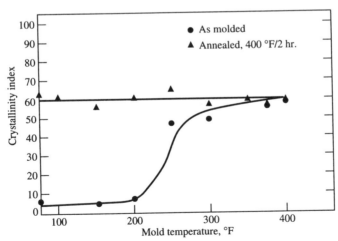

Fig. 12-25   Relation of crystallinity with molding temperatures and annealing for polyphenylene sulfide [18].

external plasticizers. Small molecules such as water can perform this function. However, external plasticizers are also purposely added in order not to change the polymer's characteristics but also to improve its processibility. External plasticizers, whether purposely added or absorbed, must be similar to the polymer in terms of polarity and intermolecular forces. Also, the external plasticizer must have a low diffusion rate from the polymer, that is, a high level of retention.

A broad spectrum of chemicals is used for external plasticizers. These include: phthalate esters (dioctyl phthalate); phosphate esters (tricresyl phosphate); low molecular weight polymers (polypropylene glycol); oleates, sebacates, adipates (dioctyl sebacate, dioctyl adipate); fatty acid esters; hydrocarbon derivatives; sulfonamides; and glycol derivatives.

The net effect of external plasticization is to increase chain flexibility and, hence, affect the polymer's properties as shown in Table 12-1. A particularly important property change is an increase in impact strength.

Polymer end properties can also be affected by adding reinforcing agents or by blending with other polymers. Although these cases are not always effects of processing, they do represent alteration of final behavior by changes in structural parameter behavior.

The usual reinforcing agents are fibers (glass, carbon, etc.) dispersed in a polymeric matrix. These reinforcements basically disperse stress through the material to inhibit failure and stop crack propagation if failure commences. The net result is an enhancement of the polymer's properties, particularly the tensile strength and flexural modulus. Table 12-4 compares some properties for unreinforced and glass-reinforced (30% glass content) plastics. As Table 12-4 indicates, property improvement is substantial (tensile strengths increased by a factor of 2 or 3 flexural modulus increased three- or fourfold).

Carbon reinforcements in thermoplastics offer even greater improvements in mechanical properties, such as tensile strength and stiffness. Other properties particularly enhanced include coefficient of expansion, which is lower, creep resistance, and toughness. Another advantage is that carbon-reinforced items are electrically conductive and hence decrease static electricity.

Reinforcing thermosets also offer great advantages. Not only do tensile and flexural strengths improve but also impact strengths. The cost-performance benefits are very attractive for reinforced thermosets.

Blending polymers essentially influences the modulus-temperature behavior. In particular, the high-modulus–high-impact strength behavior can be enhanced. This is done by blending two polymers, one of which is rubbery and the other rigid, at room temperature. Figure 12-26 shows the effect of blending polystyrene and 30/70 butadiene–styrene copolymer (curves are by weight percent of polystyrene in the blend). The use of blends represents another example of the vast array of possibilities available to the polymer practitioner.

TABLE 12-4   Comparison of Unreinforced and Reinforced Thermoplastics Properties

Polymer	Tensile Strength 10^3 psi	Flexural Modulus 10^6 psi	Deflection Temperature 164 psi (°F)	Thermal Expansion 10^{-5} in./in. °F
ABS	14.5 (6.0)	1.10 (0.32)	220 (195)	1.6 (5.3)
Acetal	19.5 (8.8)	1.40 (0.40)	325 (230)	2.2 (4.5)
Nylon 6	23.0 (11.8)	1.20 (0.40)	420 (167)	1.7 (4.6)
Nylon 66	26.0 (11.6)	1.30 (0.41)	490 (150)	1.8 (4.5)
Polycarbonate	18.5 (9.0)	1.20 (0.33)	300 (265)	1.3 (3.7)
Polyester (PBT)	19.5 (9.0)	1.40 (0.34)	430 (130)	1.2 (5.3)
Polyethylene (high-density)	10.0 (2.6)	0.90 (0.20)	260 (120)	2.07 (6.0)
Polypropylene	9.8 (4.9)	0.80 (0.18)	295 (135)	2.10 (4.0)
	13.5 (7.0)	1.30 (0.45)	215 (180)	1.9 (3.6)
	18.0 (10.0)	1.20 (0.40)	365 (340)	1.4 (3.1)

Note: Terms in parentheses are unreinforced.
Source: Ref. 25.

12.6   POLYMER PROPERTIES AND POLYMER PROCESSING

The preceding material in this chapter shows the importance of the linkage of polymer processing, polymer structural characteristics, and polymer properties. In order to get the most out of processing operations, the polymer engineer or scientist must follow logical and orderly procedures, even if, at times, they are of a qualitative or semiquantitative nature.

One very necessary aspect is to maintain careful tracking of the important structural characteristics of polymers. This means that not only must a polymer be thoroughly characterized before processing but also, if possible, during the processing and, most certainly, at the end of the process. Such characterization

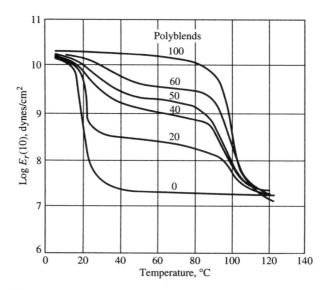

Fig. 12-26   Effect of blending polymers, polystyrene, and polystyrene–butadiene copolymer [19].

shows the impact of processing condition on the polymer's characteristics which, of course, fix end properties. Although it is not possible to quantify the effect of process variables on structural characteristics, the polymer engineer or scientist can qualitatively predict the impact of changing process variables. This is important to remember since the engineer's unfortunate response to undesirable product properties on occasion is wildly to change a number of process variables simultaneously without considering their effect on structural characteristics. A careful evaluation of the product problem in terms of structural characteristics and the required process changes needed to give them do far more good than "knob-twirling" or "mumbo-jumbo" process changes.

The entire area of the relationships between polymer processing, polymer structural characteristics, and polymer end-use properties is a very fruitful one for future study. A particularly helpful aspect is the quantification of these important relationships. Still another is the prediction of synergistic results when several structural characteristics are altered simultaneously. Overall, the goal is to be able to predict polymer end-use properties to the same degree that is possible for metal processing.

Polymer engineers and scientists who are able both to recognize the importance of the polymer-processing–polymer structural characteristic–polymer property interrelation and to apply available knowledge in an orderly and logical sequence will not only optimize their own professional performance but also master a complicated and important industrial area.

REFERENCES

1. WINDING, C.C., and HIATT, G.D., *Polymer Materials*, McGraw-Hill, New York (1961).
2. MARK, H., *Ind. Eng. Chem.* **34**, 1343 (1942).
3. BILLMEYER, F.W. Jr., *Textbook of Polymer Science*, Wiley, New York (1971).
4. GRISKEY, R.G., and LEE, B.I., *J. Appl. Polymer Sci.* **10**, 105 (1965).
5. GRISKEY, R.G., CHEN., F.C., and BEYER, G.H., *AIChE J.* **15**, 680 (1969).
6. RUBIN, I.I., *Injection Molding Theory and Practice*, Wiley, New York (1972).
7. BALLMAN, R.L., and TOOR, H.L., *Mod. Plastics* **38**, 10:113 (1960).
8. JACKSON, G.B., and BALLMAN, R.L., *S.P.E. J.* **16**, 1147 (1960).
9. SIPPEL, A., *Z. Elektrochem.* **50**, 152 (1944).
10. MATSUO, T., and KASE, S., *Sen-i-Gakkashi* **16**, 458 (1960).
11. SPRUIELL, J.E., and WHITE, J.L., *33rd Annual Technical Conference of the Society of Plastics Engineers,* **XXXI**, 188 (1975).
12. GRISKEY, R.G., and FOK, S.Y., *Proceedings of the M.E.S.D. Conference of the AIChE* **1**, 332 (1970).
13. ZIABICKI, A., *Fundamentals of Fibre Formation*, Wiley, New York (1976).
14. BELCHER, S.L., in *Plastic Blow Molding Handbook*, N. Lee, (ed.), Van Nostrand Reinhold, New York (1970), Chap. 4.
15. PLATZER, N., in *Processing of Thermoplastic Materials*, E.C. Bernhardt (ed.), Reinhold, New York (1959), Chap. 8.
16. WAGNER, A.H., YU, J.S., and KALYON, D.M., *Annual Technical Conference of the Society of Plastics Engineers*, **35**, 303 (1989).
17. ROSATO, D.V., and ROSATO, D.V., *Injection Molding Handbook*, Van Nostrand Reinhold, New York (1986).
18. HILL, H.W., JR., and BRADY, D.G., *Polymer Eng. Sci.* 16(12) 831 (1976).
19. TOBOLSKY, A.V., *Properties and Structure of Polymers*, Wiley, New York (1960).
20. KATAYAMA, K., AMANO, T., and NAKAMURA, K., *Kolloid Z.* **226**, 125 (1968).
21. TAKAHASHI, M., *Sen-i-Gakkaishi* **15**, 368 (1959).
22. SHEEHAN, W.C., and COLE, T.B., *J. Appl. Polymer Sci.* **8**, 2359 (1964).
23. CLARK, E.S., *Soc. Plast. Engr. J.* **23**, 46 (1967).
24. Ibid., *Appl. Poly. Symp.* **20**, 325 (1973).
25. *Materials Reference Issue, Machine Design,* **147**, Penton Publishing, New York (1978).

INDEX